Springer Series in Electrophysics
Volume 12

Edited by Walter Engl

Springer Series in Electrophysics

Editors: Günter Ecker   Walter Engl   Leopold B. Felsen

VLSI Technology

Fundamentals and Applications

Editor: Y. Tarui

With 377 Figures

Springer-Verlag
Berlin Heidelberg New York Tokyo

53132002

Professor Dr. Yasuo Tarui

Tokyo University of Agriculture and Technology, Nakamachi, Koganei,
Tokyo 184, Japan

Series Editors:

Professor Dr. Günter Ecker

Ruhr-Universität Bochum, Theoretische Physik, Lehrstuhl I,
Universitätsstrasse 150, D-4630 Bochum-Querenburg, Fed. Rep. of Germany

Professor Dr. Walter Engl

Institut für Theoretische Elektrotechnik, Rhein.-Westf. Technische Hochschule,
Templergraben 55, D-5100 Aachen, Fed. Rep. of Germany

Professor Leopold B. Felsen Ph.D.

Polytechnic Institute of New York, 333 Jay Street, Brooklyn, NY 11201, USA

Title of the original Japanese edition: *Cho LSI Gijutsu* ed. by Yasuo Tarui
© Yasuo Tarui 1981
Originally published by Ohmsha, Ltd., Tokyo (1981), English translation by Yasuo Tarui

ISBN 3-540-12558-2 Springer-Verlag Berlin Heidelberg New York Tokyo
ISBN 0-387-12558-2 Springer-Verlag New York Heidelberg Berlin Tokyo

Library of Congress Cataloging-in-Publication Data. Chō LSI gijutsu. English. VLSI technology. (Springer series in electrophysics ; v. 12. Translation of: Chō LSI gijutsu. 1. Integrated circuits – Very large scale integration. I. Tarui, Yasuo. II. Title. III. Series. TK 7874.C544413 1986   621.381'73 85-17178

© Springer-Verlag Berlin Heidelberg 1986
Printed in Germany

Offset printing: Beltz Offsetdruck, 6944 Hemsbach. Bookbinding: J. Schäffer OHG, 6718 Grünstadt.
2153/3130-543210

Preface

The origin of the development of integrated circuits up to VLSI is found in the invention of the transistor, which made it possible to achieve the action of a vacuum tube in a semiconducting solid. The structure of the transistor can be constructed by a manufacturing technique such as the introduction of a small amount of an impurity into a semiconductor and, in addition, most transistor characteristics can be improved by a reduction of dimensions. These are all important factors in the development. Actually, the microfabrication of the integrated circuit can be used for two purposes, namely to increase the integration density and to obtain an improved performance, e.g. a high speed. When one of these two aims is pursued, the result generally satisfies both.

We use the English translation "very large scale integration (VLSI)" for "Cho LSI" in Japanese. In the United States of America, however, similar technology is being developed under the name "very high speed integrated circuits (VHSI)". This also originated from the nature of the integrated circuit which satisfies both purposes. Fortunately, the Japanese word "Cho LSI" has a wider meaning than VLSI, so it can be used in a broader area.

However, VLSI has a larger industrial effect than VHSI. The reason lies in the economic effect of the reduction of the cost per unit function, resulting in rapid enlargement of the area of application and expansion of the scale of the industry; all these effects originate from the scale of the integration. Therefore, as industry proceeds to introduce, first, very large scale integration, improvement of performance, such as high speed, will accompany this.

This book contains the main results of research performed in the Cooperative Laboratories of the VLSI Technical Research Association, together with some additional, related technology. The Cooperative Laboratories were organized for the four years 1976-1979, with about 100 people being temporarily transferred from five computer manufacturers and the Electrotechnical Laboratory.

V

Technological Fundamental necessities and common interests have determined the selection of the research theme at the Cooperative Laboratories. Because of the principle of the selection and the limited period of 4 years, the development of the fabrication apparatus and technology was selected as the first important theme. The second important theme was the problem of the silicon crystal. In addition, fundamental and common interests on the part of the process technologies, test technologies and device technologies were selected. Thus, as the selection of the themes was made on the basis of fundamental and common interests, we expect this book will be interesting as fundamental knowledge for readers.

Finally I would like to express my sincere appreciation to people concerned in the Government, related companies, Universities, Japan Electronic Industry Development Association and the VLSI Technology Research Association headed by Mr. N. Nebashi, former executive director, for their guidance and cooperation which made this worthwhile research possible.

Tokyo, October 1985 *Yasuo Tarui*

Contents

List of Contributors

Aizaki, Naoaki	NEC Corporation
Asai, Takayuki	Hitachi Ltd.
Fujiki, Kunimitsu	NEC Corporation
Fukuyama, Toshihiko	Fujitsu Limited
Funayama, Tohru	Fujitsu Limited
Hisamoto, Yasuhide	Hitachi Ltd.
Hoh, Koichiro	Electrotechnical Laboratory
Honda, Tadayoshi	Fujitsu Limited
Iizuka, Takashi	Optoelectronics Joint Research Laboratory
Iwamatsu, Seiichi	Suwa Seikosha, Co., Ltd.
Kato, Hirohisa	Toshiba Corporation
Kishino, Seigou	Hitachi Ltd.
Kobayashi, Ikuo	Fujitsu Limited
Komiya, Hiroyoshi	Mitsubishi Electric Corporation
Komiya, Yoshio	Electrotechnical Laboratory
Koyama, Hiroshi	Mitsubishi Electric Corporation
Matsumoto, Hideo	Fujitsu Limited
Matsushita, Yoshiaki	Toshiba Corporation
Migitaka, Masatoshi	Toyota Technological Institute
Miura, Yoshio	NEC Corporation
Mori, Ichiro	Toshiba Corporation
Nagase, Masashi	Toshiba Corporation
Nakamura, Akitomo	NEC Corporation
Nakamura, Tadashi	Fujitsu Limited
Nakamura, Takuma	Mitsubishi Electric Corporation
Nakasuji, Mamoru	Toshiba Corporation
Nomura, Masataka	Hitachi Ltd.
Okada, Koichi	NEC Corporation
Oku, Taiji	Mitsubishi Electric Corporation
Onodera, Kazumasa	NEC Corporation

Ota, Kunikazu	NEC Corporation
Otsuka, Hideo	Toshiba Corporation
Sano, Syunichi	Toshiba Corporation
Sekikawa, Toshihiro	Electrotechnical Laboratory
Shibata, Hiroshi	Mitsubishi Electric Corporation
Shimizu, Kyozo	NEC Corporation
Shinozaki, Toshiaki	Toshiba Corporation
Shirai, Teruo	Fujitsu Limited
Sugiyama, Hisashi	NEC Corporation
Tajima, Michio	Electrotechnical Laboratory
Takasu, Shin	Toshiba Ceramics Co., Ltd.
Tarui, Yasuo	Tokyo University of Agriculture and Technology
Wada, Kanji	Toshiba Corporation
Yamamoto, Shinichiro	Toshiba Corporation
Yanagisawa, Shintaro	Fujitsu Limited
Yoshihiro, Hisatsugu	Hitachi Ltd.

1. Introduction

Very Large Scale Integration (VLSI) is in technical concept within the widely
expanded field of the semiconductor integrated circuits. The integrated circuit itself was invented in the process of microminiaturization of electronic
circuits. In the early stage the purpose of microminiaturization was literally
to make the electronic devices smaller and lighter, especially for use in
missiles and satellites.

Microminiaturization started in the form of a high-density assembly with
components being fabricated in small size and combined in condensed form. The
assembling of complex circuits with discrete components of variable size soon
reached a limit because handling of the components and, in particular, their
wiring became complex like a spider's net.

This limitation led to the idea of the "module" structure. Such structures
are characterized by an assembly of component units with standardized regularity and separation. In the module structure, however, it is still necessary
to handle each component one by one. This was a limitation for further miniaturization and became economically unfeasible.

The breakthrough in the difficulty of handling each component was achieved
by the concept of the integrated circuit for which components are fabricated
together and handled together. The integrated circuit is defined as "the
microstructure of many circuit elements inseparably associated on or within
a continuous substrate". In other words, an integrated circuit is handled as a
functional unit for the design, the fabrication and the test. Thus, the concept of circuit integration opened a new area of microminiaturization, but
it suffered from the difficulty of handling each component separately.

1.1 The Significance of Semiconductor Integrated Circuits

When the semiconductor integrated circuits were entering production, continuously improved by technological advances, it became clear that the merits of

integration not only lead to the original aim of smaller and lighter devices, but to an increase of high reliability, low cost and high performance.

The main reason for those merits today is the dramatic escalation of the units of electronic devices from R (resistance), C (capacitance), L (inductance) and transistors to the functions of the systems. Instead of the assembly of components one by one, only a series of photographic processes determines the locations of several millions of components, or how several millions of wires are connected. In addition, as all components are put into monolithic form, many electrical connections and encapsulations are superfluous.

Thus, the merits of integrated circuits with respect to cost, reliability and performance are recognized as generally applicable to all electronic devices. To utilize them, integration became the guiding principle in the electronic industry.

The advantages of integration are becoming more evident as the number of components on a chip is increasing. For this reason, the number of integrated devices or the "integration level" on a chip is to be increased. This increase led to an overall economic optimum, including yield considerations.

1.2 Prospects of High-Density Integration

As mentioned above, the development from integrated circuits (IC) to large-scale integration (LSI) has been achieved in the last twenty years. A quantitative expression is given by the complexity of a chip. The annual change of this number of components on a chip is commonly used to represent that tendency. Figure 1.1 shows it for memory circuits; the number of components in-

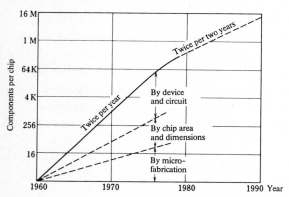

Fig.1.1. Illustrating the increase of the number of components per chip and the decomposition into different steps

creased twice per year during 1960 to 1980. The origin of the increase has been traced to three factors [1.1].

The important point is the outlook for the future. The contribution due to the development of self-aligning methods for devices and circuit-simplifications, which were the main factors so far, reached almost the limit of simplification. Therefore the main factors we may expect in the years to come are microfabrication and an increase of chip area. From this point of view it is expected that the complexity will continue to increase, but with a changed slope, namely by a factor of two per every two years with the turning point around or a little before 1980, as shown in Fig.1.1.

VLSI is the acronym used today for very high density integration. As the pattern for high-density integration advanced in fineness, it reached a limit where no more reduction of linewidth became possible by conventional photomask-fabrication technology and the contact pattern-transfer method. To overcome this difficulty, the development of a series of new technological tools and processes is necessary.

VLSI is a concept of advanced devices utilizing technology like electron beam X-ray and ion-beam techniques. These techniques are developed one by one and applied to production step by step; so the number of components on a chip is increasing gradually. Therefore, it is difficult to define VLSI. However, commonly cited figures, for numerical simplicity, are 100 components for standard IC, 10000 components for standard LSI, and 1 million components for standard VLSI.

1.3 Device Dimensions and Density of Integration

In a first-order approximation, the device dimensions are proportional to the decrease of fabrication dimensions with the advancement of microfabrication technology. On the other hand, the integration density is inversely proportional to the square of the device dimensions. Figure 1.2 shows the possible increase in the density of integration for a MOS dynamic memory with decrease in minimum feature size [1.2]. Three lines are drawn for MOS technologies: the lowest line is the one for single polysilicon-layer technology, the middle line is the one for the double polysilicon-layer technology. Those two lines were extrapolated from reported figures. If we take the average of those two curves, the integration density of about 1 Megabit/cm^2 corresponds to a minimum size of 1 μm.

One million bits per cm^2 with one micrometer channel length is only the state-of-the-art of present technology. The improvement of the memory cell

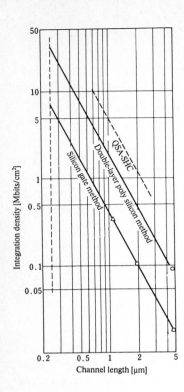

Figure axes: Integration density [Mbits/cm²] — values: 50, 10, 5, 1, 0.5, 0.1, 0.05. Channel length [µm] — values: 0.2, 0.5, 1, 2, 5. Curves labeled: Silicon gate method, Double-layer poly silicon method, QSA-SHC.

structure can increase the density by a factor of two or more. An example is shown by the dotted line in Fig.1.2. This line represents the case of a memory cell with adequate margin by QSA-SHC, to be described in Chap.8.

Therefore, appreciably high-grade VLSI will be produced with a linewidth of 1 µm. Concerning the microfabrication technology, it is necessary to produce patterns of roughly 1 µm. This range of linewidth is the aim of microfabrication technology which will be described in the following.

1.4 Outline of the Microfabrication Technology

The microfabrication technology is the most important technology within the VLSI technologies, and most efforts are directed toward it. Since this technology is essentially necessary to increase integration, a new method must be developed to break the barrier of conventional technology.

Figure 1.3 gives an overview of fine-pattern lithography methods developed or under development for VLSI and LSI; they will be described in detail in this book. For delineation from pattern data, there is a conventional optical

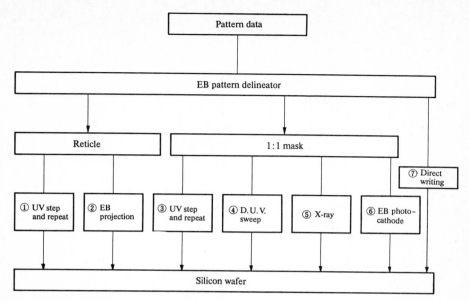

Fig.1.3. Overview of fine pattern lithography methods

method. However, the electron beam (EB) pattern delineator is superior not only with regard to the fine-line width but also to the delineation time and cost. Therefore, with the advancement of the EB delineator, the optical method will be replaced.

Direct writing shown at the far right in Fig.1.3, is the method of exposing the wafer directly to an electron beam. This method is effective for drawing fine lines and shortening the turn around time. Therefore, it will be used in the development phase and the pilot production. For mass production, however, pattern transfer is more economical as far as it is applicable. For this reason many kinds of pattern-transfer systems have been devised.

The systems 1 and 3 in Fig.1.3 use ultra-violet (UV) light. The resolution and alignment accuracy are high because of the so-called step-and-repeat method, where a part of a wafer surface is exposed in one shot and the wafer is moved after each exposure. System 4 in Fig.1.3 is the so-called deep UV projection system using a reflection optical system because glass or quartz are not transparent to deep UV.

The pattern-transfer method by X-ray (system 5) is expected to be used in production next. The electron-beam pattern-transfer method (system 2) and the optical to electron-beam pattern-transfer method with the aid of a photo-cathode (system 6) are under development for future methods.

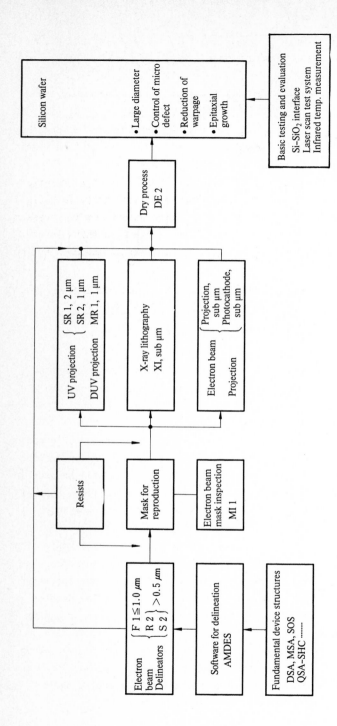

Fig.1.4. Major development steps for the basic VLSI technology in the VLSI Cooperative Laboratories

The VLSI technology described in detail in the following is a basic technology and of common interest. It was developed in the VLSI Cooperative Laboratories. Related technologies are added. The interrelation between the main technological methods is shown in Fig.1.4. The part of micropattern delineation is the same as in Fig.1.2, but in more detail. For instance, R2, S2 in Fig.1.4 are the notations used in the Cooperative Laboratories; they are shorthands for VL-R2, VL-S2, and mean the second machine of the raster-scan system and the second machine of the shaped-beam system respectively.

Following the microfabrication technology, the second subject of fundamental necessity and of common interest is the silicon wafer. Especially, in VLSI technology where micropatterns are deliniated on large wafers, it is necessary to reduce the microdefects. In addition, to delineate exact patterns on the wafer, the wrappage of the wafer must be within a limit after the process of heat treatment.

Finally, we describe dry etch systems which transfer the patterns delineated on photoresist to silicon or silicon dioxide, basic measurement and test methods, and fundamental device structures.

2. Electron Beam Lithography

To develop greater LSIs, it is necessary to increase the number of pattern elements per unit area by reducing circuit patterns. The increase of pattern elements per chip needs high-speed pattern writing. The electron-beam lithography system has stepped into the limelight as one of the systems for meeting the demand of fine and high-speed writing. The electron-beam technology utilized in such a system is immature from the viewpoint of semiconductor production. In addition, it essentially requires modern technologies, such as electrooptics, precision machinary and controlling electronics, a precision moving table, computers and software.

In this chapter we shall treat fundamental technologies and applicable software. We introduce three kinds of electron-beam lithography systems with ample experimental results developed by the Cooperative Laboratories, VLSI Research Associate, in Japan. By the understanding of the systems and their fundamental technologies, one can well appreciate the electron-beam lithography which is one of the most important technologies of VLSI.

2.1 Background

2.1.1 History of the Machine Development

Electron-beam lithography is a key technology for submicrometer dimensions of VLSI-systems. Research on electron-beam lithography for the fabrication of integrated circuits started almost at the same time (around 1967) in Japan as in other countries. The initial study was a natural extension of the scanning electron microscope (SEM) because focusing of the electron beam and high-speed deflection of it are needed. In this cradle period until 1975, several machines were developed such as JBX-2B (1968) and JBX-5A (1974) at JEOL, EBMF-1 (1973) at Cambridge Instruments, EPG-102 (1974) at Thomson-CSF and LEBES (1974) at

ETEC[1]. All these machines were designed for laboratory use only and did not have sufficiently high throughput for the production of ICs. However, in this period basic research for the production machines had started at IBM, Bell Telephone Laboratories, Texas Instruments, Thomson-CSF, JEOL, etc. Simultaneously several key technologies such as the mark detection by reflected-electron signals and the control of the stage position by a laser interferometer were developed in Japan [2.1].

Results of these basic studies have been disclosed since 1975 and the developing period started, accelerated by VLSI research projects in Japan and other countries. The most striking invention was the concept of a variable shape electron-beam system [2.2-4]. Details of the technologies developed in this period are explained in Sect.2.2.

In the 1980s, the electron-beam delineation technology became mature and the throughput for VLSI manufacturing improved. Applications to fields other than silicon integrated circuits were investigated.

In the following basic and general features of electron-beam lithography are introduced.

2.1.2 Classification of the Machines

The various electron-beam delineators hitherto developed can be classified according to the following criteria:

a) Electron-Beam Source

Thermoelectronic (thermal) emission and field-emission cathodes are used. The cathode materials are lanthanum hexaboride (LaB_6) for the former and tungsten (W) for the latter system.

b) Beam Shape

A *Gaussian beam* is an ordinary point-focused beam like in an electron microscope. The current distribution in the beam spot is Gaussian along the radius, as shown in Fig.2.1a. To expose a line of finite width by this beam, continuous parallel (e.g., 4 times) scans are necessary, as illustrated in the lower part of the figure.

A *fixed-shaped beam* is produced, as shown in Fig.2.1b, by uniformly illuminating the aperture (5) of a certain fixed shape (generally a square) and projecting its reduced image on the sample surface.

---

[1] The author is grateful to Dr. Y. Takeishi for the information on the historical progress.

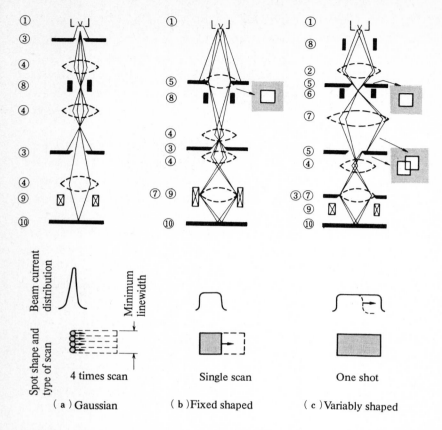

Fig. 2.1. Electron optics and beam shaping of electron-beam delineators

A *variable-shaped* beam is illustrated in Fig.2.1c. The image of the first aperture (5) is projected onto the second aperture (5') and the position of the image of the first aperture is moved horizontally by the shaping deflector (6) in the same way. The overlapping area of these (generally a rectangle) can therefore be controlled continuously. The reduced image of this rectangle is projected on the sample.

c) Beam Scanning

In *raster scanning* (Fig.2.2a), the electron beam scans the whole field independent of the presence of patterns and the beam is *on* only when it comes to the respective point on a pattern. In *vector scanning* (Fig.2.2b), the beam is steered to the point where a pattern should exist and exposes it.

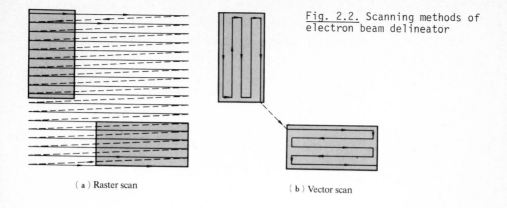

Fig. 2.2. Scanning methods of electron beam delineator

(a) Raster scan   (b) Vector scan

d) Sample Movement

As the field size of the electron-beam delineator is restricted to 2-5 mm in square due to aberration and distortion, the sample must be moved by some means to expose the entire wafer or the mask plate which have dimensions of 100-125 mm in diameter or square. One way of sample movement is the *step-and-repeat* mode illustrated by Fig.2.3a. In this mode, the sample is shifted by a field length after the exposure of one field has been completed and the exposure of the next field begins. Another way is the *continuous move mode* exhibited in Fig.2.3b, in which the sample moves continuously in one direction while the beam makes a short raster scan along the orthogonal direction of the sample movement.

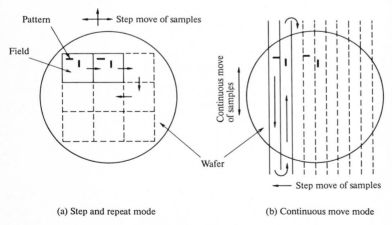

(a) Step and repeat mode   (b) Continuous move mode

Fig. 2.3. Movement of the sample (for visual convenience, the field size is exaggerated compared with the wafer size)

11

Electron-beam delineators draw a circuit pattern either on masks used in the pattern replication or directly on wafers (direct writing). Most of the machines are designed for both uses. In the direct writing, however, the alignment between successively overlaid patterns and the correction to the wafer distortion are necessary.

In Table 2.1, typical electron-beam delineators hitherto developed are classified according to the above criteria (b - d). Those machines that have been developed in the VLSI Cooperative Laboratories are marked with an asterisk and explained in detail in Sect.2.4.

Table 2.1. Classification of electron-beam delineators

Scan	Stage Move	Gaussian Beam	Shaped Beam	
			Fixed	Variable (Rectangle)
Vector	Step and Repeat	JBX-5A (JEOL) EBMF II (Cambridge Instr.) VS 1 (IBM) [2.5] EB 52 (ECL & Hitachi) [2.6] VL-F1 (Coop. Labs.)*	EBM-3, EBSP [2.7] EBM-5 (Texas Instr.)	VL-S1 (Coop. Labs.)* JBX-6A (JEOL) VL-S2 (Coop. Labs.)* ZBA-10 (Jenoptik Res. Cent.) [2.8] EB 55 (ECL & Hitachi) [2.9] EL 3 (IBM) [2.10]
	Continuous	EBES (Bell Labs.)[†] VL-R1 (Coop. Labs.)*[††]		(Bell Labs.) [2.11] VL-R2 (Coop. Labs.)
Raster	Step and Repeat		EL 1 (IBM) [2.4]	EL 2 (IBM) [2.12]

[†] Commercially available as MEBES from Perkin Elmer/Etec and EBMG-20, Ee-BES 40 from Varian/Extrion.

[††] Commercially available as EBM-105 from Toshiba

2.1.3 Factors Determining Pattern Accuracy

a) Beam Diameter

One can draw finer patterns by using electron beams with smaller diameter. However, the beam current becomes smaller and, hence, the exposure time in-

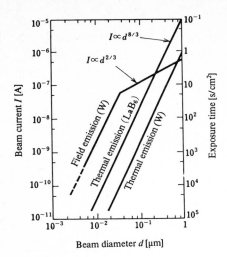

Fig. 2.4. Relationship between electron-beam diameter and beam current (exposure time is calculated by assuming a resist sensitivity of 1×10^{-6} C/cm^2)

creases (Fig.2.4). In practical delineators, the beam diameter is determined by the tradeoff between line width and throughput. In shaped beams, the beam current is proportional to the area of the beam spot.

b)*Aberrations in the Electron-Optics System*

The aberrations of electron-beam delineators are determined by the first lens in the magnification electron-optics system and by the final lens in the reduction system. The former is adopted for the delineator using field-emission electron guns (Sect.2.2.2).

Even without deflection, the projection of a point source, or one point on the edge of a shaped beam, generates a spherical aberration (d_s) and an axial chromatic aberration (d_c) at the sample surface (Fig.2.5a) [2.3,13]. In addition, the interaction aberration d_{int} due to the repulsion of electrons in a beam arises for the shaped beam because of the large beam current Sect.2.2.1). The interaction aberration is proportional to the total beam current, not the current density [2.3,4].

When the beam is deflected, 6 additional types of aberrations (deflective aberrations) occur (Fig.2.5b). Moreover, in the shaped beam which involves the projection of images with finite dimensions, 7 types of the mixed aberrations arise (Fig.2.5c). Theoretically, these deflective and mixed aberrations can be eliminated by the introduction of the MOL (moving objective lens) concept in which the objective lens is effectively moved by the superposition of magnetic fields so as to make the electron beam always pass the center of the lens after deflection [2.3,13]. In practical equipments, the

13

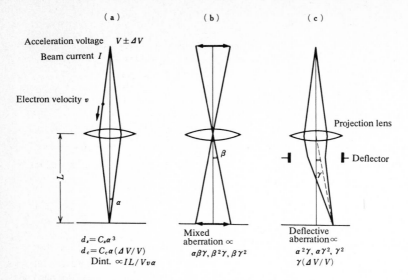

Fig. 2.5. Aberrations in electron-optical systems

dynamical focusing [2.14] and the in-lens deflection [(7) and (9) in Fig.2.1b] are adopted for the reduction of these aberrations [2.4].

c) Sample Movement, Beam-Position Control and Alignment

The movement of the sample stage along the x and y directions is measured and controlled by a laser interferometer [2.15,16] with an accuracy of the order of 10 nm. By reading the standard marks on the sample surface, the distortion in the beam deflection and the beam drift are corrected, and the pattern stitching between adjacent fields is controlled. In direct writing on the wafer, the alignment marks are fabricated on the wafer by etching and the alignment between successively overlaid patterns is made by referring to them (Sect.2.2.4).

d) Proximity Effect

The proximity effect is the distortion which appears in closely located patterns due to the scattered electrons inside the resist and the substrate. This effect has been quantitatively investigated and can be corrected in the software for the pattern delineation (Sect.2.3.2).

2.1.4 Factors Determining the Drawing Speed

When a pattern is delineated with the step-and-repeat mode on a mask or a wafer divided in N fields, the total drawing time T can be written as

$$T = N(t_e + t_0 + t_m + t_r) + T_1 \quad , \tag{2.1}$$

where t_e is the *exposure time* (per field) taken by the unit beam spot to expose the corresponding area, t_0 is the *overhead time* (per field) taken by the control system to transfer the data for the minimum element of the pattern to be drawn next (a line for the Gaussian beam and a rectangle for the shaped beam) after the completion of the first drawing, t_m is the *moving time* (per field) taken by the sample stage for its stepwise move and hold, t_r is the *registration time* (per field) taken for the mark detection and the control of the stage position, and T_1 is the *loading time* taken for the sample exchange and evacuation of the sample chamber.

The exposure time for the Gaussian beam is expressed as

$$t_e = \text{Max}(kAS/I, \; t_{step}) \quad , \tag{2.2}$$

where Max denotes the function which assumes the largest value in the parentheses, t_{step} is the time required by the deflection system to make a one-step hop of the beam in a line scanning, A is the field area, S is the resist sensitivity, I is the beam current and k is the fraction of the area the beam actually scans in a field ($k = 1$ for the raster scanning). The first term in the parentheses is the actual time necessary for the resist exposure and it must be noted that this does not depend on the beam size but only on the beam current. Therefore, the relationship $t_e = kAS/I$ also holds for the fixed-shaped beam.

For the variable rectangular beam, on the other hand,

$$t_e = SM/j = SM1_m^{\,2}/I_m \quad , \tag{2.3}$$

where j is the beam current density, M is the number of rectangles contained in a field, l_m is the maximum side length of the rectangles, I_m is the beam current for the spot with the maximum area, i.e., a square with the side lengths l_m, respectively. M is a function of l_m and when its dependence on l_m is explicitly given, we can determine the optimum value of l_m which makes t_e minimum [2.17].

The overhead time t_0 is the component which cannot be neglected in the vector scan. It greatly depends on the methods of data acquisition and transfer, the response time of the beam control system, and shape, density and arrangement of the patterns.

In the continuous-move mode, T can be expressed essentially in the same manner as in (2.1), in which the stripe that is exposed during one stroke of the stage movement is regarded as one field, and the time for the stepwise movement between stripes is accounted as $t_m + t_r$.

2.2 Components for Electron-Beam Lithography

2.2.1 Electron-Beam Source

a) Fundamentals

Features of the electron beam which determine the exposed pattern size and the exposing speed in electron-beam lithography depend on the properties of the electron-beam source, such as high brightness, low-energy spread and high stability. Therefore, to study the properties of the electron-beam source is very important.

The high brightness of the electron-beam source is necessary to obtain an electron beam with the current density for high-speed exposure. The current density J [A/cm^2] is given by the equation $J = \pi B \alpha^2$, B [A/cm$^2 \cdot$ sr] being the brightness of the electron-beam source and α [rad] the half angle of the incident beam on the target. The current density of the beam increases with the brightness of the electron-beam source.

The low energy spread is important to obtain a beam with small aberration necessary to expose the fine pattern. The beam emitted from the source enters the target after being focused by an electron lens and being deflected by the electron deflectors. The chromatic aberration d_c caused in the electron lens is given by $d_c = C c \alpha \Delta V/V$, V being the beam accelerating voltage, ΔV the voltage spread of the beam and C_c the coefficient of the axial chromatic aberration. The chromatic aberration caused by the electron deflector is proportional to the values of $\gamma(\Delta V/V)$, γ being the angle of the deflection (Fig.2.5). The low energy spread of the electron-beam source is necessary to increase the half angle of the incident beam and the angle of deflection.

The stability of the electron-beam source is important to make the exposing process practical. The change of the brightness and the movement of the emitting spot cause the unstability of the beam-current density which produces either the excess or the lack of dosing in the exposing procedure, and result in the inaccuracy of the exposed pattern.

The electron gun for the electron-beam exposing system is of the same type as used in electron microscopes [2.18-20]. There are two types of electron guns. One is the thermal emission type which utilizes the thermal electrons emitted from the metal heated up to a high temperature. The other is the field-emission type which utilizes the field-emission electrons emitted from the metal by a high electrical field. The size of the latter source is smaller than the one of the thermal electron gun. For the electron-beam exposure the triode gun with thermal emission is commonly used. The general performance of the electron gun is shown in Table 2.2.

Table 2.2. Characteristics of the electron gun

Type of the gun	Thermal emission		Field emission
Cathode	W	LaB$_6$	W
Temperature [K]	2700 – 3000	1800 – 2000	300
Brightness [A/cm$^2 \cdot$ sr]	$10^4 - 10^5$	$10^5 - 10^6$	$10^7 - 10^8$
Energy spread [eV]	~2	~2	~0.4
Vacuum [Torr]	~10^{-5}	~10^{-7}	~10^{-9}

Triode Gun. The structure of the triode gun with thermal emission is shown in Fig.2.6. The tungsten wire shaped like a hairpin with a diameter of about 0.1 mm is used for the cathode. The Wehnelt electrode has an opening with a diameter of about one to two millimeter in the center and surrounds the cathode. The top of the cathode is located at the same height or a little above the Wehnelt electrode. The anode has an opening through which the electron beam passes. The electrodes are coaxial to guarantee axial symmetry of the gun.

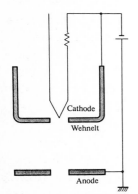

Fig. 2.6

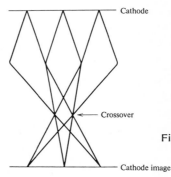

Fig. 2.7

Fig. 2.6. Structure of the triode electron gun

Fig. 2.7. Trajectories of the emitted electrons

A high positive potential is applied to the anode, and the negative potential to the Wehnelt. In general, the Wehnelt potential is supplied by the voltage produced across the bias resistor. The anode has the same grounded potential as the one of the column.

The triode gun acts as an electron lens by the potential distribution produced around the Wehnelt electrode [2.20]. By this lens action the electrons emitted from the cathode travel the trajectories shown in Fig.2.7. The electrons with Maxwellian velocity distribution emitted in various directions form a cathode image in front of the cathode. The narrowest part of the beam before the cathode image is called the crossover. The crossover can be considered to be the electron-beam source.

The theoretical maximum brightness of the electron gun is given by Langmuir's equation of $B = J_c eV/kT$, where J_c [A/cm^2] is the current density emitted from the cathode, e is the electron charge, k is the Boltzmann constant and T [K] is the temperature of the cathode. The saturation current density J_s of the metal cathode is, according to the Richardson-Dushman's equation,

$$J_s = AT \exp(-e\phi/kT) \quad , \tag{2.4}$$

where A is a constant, and ϕ is the work function of the metal. To achieve high brightness, it is necessary to increase the accelerating voltage and the saturation current density. The saturation current density is increased by heating the cathode up to a high temperature. The brightness of the tungsten electron gun is about $5 \cdot 10^4$ A/cm$^2 \cdot$ sr at a cathode temperature of 2700 K and an accelerating voltage of 20 KV.

Under the condition of a constant cathode temperature, the brightness is shown in Fig.2.8 as a function of the Wehnelt bias. The brightness increases with the applied voltage in the lower bias region and decreases in the higher bias region. The decrease of the brightness is due to the lowering of the emission-current density by the space charge effect.

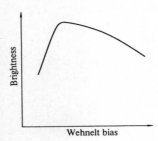

Fig. 2.8. Brightness due to charge of Wehnelt bias

The tungsten hairpin cathode can be easily operated by direct heating. A weak point of the cathode is its short life of several ten hours. Recently, the LaB$_6$ cathode has been used for its long life of several thousand hours (Sect.2.2.2b).

The Energy Spread by the Boersch Effect. The energy spread of the emitted electrons calculated from the Maxwellian-velocity distribution at the cathode temperature of 2700 K is about 0.2 eV. The observed energy spread increases with the emitted current of over 0.2 eV. This energy spread is called the Boersch effect.

18

The Boersch effect arises from the increase of the axial energy as a result of the velocity exchange by the Coulomb collisions between the beam electrons. One origin of the Boersch effect is the velocity disorder of the emitted electrons. The lateral velocity distribution observed on coordinates moving with the beam electrons is the Maxwellian velocity distribution with the cathode temperature T, while the axial velocity distribution with the temperature T_a is given by the equation of $T_a = kT/eV$ [2.21]. This equation accounts for the fact that the beam energy is proportional to the square of the beam electron velocity. The axial velocity spread becomes remarkably small because the axial temperature T_a is much lower than the lateral temperature T. However, the axial velocity spread increases as a result of Coulomb collisions; then the axial energy spread becomes large. Another origin of the Boersch effect are the random collisions of the transverse velocity components of the beam electrons, generated by the lens action of the gun. The energy spread increases with beam current because the frequencies of collisions are proportional to the beam current.

KNAUER [2.21] reported that a theoretical energy spread of 2.4 eV at the emission current of 150 μA in the triode gun is consistent with the experimental value.

Stabilization of the Gun. Instabilities of the emission characteristics are caused by fluctuations of the cathode temperature and changes of the potential distribution. Note that the shape of the cathode is deformed by evaporation of cathode materials. The work function of the single-crystal cathode like LaB_6 depends on the crystal orientation and the fluctuations of the emission characteristics on the brightness and the source position are complicated (Sect.2.1.1b). These fluctuations are corrected by adjusting the heating power, the Wehnelt bias and the alignment-coil current. Beam-current stabilization below 5% has been attained by controlling the heating power and the aligner. This automatic stabilization of the electron gun is very important because the electron-beam exposure systems are operated for many hours.

b) Lanthanum Hexaboride (LaB₆) Cathode Electron Gun

As lanthanum hexaboride (LaB_6) has a low work function and a low vapor pressure at high temperature, it has superior characteristics for cathode materials. Figure 2.9 shows the thermal emission characteristics of LaB_6 and other cathode materials. On the right-hand side are materials which give adequate current density at lower temperature. These are better cathode materials. An LaB_6 cathode is superior to a W hairpin and a treated tungsten cathode. The L-cathode, the impregnated cathode and the oxide-coated cathode have poor

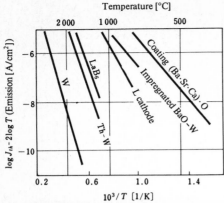

Fig. 2.9. Thermal-emission current denisty for various thermal cathodes as a function of temperature

(a)

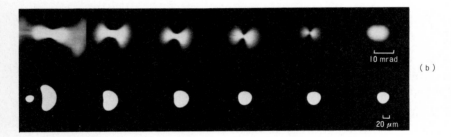

(b)

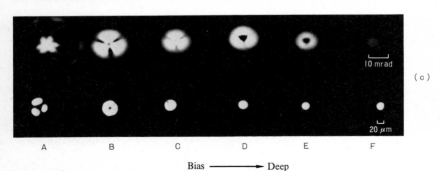

A   B   C   D   E   F

Bias ──────→ Deep

Fig. 2.10. Emission (upper) and crossover (lower) patterns for a single-crystal LaB$_6$ cathode electron gun for various bias conditions. (a): <100>, (b): <110>, (c): <111>

machinability, and have the unfavourable feature because these cathodes release a lot of CO or CO_2 gas when activation for thermal emission takes place;
thus, they contaminate the column. Therefore, LaB_6 is the best cathode material
for electron guns for an electron-beam delineator.

At first, poly-crystal LaB_6 cathodes were tried. However, their stability
was poor and lifetimes were not so long. On the other hand, single-crystal
LaB_6 cathodes were stable and they keep smooth surfaces even after being used
for a long time. When employing single-crystal LaB_6 cathodes for electron guns,
the choice of the crystal orientation is important. Work function value, facet
formation characteristics, stability for residual gas, particularly oxygen, and
tip shape should be considered in choosing the crystal orientation. Figure 2.10
shows emission and crossover patterns for <100>, <110>, and <111> oriented
single-crystal LaB_6 cathode electron guns. At the left-hand side, in the low-
bias condition, the thermal-electron emissivity difference between crystal
orientations forms a complex but symmetrical pattern. From these symmetries
it can be determined which crystal orientation has the greatest electron
emissivity. The high-bias case at the right-hand side corresponds to the space-
charge limited region, and emission and crossover patterns are simple circular
patterns, independent of crystal orientation.

LaB_6 is active and reactive to surrounding material. Therefore, the supporting mechanism and the heating method involve problems. Many designs were
tried to overcome the problems. In Vogel's mechanism, the LaB_6 tip is held by
pyrolitic graphites [2.22]. The graphite is not likely to react to LaB_6 and
can be directly heated at a low current level, because of the large electric
resistivity in the axis direction. The mechanism of this type has merits for
low heating power and when a small cathode tip is applicable. Therefore it
may become the most popular heating method.

Electron Gun for Gaussian Probes. Gaussian probes are formed from demagnified
crossover images. The electron gun has to satisfy four requirements: (i) a
large beam current for the delineating beam size, (ii) long cathode life,
(iii) small fluctuations of beam conditions, such as brightness, beam size,
beam position, etc., and (iv) a small energy distribution.

(i) As shown in Fig.2.4, when the beam diameter is larger than 0.2 μm,
LaB_6 is the best cathode material. (ii) The lifetime of LaB_6 may reach more
than 2000 hours. This performance is important for reducing the delineation
downtime. (iii) Brightness fluctuations are beam-current fluctuations; they
degrade the delineation accuracy. Although the crossover position drift
causes beam position drift, this drift is demagnified to the extent of the

crossover image demagnification factor, and there is not so big a problem. The fluctuation of the emission direction causes a beam-current fluctuation. This is because the beam current intercepted at a spray aperture fluctuates. For an LaB_6 electron gun, this fluctuation is small, and this is of little concern.

For an actual delineation system, a maximum lifetime of 4000 hours was obtained. The LaB_6 tip had a curvature of a 15 μm radius, the brightness was $8 \cdot 10^5$ A/cm$^2 \cdot$ sr, and a beam-intensity stability of 4%/two hours was obtained without servo stabilizer.

Electron Gun for a Shaped Beam. The required electron gun characteristics for a shaped beam are as follows:

(i) The electron gun must homogeneously illuminate the shaping aperture and simultaneously produce a sufficient crossover image at the final lens aperture for Köhler illumination [2.23]. For this illumination the crossover image is focused onto the plane of the entrance pupil of the final lens.

(ii) The electron gun must give optimum brightness required from beam conditioning.

(iii) Long life and brightness stability are also necessary as for a Gaussian beam.

(iv) The fluctuation of emission direction degrades the homogeneity of the beam-intensity distribution, and fluctuation of the crossover position results in fluctuations of the beam current. This is because the intercepted current at the final lens aperture fluctuates, in the case of Köhler's illumination condition. As the last two fluctuations degrade the delineation accuracy, these fluctuations must be small.

There are two imaging concepts. One is Köhler's illumination, and the other one is the critical illumination, in which the crossover is imaged on the shaping aperture. The quality factor required for an electron gun is the same for both illuminations. Therefore we consider the quality factor for Köhler's illumination condition, which is frequently adopted. If all the beam current which passes through the shaping aperture passes through the final lens aperture, one obtains

$$\phi_g \cdot \alpha_g = L_t \cdot \alpha_t \qquad (2.5)$$

where α_g [rad] is the beam-divergence semi-angle for an electron gun at 95% of the center intensity, ϕ_g [cm] is the crossover diameter of an electron gun, α_t [rad] is the beam semi-angle from the final lens to the target, and L_t is the diameter of the illuminated area at the target; for maximum beam length L_m, we have $L_t = \sqrt{2}\, L_m$. Allowing for beam misalignment, the equation becomes

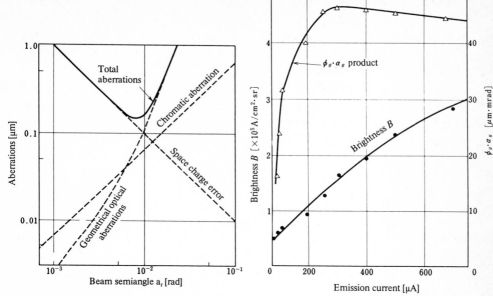

Fig. 2.11. Various aberrations as a function of beam semiangle

Fig. 2.12. Electron-gun characteristics for a variable-sized beam

an inequality, namely

$$\alpha_g \cdot \alpha_g > L_t \cdot \alpha t \quad . \tag{2.6}$$

Next, the optimum brightness is explained by an example. As shown in Fig.2.11, the axial beam resolution is determined from the geometrical aberration, chromatic aberration, and space charge error. The optimum convergence semi-angle α_{to} minimizes the total aberration; for Fig.2.11, $\alpha_{to} = 8$ m rad. Then, if the current density J of 50 A/cm^2 is desired, we obtain from

$$J = \pi\alpha_t^2 B \tag{2.7}$$

the optimum brightness B_0 as

$$B_0 = 2.5 \cdot 10^5 \text{ A/cm}^2 \cdot \text{sr} \quad . \tag{2.8}$$

We study electron-gun characteristics for the delineator VL-R2 (Sect.2.2.1a) using variable-shaped beams. The aberrations for the final lens are, as shown in Fig.2.11, $\alpha_{t0} = 8$ mm rad, $L_t = 4$ μm, and J = 50 A/cm^2. The fundamental characteristics are

$$\phi_g \cdot \alpha_g > 32 \text{ μm} \cdot \text{m rad} \quad , \tag{2.9}$$

$$B_0 = 2.5 \cdot 10^5 \text{ A/cm}^2 \cdot \text{sr} \quad . \tag{2.10}$$

23

Therefore, in comparison with characteristics for a Gaussian beam ($\phi_g \cdot \alpha_g \simeq$ 5 μm mrad, $B \simeq 8 \cdot 10^5$ A/cm$^2 \cdot$ sr), more than 6 times the $\phi_g \cdot \alpha_g$ product and less than 1/3 of the brightness are necessary. Distinctive features of the developed electron gun are characterized by the cathode surface being a 480 μm radius concave sphere, and the crystal orientation is <310>. It should be noted that a large cathode radius results in an increased electron-emission area, i.e., source area. Therefore, an increase in the $\phi_g \cdot \alpha_g$ product and a decrease in the cathode current density decreases the brightness.

The electron-gun characteristics are shown in Fig.2.12. When the emission current is 590 μA, the quality as given by (2.9,10) is fulfilled. A maximum cathode lifetime of 2700 hours was obtained for an actual delineator.

c) Field-Emission (FE) Electron Gun

Placing electrodes in front of a pointed tip (Fig.2.13) concentrates the applied electric field in the vicinity of the pointed tip. When the electric field at a metal surface exceeds 10^7 V/cm, the potential barrier at the surface becomes very thin due to the strong electric field and electrons are emitted from the metal into the vacuum by the tunnel effect. In this way, one can obtain a high-brightness (10^8 A/cm$^2 \cdot$ sr) point electron source (~30 Å) [2.24]. This phenomenon is called *field emission (FE)*. The current density at zero absolute temperature can be expressed by [2.25]

$$J = 1.54 \cdot 10^6 \frac{F^2}{\phi} \exp(-6.83 \cdot 10^7 \frac{\phi^{3/2}}{F}) \quad [A/cm^2] \tag{2.11}$$

where F is the electric field [V/cm] and ϕ is the work function of the tip surface material [eV]. The effect of temperature on the current is very small up to temperatures of 1000° C; at a high field it can practically be neglected. The current, however, depends strongly on both the field and the work function of the tip surface material, as seen from the above equation. Figure 2.14 [2.18] illustrates the relation between field and FE current density for tips with different work functions. From the figure it can be seen that the current

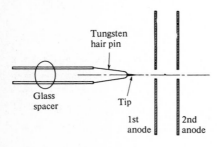

Fig. 2.13. Basic structure of a field-emission electron gun

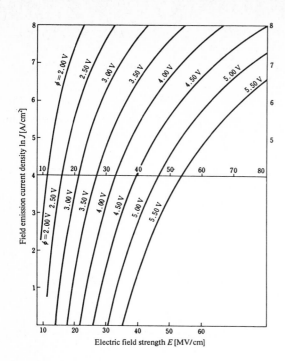

Fig. 2.14. Relations between field strength and field-emission current density

density changes by two orders of magnitude for a work-function change of 0.5 eV.

In order to produce field emission from the tip, a high voltage must be applied to the electrode with a hole in front of the tip. Electrons are taken out through the hole, which acts as an electrostatic lens (Sect.2.2.2c). A single electrode corresponds to a concave lens and diverges the electrons; the second electrode shown in Fig.2.13 is necessary in order to converge the electrons. Electrostatic lens action by the electrodes depends upon the tip's radius of curvature, ρ, the distance d_1 between the tip and the first elec-trode, the distance d_2 between the first and the second electrode, and the voltages V_1, V_2 applied to the first and the second electrode, respectively. For example, a convergent lens is formed when $d_2 = 6d_1$, $\rho/2d_1 = 10^{-3}$ and $V_2/V_1 \geq 3$.

Next, we investigate how the tungsten tip is used in practice. A 1 mm piece of tungsten single-crystal wire with (310) crystal orientation is soldered on a tungsten hair pin, as shown in Fig.2.13. The free end of the tungsten wire is immersed in a potassium hydroxide solution and etched to form a tip with a radius of about 0.1 μm by an AC current of about 10 mA. After placing the tip in the gun with a vacuum of 10^{-9} Torr, the tip is heated by a pulsed current

25

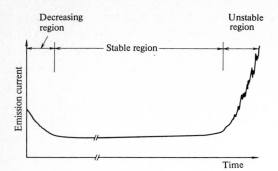

Decreasing region

Unstable region

Stable region

Emission current

Time

Fig. 2.15. Time variation of field-emission current

flowing through the filament under a high voltage at the tip. By this, the tip is cleaned and deformed so that a low work function (310) plane becomes perpendicular to its axis. After confirming normal operation under low applied voltage, normal voltage is applied to the first and the second electrode, and after an initial decrease a stable emission current is obtained (Fig.2.15). The current decrease is attributed to a work-function increase of the tip surface, caused by gas absorption in the operating vacuum. After the decrease, the emission current flows constantly, determined by the current and the vacuum. Operating periods as long as 70 h have been reported [2.26]. After the stable mode the current increases again and becomes unstable. This is due to tip-surface contamination or to surface roughness through ion sputtering during operation. In this situation, operation must be stopped at once, or the tip will be destroyed by arc discharge. If the operation is stopped before the discharge, one can restore it to the former condition by the same way of tip heating as initially. This is called *flashing*.

As mentioned above, the tip of an FE gun emits electrons with high current density ($10^6 - 10^7$ A/cm^2) under the influence of a strong force (10 ton (10 ton/cm^2) due to the high electric field and ion bombardment. Therefore, in order to get a stable current from the FE gun, the vacuum surrounding the tip must be kept high, and the tip should be made of the material which bears the severe operating condition.

Tungsten is one of the best materials because of its (i) high melting point, (ii) low vapor pressure, (iii) high thermal and electrical conductivity, (iv) high mechanical strength, (v) very pure single crystals obtainable, and (vi) moderate and constant work function. Some experiments [2.27] have been reported on LaB$_6$ and carbon, which have a small work function and strong ion-bombardment resistance, but no material except tungsten has been used in practice. When the tip is heated to temperatures for which thermionic emission does not dominate, absorbed gas molecules do not stay long on the tip, and its surface will be kept clean. At such temperatures tungsten atoms on the tip can move by surface tension, thus the rough surface of the tip caused by ion bombardment is smoothed. On the other hand, the electric field acts to move tungsten atoms to the end of the tip against the surface tension, it is, therefore, possible [2.28] to keep the tip sharp and smooth through the reciprocal action of electric field and temperature, when these parameters are well chosen. By this method, a large current can be obtained from the gun with relatively low vacuum environment. It is, however, difficult to find a stable operating condition at present. This is called *thermal field emission*. It was also reported [2.29] that a stable surface layer is formed by a pretreatment of the tip in gases such as O_2, N_2 and so on.

Butler Type Electron Gun. The electrostatic lens which consists of a metal plate with a hole (Fig.2.13) has a large aberration coefficient. This deteriorates some eminent characteristics of the FE gun such as the small size and the high current density. The aberration coefficient becomes large when the electrostatic lens of a hole is strong. As mentioned in Sect.2.2.2c, lens action of the hole becomes stronger as the field-strength difference on both sides of the hole gets larger, and an electrode structure reducing the field strength around the hole makes the aberration smaller. BUTLER [2.30] determined the optimum electrode structure by computer simulation so as to make the field around the hole zero. Figure 2.16 shows the sectional structure of this electrode, and characteristics of the FE gun with this electrode.

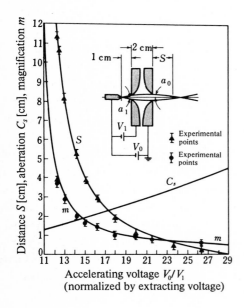

Fig. 2.16. Structure and characteristics of the Butler lens

The inner curves that mark the sections of the two electrodes are such that their curvature becomes infinite at the holes. The curves shown were obtained by calculation for the electrodes shown in Fig.2.16 when the distance between the first electrode and the tip, and the distance between the far sides of the electrodes are 1 cm and 2 cm, respectively. The figure also displays the experimental results obtained for the given electrode. In Fig.2.16, curve m gives the magnification, curve S the distance between the image point and the back side of the second electrode, and curve Cs the spherical aberration. The horizontal axis shows the voltage ratio of the second electrode voltage to the first one. It is seen that the voltage ratio should be small in order to make the spherical aberration small.

If the accelerating voltage V_0 is constant, a large current is obtained by sharpening the tip by remolding rather than increasing V_1, without increasing the spherical aberration. When a strong negative field is applied to the tip at temperatures of 1600~1800 K, a protrusion is formed of the crystal with a specific orientation. This process is called *remolding* [2.18]

Other Electron Guns. For the Butler-type lens mentioned above, the characteristics of the lens as well as the aberrations change with the voltage ratio V_0/V_1 (Fig.2.16). Therefore, V_1 cannot be changed freely to adjust the FE current. The spherical aberration becomes large as V_0/V_1 gets large. In order to reduce this drawback, some experiments have been reported on the application of an Einzel lens [2.31] without a focus change due to the accelerating voltage [2.32] (Sect.2.2.2c), and experiments on the application of both a weak electrostatic lens (with a small aberration coefficient) and a magnetic lens [2.33].

2.2.2 Electron Optical Column

Figure 2.17 illustrates an example of an electron-optical column of an electron-beam (EB) lithography system. The electron-optical column is essentially the same as that of a scanning electron microscope (SEM). The SEM is used to observe a magnified image of a tiny object, while the EB lithography system is used to draw LSI chips with fine patterns fast and precisely. In the EB lithography system, a beam spot radius of 1000 Å is enough, and a large current with long-term stability and a long focal length of the objective lens are necessary to scan a large field without distortion. Figure 2.17 schematically shows an electron gun, the first aperture, the first magnetic lens, a blanking plate, the second aperture, the second magnetic lens, and the third magnetic

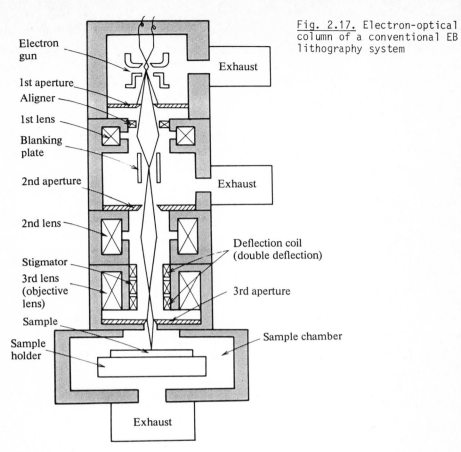

Electron gun

1st aperture
Aligner

1st lens

Blanking plate

2nd aperture

2nd lens

Stigmator

3rd lens (objective lens)

Sample

Sample holder

Exhaust

Exhaust

Exhaust

Deflection coil (double deflection)

3rd aperture

Sample chamber

Fig. 2.17. Electron-optical column of a conventional EB lithography system

lens (an objective lens) with deflection coils, stigmator coils, and the third aperture from the top. This electron-optical column has a sample holder at its bottom, while there are other column systems [2.34] with a sample holder at the top.

In the following, we shall discuss components of the electron-optical column with the exception of the electron gun which was treated in Sect.2.2.1. The first aperture limits the electron beam in entering the column, the second one cuts out the beam deflected by the blanking plates. The third aperture limits the angle of beam incident into the third lens, thus eliminating various aberrations due to the large aperture angle mentioned below. The aperture is made of a metal foil (such as platinum of ~100 µm thickness) with a hole of 100 µm in diameter. The blanking plates consist of two metal plates in parallel; they shut off the electron beam with the help of the second aperture by electrostatically deflecting it.

The essential lens of the electron optical column is not absolutely cylindrically symmetric, thus not all electrons from an ideal point source at
infinity converge to a point even if a lens without aberration (Sect.2.1.3)
is used. This is called *astigmatism*; it is caused by the unsymmetric nature
of the lens and is compensated by a stigmator coil [2.20]. The astigmatism
is usually axially symmetric, an elliptical image is converted into a round
point by a 4-pole stigmator with 4 electromagnets (Fig.2.18). When the long
axis of the image coincides with the x-axis of the stigmator in Fig.2.18,
electrons are suppressed in the x-direction and pulled in the y-direction,
thus resulting in the compensation of the astigmatism. If the stigmator in
Fig.2.18 can revolve up to 45°, it treats all astigmatism. An 8-pole stigmator,
having an additional pole between two poles in Fig.2.18, treats all astigmatism
by changing the exciting current of each pole without revolution. A stigmator
can also be made by using electrostatic devices.

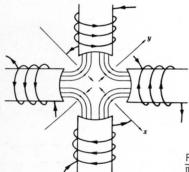

Fig. 2.18. Schematic representation of a stigmator

The EB lithography system writes patterns on a sample by scanning a spot
with beam deflection. The deflection is performed by two pairs of deflection
coils for the x- and y-axes. Two pairs of blanking plates shown in Fig.2.17
are also used for this purpose, placed perpendicular to each other. The electric deflector causes larger focus deviation than a magnetic one [2.35], the
deflection sensitivity of the former is inversely proportional to the accelerating voltage of the electrons (that of the latter is inversely proportional
to the root of the voltage [2.18]), and its performance is influenced by
sticking dust particles. Therefore, precise large-angle deflection is usually
done by a magnetic deflector. The column shown in Fig.2.17 uses two pairs of
deflector coils for each axis; the upper pair deflects the beam out of the
column center and the lower one deflects it back to the lens center and passes
the beam through the third aperture. By this method, the electron beam al-

ways goes through the center of the objective lens in order to suppress increased objective-lens aberration by deflection.

a) Electromagnetic Lens

The EB lithography system uses 2 to 3 electromagnetic lenses (Fig.2.17). Such a lens as the second one, whose sectional view is shown in Fig.2.19, has a structure with cylindrical symmetry with respect to the beam axis (Fig.2.19). The coil is surrounded by a magnetic bypass made of an iron core, through which a magnetic flux flows when the coil is excited by a current. The flux passing through the core is disturbed at the gap of the lens, and thus appears in the pass of the electrons. This flux is cylindrically symmetric, and has both radial and axial components in the plane perpendicular to the column axis (Fig.2.20).

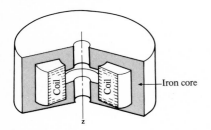

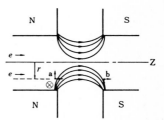

Fig. 2.19. Sectional view of a magnetic lens

Fig. 2.20. Model of flux produced by a magnetic lens

Now, let us consider qualitatively the force acting on the electrons coming to the lens parallel to the axis. The electrons, coming through the center of the column, pass through the lens without any change because the flux in the lens center is parallel to the axis. The electrons, at distance r from the axis (at a in Fig.2.20), are forced to move normally to the paper plane by the r-component of the flux, and begin to turn clockwise to the direction of travel. The r-component of the flux changes its sign at the center of the lens, and the rotational component of the electron becomes zero at the end of the lens. When the electron turns, its movement has a rotational component so that the lens action moves the electron to the axis through its interaction with the axial component of the flux. This action reaches a maximum at the lens center where the electron revolution and the magnetic field take their peak values. After passing through the center of the lens, the electron is influenced by the lens action as long as it turns, and outside the lens it flies straight to the focal point on the axis. The angle of deflection becomes larger as the electron passes the lens farther away from the center.

Thus, the electrons flying near and parallel to the axis converge to a point
(focus of the lens) outside the lens. In this case, the lens acts as a convex lens. The magnetic lens excited by a current of opposite direction still
acts as a convex lens, because the flux in the opposite direction turns the
electron in the opposite direction, thus resulting in convergent-lens action.
Therefore, the magnetic lens never acts as a concave lens [2.20].

The objective lens shown in Fig.2.17, unlike the other lenses, has different diameters above and below the magnetic pole. In this lens, whose action
is the same as that of conventional lenses, the axial magnetic-field distribution near the gap is asymmetric to the gap center; the maximum field position
(lens center) shifts toward the pole with smaller diameter. In the EB lithography system, a column with a long lens-to-sample distance, which eases detector installation, as well as long working distance (distance between the
lens center and the sample plane) is essential in order to draw patterns without distortion. For this purpose, the focal length (f) of the lens should be
long. A lens with long f and small spherical aberration is realized by a strong
excitation of a lens with large pole diameter and pole spacing [2.36]. Such a
large lens takes up large space if the lens is of equal pole diameter. In the
case of a lens with different pole diameters (Fig.2.17), the lens-to-sample
spacing can be widened because the lens center shifts downwards. This lens
has another advantage, namely that its magnetic flux scarcely leaks to the
sample because the pole diameter of this side is small.

b) Electrostatic Lens

An electrostatic lens is formed when the electric field on both sides of a
metal plate with a round hole is different. Its lens action is proportional
to the field difference. As an example, let us consider the case where an
electron is accelerated through the hole of a metal plate placed parallel
to an equipotential plane and positively biased. In this case, the equipotential lines are those shown in Fig.2.21 when we cut the equipotential surfaces by the plane passing through the center of the hole and being perpendicular to the metal plate. An electron moving on the axis from left to
right goes straight, being accelerated by crossing the equipotential lines
perpendicular to it. An electron, moving along a line away from the axis by r,
crosses the equipotential line diagonally at a (Fig.2.21). It turns upwards
by the force perpendicular to the axis. Then, the electron passing through
the hole diverges as if it comes from a virtual focus point f; the lens acts
as a concave lens. When a negative potential is applied to the plate, the hole
acts on electrons in an inverse manner; the electrons converge and the lens

32

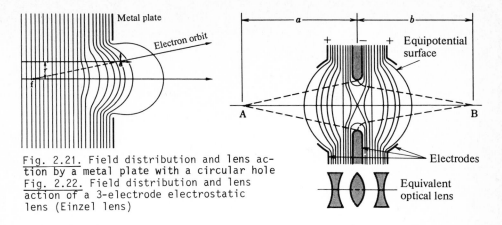

Fig. 2.21. Field distribution and lens action by a metal plate with a circular hole
Fig. 2.22. Field distribution and lens action of a 3-electrode electrostatic lens (Einzel lens)

acts as a convex lens. The magnetic lens only performs as a convex lens (Sect.2.2.2a), whereas the electrostatic lens acts as both a convex and a concave lens according to the polarity of the voltage applied to the metal plate.

Next, we shall describe an einzel lens [2.37] with 3 electrodes, as shown in Fig.2.22. When the same potential is applied to the first and third electrode, and a lower potential to the second electrode, two concave lenses are formed outside the einzel lens, and a convex lens is formed at its center similarly to that shown in Fig.2.21. The einzel lens has the following features: the electron velocity is the same in front and behind the lens because the potential distribution is symmetric to the second electrode, the focal length is kept constant when the potential on the second electrode varies proportional to that on the other electrodes.

The electrostatic lens has a larger aberration than the magnetic lens, and the dust particles attracted onto the lens electrode by the strong field in the lens prevent a reproducible operation of the lens. Therefore, the electrostatic lens is hardly used in an electron-optical column. The special application to the FE gun is mentioned in Sect.2.2.1.

c) Demagnifying Electron-Optical Column

The crossover size of the thermal electron emission gun is around several 10 μm (Sect.2.2.1), the crossover should be demagnified by a lens to get a spot size of ~0.1 μm needed in an electron-beam lithography system. The demagnification is usually done by a demagnifying electron-optical column with 3 magnetic lenses, as shown in Fig.2.17; the demagnification is several hundredfold. A lens for the demagnifying column with a demagnification of 1/27 is shown in

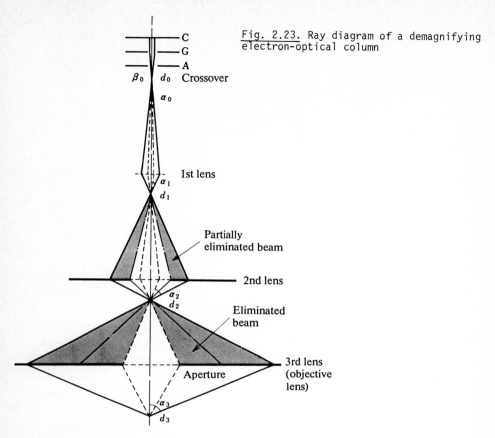

Fig. 2.23. Ray diagram of a demagnifying electron-optical column

Fig.2.23, where β_0, d_0 and α_0 are brightness, diameter of crossover, and semi-divergence angle, respectively; α_1 and d_1, α_2 and d_2, α_3 and d_3 are the semi-convergence angle and image diameter of the first, second and third lens, respectively. In the Helmholtz-Lagrange theory of optics [2.19]

$$d_0 \alpha_0 \sqrt{V_0} = d_3 \alpha_3 \sqrt{V_3} \quad . \tag{2.12}$$

V_0 and V_3 are the potentials of the cathode at the crossover point and the sample plane, respectively; they are usually equal. From (2.12), the angle is proportional to the demagnification factor. Aberration is most important in the third lens, while it is less important in the other lenses because the aberration increases with α, as shown in Fig.2.5; when the beam is limited at the third lens by its aperture in order to keep the aberration below the allowable maximum, the beam with large convergence angle is eliminated at the third lens, and the convergence angle for the first and second lens with request to the available beam becomes very small.

d) Magnifying Electron-Optical Column

The FE gun mentioned in Sect.2.2.1c has a source diameter of ~ 100 Å, which is smaller than that needed by an EB lithography system, the electron source is usually magnified by an order of magnitude to draw large current from the gun. A Scanning Electron Microscope (SEM) usually uses a spot diameter less than 100 Å, the magnifying electron-optical column is peculiar to the EB lithography system.

Contrary to the discussion done with Fig.2.23, in a magnifying optical column, the semi-convergence angle is the larger the closer the lens is to the gun. In the column shown in Fig.2.24, the first and second lens (objective lens) provide equal contribution to the aberration because the aberration of the first lens could be smaller than that of the second one by strong excitation. Moreover, electrons are extracted from the gun by a large semi-divergence angle (Fig.2.24), the aberration of the static lens in the gun strongly affects the characteristics of the column. Therefore, one should design all the lenses to have equal contribution to the aberration of the column. The column in Fig.2.24, in contrast to that of Fig.2.23, uses a weakly excited (long focal length) objective lens, which could be omitted because the working distance can be long in a magnifying lens. The electrons scattered onto the sample are few in the column in Fig.2.24, as stray electrons are easily removed by the aperture at the first lens.

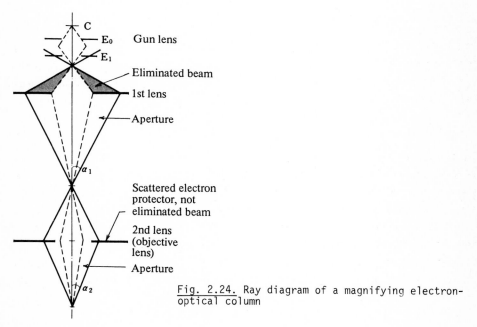

C

E_0 Gun lens

E_1

Eliminated beam

1st lens

Aperture

α_1

Scattered electron protector, not eliminated beam

2nd lens (objective lens)

Aperture

α_2

Fig. 2.24. Ray diagram of a magnifying electron-optical column

2.2.3 Electron-Beam Delineation

In a TV set, various figures or characters appear on the fluorescent screen
of a Braun tube through scanning it with an intensity-modulated focused electron beam. In the case of delineating IC-device patterns, the latent images
are formed in the photoresist through basically the same process of controlling
an electron beam like in a TV receiver. However, both size and location of all
figures must be controllable in any layer and at any place of a wafer to make
the device work as designed (Sect.2.2.4). From the IC manufacturer's cost point
of view a higher throughput rate is required for the direct exposure-on-the-
wafer system, which is comparable to a conventional UV replicator (within few
minutes per wafer). The rate requirement by the system for mask or reticle
making is not so critical because of its off-line process.

For the delineation of a simple figure, the method used is to scan out the
inside of the pattern area with a small round spot of a focused beam (Fig.2.1a).
To produce an apparently linear edge figure with a series of spots it is known
by experience that at least four beam spots must be in line along the edge of
the pattern. In such a machine the beam moves either linearly parallel to one
direction, or in two directions or in a vortex for scanning to make a comparatively small simple figure. The continuous scan along the fringe of a figure
is good for getting a sharply exposed pattern.

The minimum spot size possible can be about several ten $\mathring{A}$. For delineating
VLSI patterns at an adequate rate, a size between 500 and 2500 $\mathring{A}$ for the half-
value width of the beam profile is adopted to obtain accurately exposed device
patterns up to about 1 μm on the photoresist of several thousand $\mathring{A}$ to 1 μm
thick, or larger for making a reticle. When a beam of relatively large current
density is focused, the sharpness of the spot is degraded by the space-charge
effect and the Boersch effect (Sect.2.2.2), as well as by various aberrations of the optics. These effects can be reduced through shortening the
work distance between lens and wafer, and widening the semi-angle of the
beam, which may, however, increase blur and distortion on deflection. Consequently, the work distance or focal length of the projection lens needs
to be optimized.

To prevent large deformation of a figure due to the intraproximity effect
which occurs on close exposures inside the figure, the small spot beam can be
controlled finely either in the scanning speed or in current density (Sec.2.3).

In a practical electron-beam delineator the pattern design [2.38] as shown
in Fig.2.25 from which each subpattern with a proper scan according to the
figure and size can be selected, or self-scanning within the pattern area
providing only an input to determine beam-on and -off points [2.39] is effec-

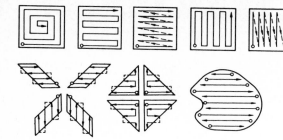

Fig. 2.25. Pattern menu for small-spot scanning

o	Starting point of exposure
----	Beam spot path under blanking
——	Beam spot path under blanking-off (exposing)

tively adopted to increase the delineation rate by eliminating excessive control and transmission data.

Most IC device patterns are constructed by combining rectangles. There is an exposure system called the fixed-size-shaped electron-beam exposure system [2.40]. In this system the projected spot of the beam image is shaped into a regular square, and the pattern is divided into a combination of the squares for mosaic-like exposure (Fig.2.1b). The 16 exposures to form the smallest square by round beam spots are replaced by one in this system. However, the current density of the shaped beam cannot be so high as that of the small spot, because the beam radiated from the cathode chip is widened to cut out the fringe with a shaping aperture. This keeps the current-density distribution uniform inside the square spot; however, the density decreases, which lengthens the exposure time for a shot. As a result, the throughput rate cannot be reduced in proportion with the decrease of the number of exposure shots.

Regarding the range of the variably-shaped beam size [2.40], the wider is not always the better, because the current distribution in the beam spot image must be flat even in the largest square (Fig.2.1c). The coverage of the beam size must be properly chosen [2.41].

Extending the beam-shaping technology, the character projection of unit figures is effective in cases where many repitious exposures are required like the chevron or T-I bars in bubble devices [2.42,43]. In such a system a figure selected through deflecting the beam to one of the unit-figure apertures provided around the axis of the beam path is projected onto the surface of a work piece at a reduced size.

To obtain sharp figures, it is necessary to properly design and manufacture the electro-optical system. In addition, the inside of the column must be kept clean. The inside is likely contaminated during exposure to the air for loading or unloading the work pieces and by chemical reaction of the photo-polymer excited by the beam. Non-conductive contaminants on the column components, especially those which have contact with the beam such as the shaping apertures, may cause harmful disturbance of the electric field by accumulation of charges.

a) Delineation of Spatially Separated Figures

In the case of many figures spread over an area, or in the case of a large or complex figure which cannot be exposed with a simple scan of a small beam spot or with a single shot of a shaped beam, the electron beam must be deflected widely to expose all figures. The deflection is done by a raster scan over the field even though there does not exist a figure like in TV, or by hopping from one figure to the other. The latter is called the vector scan (Fig.2.2) [2.44].

In order to deflect the beam to a given location, the input of digital data is changed to analog values by a D-A converter which supplies the deflection voltage or the deflection current. To determine the location of the beam spot, the data can be transferred through the same controller to scan the inside of the figure by a small spot machine, while a separate control path for a shaped beam or a machine with pattern designs is used. The accuracy of the location is determined by the resolution of the D-A converter, the full range of which is almost equal to the size of the deflection field. Since the resolution and the work speed of the D-A converter are incompatible, an optimization is needed: usually a 12 - 16 bit D-A converter is employed in consequence (resolution: 1/4096 - 1/65536 of the deflection field).

When the beam deflection is large, the spot on the surface of a work piece gets blurred and its location becomes less defined. The former may be minimized by a dynamic angle-dependent correction [2.14]. The latter can practically be compensated during exposure by previously measured distortion values at every site in the scanning field. Even with such compensations and permitting a residual distortion of some 0.2 μm, the maximum size of a scanning field may be restricted to 0.2 - 2 mm (up to about 5 mm when requirements on blur etc. are somewhat relaxed) because of increased aberration. Therefore, the size of most IC chips exceeds the field size.

In the vector scan machine, the beam spot flies to another, spatially separated figure under blanking after exposure. At this moment, an oscillation or a time delay often occurs in the control system, which requires proper design of the time constant of the deflection system and the cable length. The gitter which appears in the D-A converter under such a transient condition

causes disturbance of the figure. It is suggested that the exposure begins
when such a phenomenon has died away or is decreased to an allowable level,
even in a high-speed delineator equipped with a transient suppression device.
In magnetic deflection hysteresis of the component materials surrounding the
beam path may be responsible for losing the location. This can be harmless
when the beam spot scans the same path without a figure, i.e., in a raster
scan over the area [2.45].

b) Delineation of Figures Over the Surface of a Large Work Piece
Refined manufacturing has recently led to larger diameters of the Si wafer
(100 - 125 mm). To spread device figures over the surface of such work pieces,
they must be moved mechanically to traverse the exposure area which is restricted by the limited scanning-field size. The motion can be accomplished
as follows:

(i) The stage moves continuously in one direction, as the electron beam
scans vertically to the motion. When it reaches an edge, the stage shifts to
the next stripe and moves back along the stripe with beam scan. This is called
"continuous stage motion" [2.46] (Fig.2.3b).

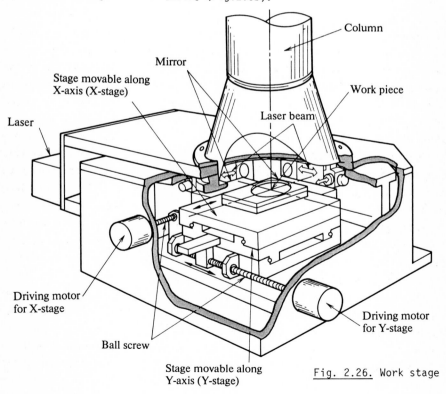

Fig. 2.26. Work stage

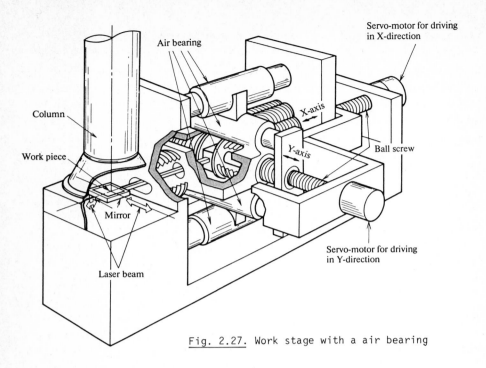

Fig. 2.27. Work stage with a air bearing

(ii) The stage stops during exposure, and then moves rapidly to the next field for exposure. This is called "step-and-repeat stage motion" (Fig.2.3a).

Any shape of the beam spot is applicable to these concepts. In the continuous-motion system the requirements on the electron optics and its control is remarkably simpler because the deflection is only in one direction, and its range is as narrow as 125 - 250 μm [2.46].

The work stage consists essentially of two platform. Each platform is moved by a high precision ball-screw with a pulse-motor or a servo-motor along a precisely straight guide-rail suppressing rolling or pitching motion to within several μm (Fig.2.26). The air-bearing suspension [2.46] makes the platform motion very smooth by keeping the motion in extremely high stiffness to the vertical force (Fig.2.27). In order to adjust the control-loop characteristics so that quick start-and-stop response and highly accurate stop positions are realized, free from the external low-frequency vibration by air suspension, the stage must be constructed with the light-weight and high-stiffness material for obtaining high eigenfrequencies.

Even with such systems, the accuracy of the stage location is not sufficient to butt fine patterns between the scanning fields and superposing patterns on layers. The stage location error measured by a laser interferometer [2.16]

40

is utilized to the offset of the electron beam scanning [2.47] and can thus
be compensated within 0.2 - 0.3 μm. Preceding this process the location scale
of the deflecting beam must be corrected by reading the locations of the stage
marks simultaneously with the laser and the electron beam. For this purpose,
it is important that temperature changes or vibrations do not affect the
relative position between electron-optical column and stage by diverging
from the common origin. To detect the previously set marks in direct writing,
more than three are required on a chip or in a field of the electron beam to
determine the axes and the scales. The patterns of each field or of each layer
are delineated with small butting or overlay error. This method is very effective when overlaying the patterns formed by separate machines or when
patterning on a distorted wafer by thermal processing (Sect.2.2.4).

When delineating IC-device patterns on a wafer, the above-mentioned processes concerning the pattern itself, scanning of the field and the wafer
level are executed according to the flow chart shown in Figs.2.28 and 29.
The data of the pattern, which converted to the delineator format, are received on magnetic tape (MT), introduced into the delineation program according to the schedular, and stored on a magnetic disc (DSK). These data are
transmitted to the driving unit of the electron-beam machine. In some machines
the transmitted data are divided into groups so that the data for one field
or one stripe can be stored in a high-speed IC buffer memory. The memory content is converted to another field or another stripe data during the stage
motion; this cuts the idle time of delineation. In the shaped-beam machine,
as processing of the data for a pattern is done simultaneously in one shot,
the serial input data must be converted into parallel input data in the
registers.

The input data must include at least the location of the pattern (the sum
of the X and Y coordinates is around 32 bit). the size of the pattern (the
sum of both sides is around 20 bit), blanking (about 2 bit) and the selection
of the control items (when selecting 1 from 6 items around 10 bit). Then, one
unit square pattern requires 60 - 70 bit. Therefore, the sum of the number of
unit squares for VLSI patterns is supposed to be 10^8, it requires $10^9 - 10^{10}$
bit data. This is the reason why the compression and parallel processing of
the data should be introduced into the control computer of the electron-beam
delineator although it is equipped with a high-speed large-capacity memory
like a usual large computer. Data compression is performed through reducing
the word length or utilizing the regularity of the delineation sequence or
repeating the same device pattern in a chip and then repeating the same chip
pattern on a wafer. Compensation of the field distortion or the beam location

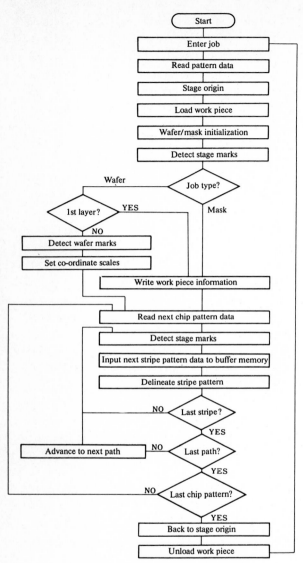

Fig. 2.28. Flow chart for continuous stage motion

can be executed locally so that such data need not pass through the main data bus.

On the other hand, most of the data formats for patterns are standardized so that conventional X-Y plotters or optical pattern generators at hand can be utilized in the design phase. Thus, the data must be rearranged into the format and word order suitable to the electron-beam delineator. Some specific

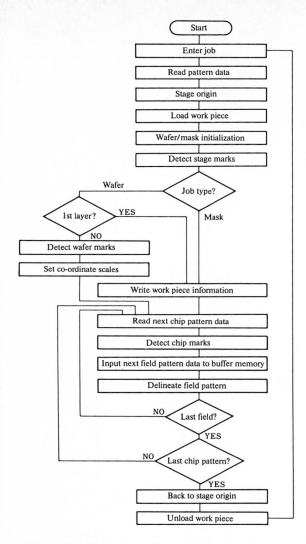

Fig. 2.29. Flow chart for step-and-repeat stage motion

problems still need to be solved, like the multiple exposure hazard in converting data of an aperture projection system into the optical exposure. The amount of the data to be processed in a short time is so extensive that a large computer is required. Therefore, the electron-beam delineator should be designed with a system architecture suitable to the particular IC manufacturing process by use of the matched software system.

2.2.4 Alignment

In electron-beam writing on wafers, which differs from mask-making, high precision alignment is necessary at each writing level. After writing the first-level patterns at the standard position and processing the wafer, the wafer cannot be positioned again at absolutely the same place as it was before. Therefore, the second-level pattern can be written only with due regard to compensation for position, angle and deflection amplitude discrepancies.

Alignment accuracy should be less than a few tenths of the minimum pattern line width. For VLSI patterns, in which the minimum pattern line width is 0.5 - 1.0 µm, 0.1 - 0.2 µm alignment accuracy is needed.

Usually, the following procedures are taken to realize the necessary accuracy. The electron-beam scan around alignment marks the wafer, and the mark position shifts are decided upon in comparison with standard mark positions. Then, the pattern writing positions are modulated according to these shifts.

These procedures are suited to the electron-beam delineator, because a high mark detection accuracy is possible, and the position-deciding tool is the same electron beam as the pattern-writing tool.

For highly precise alignment, the alignment-mark structure, the mark-position-deciding method and the writing-position compensation method must be optimized. These are discussed in the following.

a) Alignment Mark Structure

The alignment marks cannot be damaged during the wafer manufacturing process like etching or chemical vapour deposition (CVD), to obtain a sufficient signal-to-noise (SN) ratio for the position-deciding signal.

Silicon steps, silicon-dioxide steps, V-shaped grooves and metal have been used as the alignment mark. Each of these marks has particular advantages and should be selected according to the wafer process and the wafer surface direction involved. Characteristics of these marks are listed in Table 2.3.

These marks have been determined on the basis of experiments and simulations to find optimum shape and dimensions. Simulation results for silicon steps, as a function of step height and inclination angle, are shown in Fig.2.30 and those for V-shaped grooves as a function of groove depth in Fig.2.31. These results reveal that the step height or the grooved depth should be greater than 2 µm.

b) Mark Position Deciding Method

Among the various kinds of signals obtained by electron-beam scanning around the alignment mark, four different kinds of current have been studied:

Table 2.3. Alignment mark comparison

	Preparation	Damage in Processes	Other Factors
Silicon steps	easy by ordinary etchant	somewhat easily damaged	
Silicon-dioxide step	easy by ordinary etchant	easily damaged	
V-shaped groove	special etchant is necessary	somewhat easily damaged	highly precise pattern
Metal	metal species are limited	hard to damage	large S/N ratio

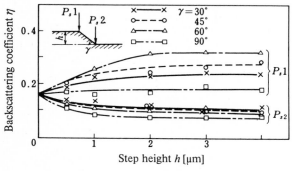

Fig. 2.30. Calculated backscattering-coefficient dependence on step height and inclination angle of the mark

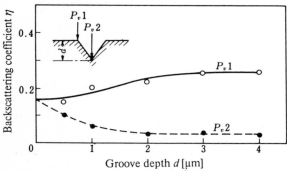

Fig. 2.31. Calculated backscattering-coefficient dependence on V-shaped groove depth

(i) backscattered electron current; (ii) secondary electron current; (iii) absorbed electron current; (iv) pn junction voltaic current.

Backscattered electron current and secondary electron current are frequently used. An electrical attachment to the wafer is needed in the case of absorbed electron current or pn junction voltaic current. Moreover, the pn junction must have been formed in the wafer to utilize pn junction voltaic current.

In the case of backscattered or secondary electron current, pn junction detectors or scintillation detectors are used. When pn junction detectors are employed, the amplification factor for high-energy electrons is large and setting position restrictions are small. When scintillation detectors are used, a large amplification factor can be easily obtained with a photomultiplier. Comparing backscattered and secondary electron currents for a V-shaped groove mark, the SN ratio for backscattered electron current becomes larger and that for secondary electron current smaller, when the incident electron energy varies from 10 to 20 keV. If the secondary electron current is utilized, there are disadvantages in that the signal current is likely varied by sample contamination or resist coating, and that the writing electron beam position is affected by the accelerating electric field of secondary electrons.

For these reasons, the backscattered electron current has most frequently been utilized up to now. A typical sequence in processing the obtained signal and deciding the mark position is shown in Fig.2.32. The detected signal results from the backscattered electron current obtained by electron beam scanning on the step-type mark. After being amplified, limited for noise reduction, shaped to standard pulse size and passed through the AND circuit together with the sampling pulse, the signal is used to calculate the mark position.

Besides these analog-signal processing method, there is one digital processing method using a waveform memory circuit. To improve the detection accuracy, an averaging method for forward and backward scans or multiple scans is frequently applied.

The analog processing method has the advantage of high-speed processing, and the digital processing method has an advantage in that the averaging can be done at the first processing level.

The procedure to decide on the center position for an inclined L-shaped mark is as follows. Several line scans are made in the X and Y directions, and the most approximate lines for each arm of the mark are decided upon.

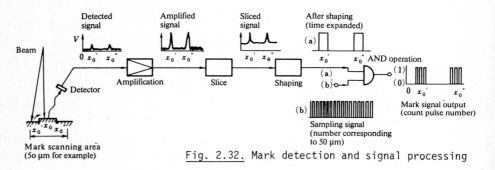

Fig. 2.32. Mark detection and signal processing

The relative mark position is calculated as the crossing point of these lines and the absolute mark position is determined by addition of the table position measured by a laser interferometer. Usually, the accuracy of the thus calculated mark position is as small as or smaller than 0.05 μm; this is sufficient for alignment.

The compensation angle should be as small as possible, to fully utilize the electro-optical system. Therefore, the wafer must be prealigned within a certain small angle deviation from the standard angle. The prealigned angle tolerance is 10^{-4} to 10^{-3} rad, depending upon the system used. Generally speaking, raster-scan systems have a small tolerance because of their delineation characteristics.

In practice, the wafer prealignment marks are prepared and the wafer-to-wafer cassette angle is prealigned with an optical microscope. After setting the wafer cassette on the moving table, at first, wafer marks and then the alignment marks for each die or a certain area are searched for.

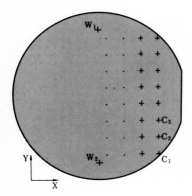

Fig. 2.33. Layout example for wafer marks (W1, W2) and die marks (C1,C2, ---)

For example, in the case of Fig.2.33, where the standard positions for the first wafer mark W1, the second wafer mark W2, and one of the die marks C1 are (X_{W1}, Y_{W1}), (X_{W2}, Y_{W2}) and (X_{C1}, Y_{C1}) and the measured positions for the wafer marks are (X'_{W1}, Y'_{W1}) and (X'_{W2}, Y'_{W2}), the die mark C1 displacement $(\Delta X_{C1}, \Delta Y_{C1})$ is given by

$$\Delta X_{C1} = (X'_{W2} - X_{W2}) - d(Y_{C1} - Y_{W2})$$
$$\Delta Y_{C1} = (Y'_{W2} - Y_{W2}) + d(X_{C1} - X_{W2}) \tag{2.13}$$

with $d = -(X'_{W1} - X'_{W2})/(Y'_{W1} - Y'_{W2})$. Consequently, the actual C1 position is given by

$$X'_{C1} = X_{C1} + \Delta X_{C1} \qquad Y'_{C1} = Y_{C1} + \Delta Y_{C1} \quad . \tag{2.14}$$

Using these equations, the compensated die mark positions are determined and the compensated writing positions can be selected upon.

c) Writing Position Compensation Method

A suitable compensation method for the X-Y displacement, the setting-angle change and the lateral distortion in the wafer, determined by mark detection, depends on whether the delineating method is a vector-scan method or a raster-scan method. For the vector-scan method, the writing position compensation can be accomplished similarly to the deflection-distortion correction. Actually, there are only higher orders of deflection distortion. Therefore, the first-order term (displacement, angle and deflection amplitude) and the second-order term (trapezoid) of distortion must, in addition, be given for the writing-position compensation. On the other hand, for the raster-scan method, the writing deflection is essentially restricted to the extent that the deflection distortion can be neglected. Therefore, there is usually no such hardware for higher-order deflection distortion as in the vector-scan method. Instead, a specific compensation method is devised to utilize such raster-scan method characteristics for which the writing scan is accomplished as the writing area stripe.

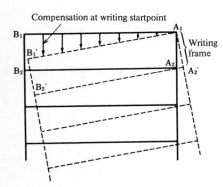

Fig. 2.34. Writing position compensation method for a raster scan system

As in Fig.2.34, the starting-point for writing is changed according to the angle compensation value. In the next writing frame, the first starting point is determined by the software, and the following starting point for each writing line is gradually displaced by hardware. The displacement value for this starting point can be set as a certain value for each certain small area. Therefore, the lateral distortion of the wafer, which might take place during high-temperature treatment, can be compensated for successfully. (Wafer warpage correction will be treated in Sect.2.3.4).

The time used for the above-mentioned alignment depends on the system and is several hundred milliseconds per die for a typical vector-scan system or a few minutes per wafer for a typical raster-scan system.

d) Comments

To increase the SN ratio of the mark-scanning signal, the shaped beam is used with advantage. To apply a beam elongated in the direction perpendicular to the scanning direction means the addition of a signal at the scanning elec-tron-beam level; it results in high accuracy and shortening of the mark-scanning time.

The signal SN ratio is influenced by the processed wafer structure, such as the coated resist profile or the CVD layer profile. The mark structure and the signal processing method should be optimized according to the specific wafer structure processed. The SN ratio for the mark-scan signal increases proportionally to the electron-beam intensity. However, excessive beam in-tensity must be carefully avoided, since it causes problems, such as resist melting or evaporation.

When the same registration mark is used repeatedly in different wafer pro-cesses, it must be examined for mark demage or deformation.

Usually, multiple detectors are set up, and the signals by these detec-tors are utilized in sum or difference form, depending on the mark structure.

The above-mentioned alignment techniques represent only fundamental prin-ciples. In actual equipment, the electron-beam position stability and the wafer cassette position stability are very important problems. Especially in the raster-scan method, high stability is needed because the specimen table is always moving at high speed.

2.2.5 Radiation Damage

High-energy electron beams, like other kinds of ionizing radiation, can affect the molecular weight of polymers in two ways. They can increase it by linking molecules together (cross-linking) or they can decrease it by inducing main-chain degradation. Thus, chemical properties, such as solubility, of the ir-radiated polymer change thus making the electron-beam lithography possible.

Electron-beam lithography can be applied to maskless patterning directly on a resist coated wafer, as well as to making, for example, X-ray lithography masks. As far as the latter is concerned, the radiation-damage effects are negligible. In the former case, however, those device parameters must be taken into consideration. The electrons which have past through the resist to reach the silicon substrate and/or surface films such as SiO_2, often have sufficient

energy to induce radiation damage in them. Resolution considerations often
lead to appropriate incident energy levels.

The nature and extent of damage in the silicon dioxide and silicon itself
depend on the nature of the ionizing radiation employed. In the case of electrons with energy up to several hundred keV, radiation damage is primarily
caused by (i) excitation and ionization, and to a lesser extent by (ii)
electron nucleus collision that causes atom displacements (lattice defects)
[2.48].

Ionizing effects in oxide films are important from the device view-point.
Some of the major effects, summarized by SNOW et al. [2.49], are shown in
Table 2.4. The space charge developed during irradiation is positive and can
be neutralized by thermal treatment at temperatures below 400°C, or by photo-
electron injection from metals or silicon. The fast surface states can be annealed thermally in the same temperature range, too.

Table 2.4. Summary of the effects of the radiation-induced changes in the oxide on device parameters.
Serious effects are shown in capitals

Phenomenon \ Device	p-n junction diode	bipolar transistor	MOS transistor	junction field-effect transistor
Space charge build-up in the oxide	REVERSE CURRENT DEGRADATION	REVERSE CURRENT DEGRADATION	REVERSE CURRENT DEGRADATION	REVERSE CURRENT DEGRADATION
	breakdown voltage change	breakdown voltage change	breakdown voltage change	breakdown voltage change
		h_{FE} DECREASE AT LOW-CURRENT LEVELS	THRESHOLD VOLTAGE CHANGE	
Fast surface states creation at the Si-SiO$_2$ interface	reverse current increase	reverse current increase	reverse current increase	reverse current increase
		h_{FE} DECREASE AT LOW-CURRENT LEVELS	g_m decrease	

These, however, are not all of the ionization effects in the oxide. Recent,
more sophisticated studies, prompted by the increasing use of ionizing radiation in device fabriaction (electron-beam lithography, reactive-ion etching,
ion implantation, etc.) [2.50], show that it also generates damage associated
with neutral electron traps [2.51-53]. These traps have electron-capture cross-
sections of 10^{-15} to 10^{-18} cm^2 and densities of 10^{11} to 10^{12} cm^{-2}. Even at
500°C, a 30-minute annealing does not reduce the trap density to the level of
the unirradiated case.

The nature of the neutral damage is unknown. However, compaction measurements show that structural change is created by electron energy deposited

in ionization [2.54]. The change has been attributed to broken bonds in SiO_2 tetrahedra, and anneals in a well-defined stage centered at 650°C; this suggests that they may be the same defects as the neutral trapping sites.

Effects of the electron-beam damage in an earlier stage of device fabrication is less important. They can easily be removed by a subsequent high-temperature thermal treatment. On the other hand, the damage produced in the final metallization process is more disastrous because it is hardly removed by available temperature treatment. One of the more critical cases is aluminum metallization. A heat treatment of about 550°C to anneal out the neutral traps readily causes aluminum to react with the oxide and the silicon, thus failing the device. RF plasma annealing and other means are being tried to anneal out the damage at a lower temperature.

Collision with atoms forming the lattice is another form of high-energy electron interaction with crystalline solids. In this case, if sufficient energy is imparted onto the colliding atom, it is displaced (recoiled) to a neighbouring interstitial site, leaving a vacant lattice site (vacancy) behind. This type of event is called a displacement event, and the interstitial atom-vacancy pair (Frenkel defect) is the simplest form of the lattice defects produced by irradiation of ionizing radiation. Such simple defects and some of the more complicated ones, e.g., divacancy, can be produced directly in the damage process. These defects, at an appropriate temperature, can migrate and form more complex defects with other defects or with impurity atoms. Their physical properties and effects on device parameters have been studied extensively [2.55-57].

It should be noted, however, that electrons with sufficiently high energy (200 to 300 keV for silicon) produce recoil atoms only. Thus, in a conventional electron-beam lithography system with the energy of up to several tens of keV's, the effects by such defects can be neglected. They are more important in ion implantation, where the mass of incident particles is comparable to that of the target material [2.58].

2.3 Software for Electron-Beam Lithography

The production sequence for an LSI chip is shown in Fig.2.35. The software prepares data necessary for electron-beam lithography to produce the mask or to expose an electron-beam pattern directly on the wafer.

Figure 2.36 illustrates the design of a software system. The data input process includes the development of multi-layer structure design data into single-layer data for each separated layer level and the transformation of

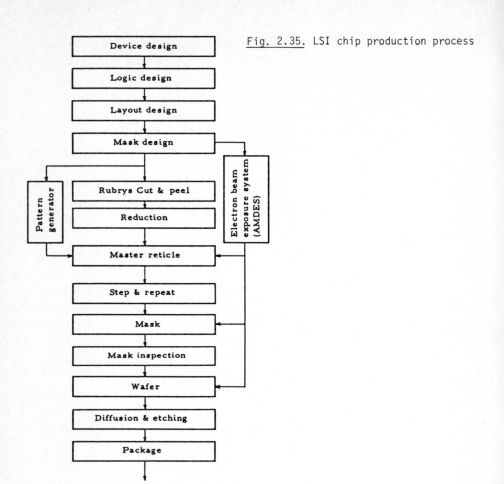

Fig. 2.35. LSI chip production process

Device design
Logic design
Layout design
Mask design
Rubrys Cut & peel
Reduction
Master reticle
Step & repeat
Mask
Mask inspection
Wafer
Diffusion & etching
Package

Pattern generator

Electron beam exposure system (AMDES)

circular data into approximated polygon data. The region partitioning process
divides the design region into subdivisions. All the design patterns are
grouped according to the subregion to which they belong, and partitioned into
more than one subpattern if necessary; the group codes are given. Each pattern
is indexed. These indices are based upon pattern location and size. Patterns
grouped into subregions and indexed are collected into a figure master file
to be sorted by the group code and indices.

The flow path is then separated into two processes, positive pattern processing and negative pattern processing, the appropriate one is chosen by
the user.

Both pattern processings include elimination of multiple exposure by eliminating overlapping patterns, or their complement areas in the case of negative

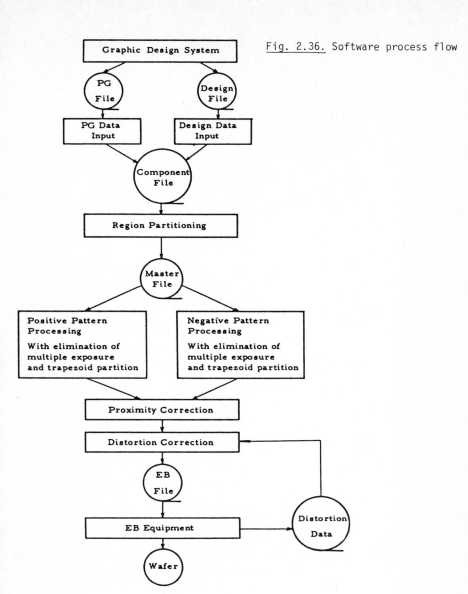

Fig. 2.36. Software process flow

processing, and by partitioning of the resulting patterns into trapezoidal sub-patterns to fit the input pattern format of the electron-beam equipment.

Output patterns of the pattern processing subsystem are then submitted to error corrections if the user feels it is necessary. These are proximity-effect correction, distortion correction and wafer-warpage correction. The proximity-effect correction is accomplished by controlling each pattern's exposure in-

53

tensity. The distortion correction is done by properly deforming the shape of input patterns.

The amount of distortion is obtained by measuring the difference between test patterns beforehand and the superimposed pattern emerging from the electron-beam column. The measuring data are stored on magnetic tape and utilized later to match the distortion function in fitting parameter calculation.

Wafer warpage is caused by heat-cylce processing. Together with the increase in wafer diameter, the warpage resulting from each heat-processing cycle produces pattern distortions which cannot be ignored. It is necessary to select the number of registration marks needed and their locations on the warpaged wafer so as to satisfy the required precision specifications.

The electron-beam file contains input patterns to the electron-beam equipment thus processed, together with additional information to control the electron-beam property and the move of the wafer stage. A typical hardware construction for data processing is illustrated in Fig.2.37 [2.59]. Two electron-beam exposure equipments are used here. One is the JEOL-5AR, manufactured by JEOL Ltd, and the other is the EBMF2, manufactured by Cambridge Instruments. Both are operated in the vector-scan mode.

These equipments are connected on-line with the 50-K band data communication line via a master-slave link (MSL) located at both ends, to the mask pattern designing equipment (based on the AGS 860-V system manufactured by Applicon, Inc.). This provides interactive input and a modification function for the design engineer of the mask patterns.

A disk storage device, with 100-M word capacity, is attached to the mask pattern designing equipment as a pattern data bulk storage. A CRT display unit operating in the raster-scan mode provides a bulky pattern data rapid display method.

The software package used on a large-scale digital computer (ACOS 700 system)[2] includes those functions for removing overlapping patterns, distortion correction, proximity-effect correction, and wafer-warpage correction, to perform high-speed and high-accuracy processing of a large volume of data. The computer is connected to the designing equipment with a data communication line to help efficient transfer of design data among component equipments.

---

[2] developed by Nippon Electric Co., Ltd.

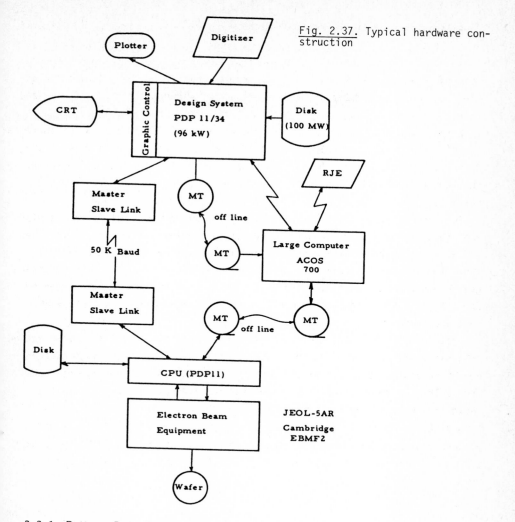

Fig. 2.37. Typical hardware construction

2.3.1 Pattern Data Processing

a) Input Files

The mask-pattern design-data file is generally produced by an LSI interactive
graphic system. Fundamental patterns given by the design engineer are em-
bedded in the hierarchical design-data structure and are expanded to their
proper locations within the design field by specified mapping and transformation information. The fundamental pattern data include information on the
sandwich layer the pattern occupies. Thus, transformation into single-layer
data is necessary. The following fundamental design patterns are permitted

(data specification includes level data):

- Fixed rectangle: rectangle with fixed size, parallel to the X or Y axis.
- Extendable rectangle: rectangle with edge(s) extendable by mapping into higher-level data structures, parallel to the X or Y axis.
- Polygon: arbitrarily shaped polygon with straight edges.
- Path: represents a circuit line, node points on the line path are specified together with line width.
- Arc: a circle or a sector

Among these, the first two can be specified at atomic level, the others as primitives in the design data hierarchial structure described in the following:

- Atomic level: lowest structure level. The atomic level consists of a single rectangle, either fixed or extendable.
- Primitive level: either a collected group of atomic levels or any one of the latter three patterns cited above. Patterns included in a primitive level are referred to by referencing the name of the primitive level. The primitive level has a pair of reference points which are to be used for transforming during expansion.
- Cell level: collection of primitive levels. Specified edges of expandable rectangles, included in a primitive level, are moved such that two reference points specified at the primitive level are mapped into two reference points specified in the cell. Rotation and reflection may be performed at the same time. Pattern data included in the cell are referred to by referencing the name of the cell.

The data are expanded into the design region by referencing primitive levels and/or cell levels, thus defined by their name and specifying the necessary transformation. Primitive levels and/or cell levels can be transformed with respect to reference points; rotation and reflection are also applicable.

Design data files are constructed as follows:

- Directory file: specifies identifying information, scaling information, the names of the cells and other data.
- Primitive file: specifies the names of the primitive levels, component atomic level information and reference points.
- Cell file: specifies a cell level name, information concerning reference to and constituent primitive-level transformation together with cell-level reference points. One cell level file is created for each cell level.
- Drawing file: specifies information regarding reference and transformation of primitive level and/or cell levels to be actually mapped into the design region.

Design data file processing is divided into two phases:

- Phase 1: analysing component patterns specified by the design-data structure. Component pattern coordinate data for each primitive level and cell
level are arranged into intermediate files.
- Phase 2: analysing the drawing file, specified data expansion into the
design region and coordinate data collection into the component file, which
is used in subsequent procedures.

The data processing system AMDES [2.60] developed to draw submicron patterns for integrated circuits is compatible with existing data systems designed for conventional optical-exposure equipment.

Source-data files generated by the PG 3000 system are acceptable to this
system.

PG 3000 source data specify only rectangle patterns. A rectangle pattern
is defined by specifying the center point, the X and Y coordinates, the width
and height of the rectangle, and the rotation angle with respect to the X
axis in the counter-clockwise direction.

Arbitrarily shaped polygon data have to be defined by a collection of
rectangles, connected together and overlapping. PG file processing includes
decoding specified rectangle parameters and converting them into node-point
coordinate data. The converted information is collected in the component file,
which serves as an input file to the subsequent process.

b) Fundamental Pattern Data Processing Operations

The main purpose of AMDES is to convert design-pattern information for manufacturing VLSI chips into information compatible with electron-beam exposure
equipment. To accomplish this, the system must be able to manipulate several
kinds of pattern data.

Fundamental patterns used in the design phase are rectangles, polygons,
connecting lines or paths, circles and sectors. AMDES handles pattern data
expressed as "simple polygons" in the computer memory. A simple polygon is defined as a polygon with a limited number of nodes; straight edges do not cross
or touch other edges.

Finally, pattern shapes acceptable to the electron-beam exposing equipment
are limited, in the JEOL-5AR system, to triangles or trapezoids, with lateral
edge(s) parallel to the X axis. The following are pattern-data manipulation
categories that AMDES performs:
- Replacing design data with equivalent computing data (Fig.2.38). Circles
  and sectors are replaced by approximated polygons. Replacement methods for
  other patterns are selfexplanatory.

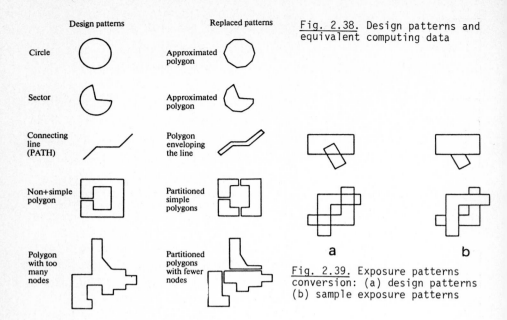

Fig. 2.38. Design patterns and equivalent computing data

Design patterns — Replaced patterns

Circle — Approximated polygon

Sector — Approximated polygon

Connecting line (PATH) — Polygon enveloping the line

Non+simple polygon — Partitioned simple polygons

Polygon with too many nodes — Partitioned polygons with fewer nodes

a   b

Fig. 2.39. Exposure patterns conversion: (a) design patterns (b) sample exposure patterns

- Design-pattern conversion to exposure systems (Fig.2.39). Because multiple exposures with an electron-beam cause an undesirable swelling of the pattern contour, overlapping of two or more design patterns should be avoided. There are many equivalent conversion possibilities; several of them result in fewer partitioned patterns, others result in patterns with simpler shapes. For electron-beam-exposure equipment, reducing the total pattern count may reduce the exposure time. Consideration of optimization versus some measure, i.e., output pattern count, is left for future problem solving. No optimization of this kind has been included in this system up to now, because the main exposure time depends on the total area of the exposed patterns.
- Design pattern conversion to negative polarity patterns (Fig.2.40). Negative polarity means exposing the background field, while designed pattern shapes are left unexposed. This kind of operation is required in LSI sandwich production.
- Partitioning exposure patterns to those acceptable by the electron-beam equipment (Fig.2.41). The exposure pattern shown on the left-hand side of Fig.2.41 should thus be divided into four subpatterns that are input to the electron-beam equipment.
- Exposure-field partitioning (Fig.2.42). For large chips, the usual practice is to limit the electron-beam scan field to avoid undesirable deflection

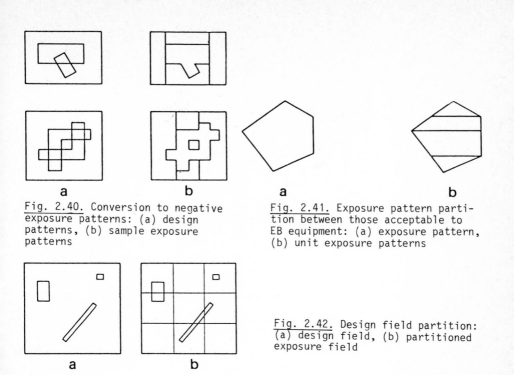

a   **b**   **a**   **b**

Fig. 2.40. Conversion to negative exposure patterns: (a) design patterns, (b) sample exposure patterns

Fig. 2.41. Exposure pattern partition between those acceptable to EB equipment: (a) exposure pattern, (b) unit exposure patterns

Fig. 2.42. Design field partition: (a) design field, (b) partitioned exposure field

a   **b**

distortion. The chip area field is then subdivided into smaller rectangular subfields. Exposure is performed by moving the stage of the wafer bed successively from subfield to subfield.

c) Algorithms

This section deals with the information format and operational algorithms for processing pattern data by computer. Pattern A is expressed as a polygon with n nodes and n straight edges. Pattern data A is a coordinate sequence of node points P_1, ..., P_n with edges $e_1 = (P_1, P_2)$, $e_2 = (P_2, P_3)$, ..., $e_n = (P_n, P_1)$. The arrangement of points may either be clockwise or counter-clockwise along the contour of the pattern. The pattern data area A is defined positive if it is clockwise. Any point among A node points may be chosen as starting point P_1, provided that the sequence is properly retained. Figure 2.43 illustrates the method of chaining and the pattern data polarity.

Assume patterns A and B overlap each other (Fig.2.43c). After these patterns have been developed in the pattern processing buffer (main working buffer), all A points on the arbitrary edge intersecting with the arbitrary edge of B are obtained. Those newly found points are included in the pattern data node-edge chaining (Fig.2.43d).

59

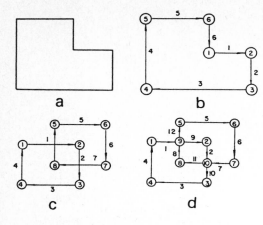

Fig. 2.43. Patterns and pattern data: (a) pattern, (b) pattern data with negative area, (c) overlapping patterns, (d) introducing intersection points and rechaining

The difference B-A for simple patterns A and B is divided into the following four cases (Fig.2.44): (i) A < B, (ii) A ≥ B, (iii) A ∧ B = 0, and (iv) A ∧ B ≠ 0 and A ⫫ B. Hatched areas in Fig.2.40 show differential patterns for the corresponding cases. In the first case, where pattern A is in pattern B, the resultant will not be a simple pattern. In this case, it is necessary to divide the resultant pattern into two or more simple patterns with equivalent contours (Fig.2.45). The second and the third case are selfexplanatory. Operations for the fourth case are: Develop pattern A in the negative direction and pattern B in the positive direction into the main working buffer. Obtain data on all edge intersection points and establish rechaining similar to Fig.2.43d. Rearrange node and edge chaining such that the pattern-A sequence is followed by pattern B, and vice versa, at each intersection node. Find the loop(s) with positive areas. These pattern(s) then represent elements of the difference B-A (Fig.2.46).

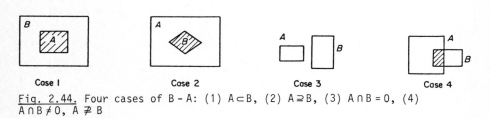

Fig. 2.44. Four cases of B - A: (1) A ⊂ B, (2) A ⊇ B, (3) A ∩ B = 0, (4) A ∩ B ≠ 0, A ≢ B

Fig. 2.45. Nonsimple pattern decomposition: (a) nonsimple pattern B - A for case 1 (b) breakdown into two or more simple patterns

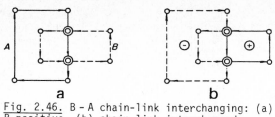

Fig. 2.46. B - A chain-link interchanging: (a) patterns A and B, A negative, B positive, (b) chain-link interchanged

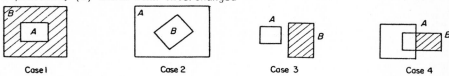

Case 1   Case 2   Case 3   Case 4

Fig. 2.47. Four A and B cases: (1) $A \subset B$, (2) $A \supseteq B$, (3) $A \cap B$, (4) $A \cap B \neq 0$, $A \not\supseteq B$

For intersecting (ANDing) two simple patterns A and B, there are four corresponding cases, as seen in Fig.2.47. Here Cases 1 to 3 are trivial and no discussion is necessary.

Operations for Case 4 are: Develop both patterns A and B in the positive direction. Obtain intersection points, rechain and interchange the node-link relation in a way similar to the previous case. Find the loop with the largest area and exclude it [this loop represents the union of A and B (AUB)]. Find other loop(s). They represent (elements of) the A and B intersection. This operation is illustrated in Fig.2.48.

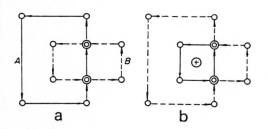

Fig. 2.48. Chain-link interchanging for $A \cap B$: (a) patterns A and B, both positive, (b) chain-link interchanged

The pattern-data structure has been designed to achieve highly effiecient data handling, efficient memory use and high-speed processing.

The main purpose of pattern processing for positive lithography is to eliminate multiple exposures. Assume that patterns P_1, P_2 and P_3 (shown in Fig.2.49a) are input to the buffer (this is the figure queue, explained in subsequent sections).

First, take P_1 and eliminate overlapping portions from P_2 and P_3. To do this, make new patterns $P_4 = P_2 - P_1$ and $P_5 = P_3 - P_1$. Next, take P_4 and eliminate overlapping portions with P_5, i.e., make the new pattern $P_6 = P_5 - P_4$. Output

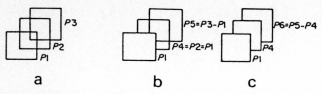

Fig. 2.49. Positive pattern processing: (a) input patterns, (b) eliminate overlapped portion with P1, (c) output patterns

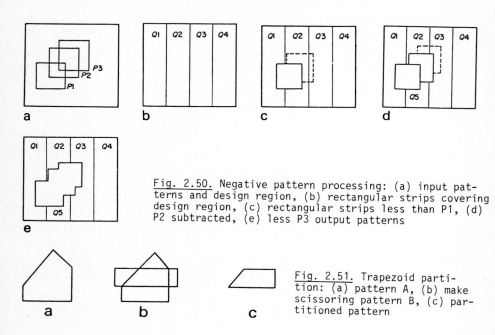

Fig. 2.50. Negative pattern processing: (a) input patterns and design region, (b) rectangular strips covering design region, (c) rectangular strips less than P1, (d) P2 subtracted, (e) less P3 output patterns

Fig. 2.51. Trapezoid partition: (a) pattern A, (b) make scissoring pattern B, (c) partitioned pattern

patterns should then be P_1, P_4 and P_6. Negative processing is carried out by subtracting each input pattern from a rectangle area representing the whole region. For practical use, the whole region is subdivided into smaller rectangles. Again, assume P_1, P_2 and P_3 are input patterns and the whole region is covered by rectangles Q_1 to Q_4 (Fig.2.50a,b). First, using P_1, obtain the differences $Q_1 - P_1$ and $Q_2 - P_1$. Further procedures are selfexplanatory (Fig.2.50). Output patterns have to be subdivided into electron-beam patterns. This procedure is illustrated in Fig.2.51.

Y-coordingate values for output-pattern nodes are sorted in ascending order. Produce data of a rectangle, parallel to the X axis, with Y coordinates being two adjacent values thus sorted. This pattern acts as the scissoring pattern for the output pattern to partition it into the required trapezoid.

If an input pattern occupies more than one subdivided design field, the pattern should be partitioned into subpatterns which belong to corresponding subregions. This can be accomplished by scissoring the input pattern with the pattern representing the whole subfield. Figure 2.52 illustrates this operation.

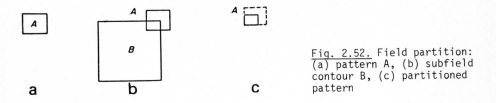

a   b   c

Fig. 2.52. Field partition: (a) pattern A, (b) subfield contour B, (c) partitioned pattern

2.3.2 Correction of Distortion

Patterns exposed by the electron-beam equipment are necessarily in error due to beam deflection distortion (Fig.2.53). When the beam has not deflected, it hits the center of the wafer, say at the point $(0,0)$. However, when the beam is deflected to impinge upon a point of the standard pattern, say point 1 with coordinates (x,y), it actually misses the point 1 by the increments Δx and Δy.

Let the coordinates of the input signal be $X = (x,y)$ and those of the point to be exposed by $U = (u,v)$. Then, the relationship between X and U can be expressed as $U = G(X)$. $G(X)$ is called the mapping function.

In order to expose point X, it is necessary to input the signal for exposure of point $\bar{X}$, expressed as

$$\bar{X} = G^{-1}(X) \quad . \tag{2.15}$$

For practical convenience, one considers a mapping function in the form of a sum of gain, trapezoidal and pin-cushion distortions:

$$\begin{bmatrix} u \\ v \end{bmatrix} = \begin{bmatrix} G_1(x,y) \\ G_2(x,y) \end{bmatrix}$$

$$G_1(x,y) = b_1 x + b_2 xy + b_3(x^3 + xy^2)$$

$$G_2(x,y) = b_4 y + b_5 xy + b_6(x^2 y + y^2) \quad . \tag{2.16}$$

The inverse mapping function $G^{-1}(X)$ is calculated by the Newton-Raphson iteration method. Some results of numerical calculation are discussed in the following paragraphs.

The first example is that of unrealistically large distortions. Let the length of the field side be 16000 points and assume that deflection errors

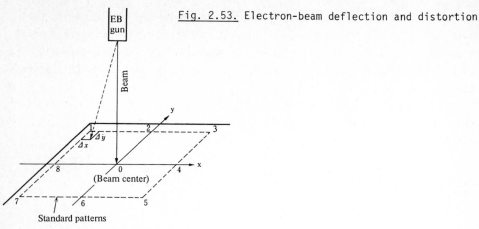

Fig. 2.53. Electron-beam deflection and distortion

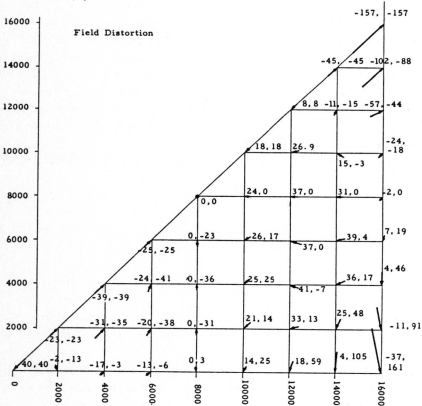

Fig. 2.54. Sample distortion distribution analysis (arrow shows distortion vector direction with relative distortion values)

on four corner points (points 1, 3, 5 and 7 in Fig.2.53) are about 100 points outside the target points, with some amount of trapezoidal distortion. Figure 2.54 shows the direction of distortion thus calculated at various points in the field. Note that arrow length exhibiting the amount of distortion and the direction are overemphasized.

The second example is that of a rather realistic amount of distortion. In this case, the amount of deflection error on four corner points is 5 points outside the target.

Table 2.5. Amount of distortion and correction (5-point distortion at corner points

ζ	η	amount of distortion		amount to be corrected	
		x	y	x	y
0	0	-5.04	-5.04	5	5
0	2000	-2.84	-2.13	3	2
0	4000	-1.27	-0.64	1	1
0	6000	-0.33	-0.08	0	0
0	8000	-0.02	0	0	0
2000	2000	-0.48	-0.48	0	0
2000	4000	0.69	0.46	-1	0
2000	6000	1.40	0.47	-1	0
2000	8000	1.63	0	-2	0
4000	4000	1.25	1.25	-1	-1
4000	6000	1.72	0.86	-2	-1
4000	8000	1.87	0	-2	0
6000	6000	1.09	1.09	-1	-1
6000	8000	1.17	0	-1	0
8000	8000	0	0	0	0

linear inverse interpolation

Table 2.5 gives magnitudes of distortions and amounts to be corrected at some points within the field. Coordinates are expressed in data points. No trapezoidal distortion was assumed. The corrections were calculated by (2.15) and rounded off to integer values. This is because the control signal to the electron-beam equipment is measured in integral data. It can be concluded from the results of Table 2.5 that linear inverse interpolation gives sufficient accuracy when distortions are relatively small. For example, even when the distortion is corrected only at the corner of a trapezoid pattern in a practical integrated circuit, the maximum error arising at the center location, far from corners, is less than about 0.01 μm. This error is negligibly small.

2.3.3 Correction of the Proximity Effect

The contour of the pattern exposed by the electron beam has a tendency of swelling due to the electron back-scattered inside the wafer material. This phenomenon causes a deterioration of the resolution of fine patterns, especially when they are located close to one another. This is called the *proximity effect*. To maintain required resolution and to obtain finer patterns with high fidelity, some kind of correction mechanism should be employed when fine patterns close together are to be exposed. Possible mechanisms are the method of controlling the exposure intensity, the method of modifying the shape of contours, or a method in which these two are intermixed, as shown in Fig.2.55.

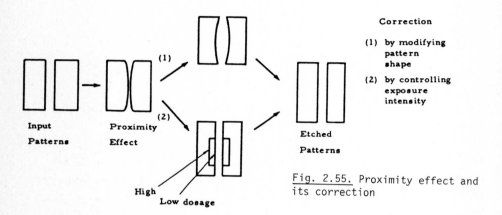

Correction

(1) by modifying pattern shape

(2) by controlling exposure intensity

Input Patterns   Proximity Effect   Etched Patterns

High   Low dosage

<u>Fig. 2.55.</u> Proximity effect and its correction

First, we shall discuss a method for controlling each pattern's exposure intensity in a first-order approximation, such that every pattern is separable from all others. Other methods, for instance exposure intensity control with area partition of each pattern and shape modification, will be discussed in the following subsection.

The most straightforward way to obtain the exposure at an arbitrary point is to integrate the exposure intensity distribution (EID) function [2.61]; i.e., at point x

$$E(x) = \int\int I(y)F(//x - y//)d^2y \quad . \tag{2.17}$$

F and I(y) are the EID function for unit exposure intensity and the exposure intensity at point y, respectively, and // // denotes the norm. The function F(r) is obtained by a least-square fitting to experimental data (Fig.2.56). Point x will be etched if the value E(x) exceeds a given threshold value.

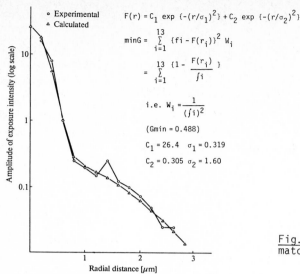

F(r) = C_1 exp {-(r/σ_1)^2} + C_2 exp {-(r/σ_2)^2}

$$minG = \sum_{i=1}^{13} \{fi - F(r_i)\}^2 W_i$$

$$= \sum_{i=1}^{13} \{1 - \frac{F(r_i)}{f_i}\}$$

i.e. $W_i = \frac{1}{(f_i)^2}$

(Gmin = 0.488)

$C_1 = 26.4$ $σ_1 = 0.319$

$C_2 = 0.305$ $σ_2 = 1.60$

Fig. 2.56. Sample EID function matching

Thus, if one can control I(y), it is possible to control whether point x is etched or not, and eventually to correct the proximity effect.

In order to bring this method to reality, it is necessary to devise a way of reducing the number of points at which the exposure E(x) is to be calculated. For this purpose, so-called "representative points" are selected, where the possibility of a density exceeding the threshold value is expected to be high, by evaluating some criteria explained below. In this way, a practical algorithm has been developed to correct the proximity effect.

a) Fitting the EID Function

Assume that the EID function is of the form

$$F(r) = C_1 \exp[-(r/\sigma_1)^2] + C_2 \exp[-(r/\sigma_2)^2] \quad . \tag{2.18}$$

Let the sum of squared residuals be the function

$$G(C_1, C_2, \sigma_1, \sigma_2) = \sum_{i=1}^{N} W_i [F(r_i) - f_i]^2 \quad , \tag{2.19}$$

where N is the number of observations, f_i is the observed intensity at point r_i (i = 1, ..., N), and W_i is the given weight at point r_i.

Next, we minimize the sum of the squares

$$sum = \min\{G(C_1, C_2, \sigma_1, \sigma_2)\} \tag{2.20}$$

67

with respect to its parameters C_1, C_2, σ_1 and σ_2. Then a least-square fitting of the function $F(r)$ to a set of observed data is obtained.

Minimization of (2.20) is performed by a special multi-dimensional minimization algorithm. Figure 2.56 illustrates an example for the fitted function together with observation data for the sake of comparison. The result is seemingly satisfactory and it can be concluded that (2.18) can adequately be adopted as a general procedure for obtaining the EID function.

b) Dosage Calculation and Representative Points

Let a sequence of exposure intensities be $I_0 > I_1 > I_2 > \ldots > I_s$, and a set of patterns to be exposed be A_1, A_2, $\ldots$, A_n. Denote the beam intensity I_j at pattern A_i by $\ell(i) = j$. The purpose of the algorithm is to assign the exposure intensity I_j to pattern A_i such that no serious deformation due to the proximity effect is expected. The algorithm is carried out in the following steps. First, assign an equal level of intensity [say, highest, $\ell(i) = 0$] to all the patterns A_i ($i = 1, 2, \ldots, n$). Next, the level of intensity is modified by repeating the following steps:

Step 1: Pick out any pattern A_i. Determine all patterns A_{i_1}, $\ldots$, A_{i_m} which are adjacent to A_i. A practical method will be shown below. Let ε be a "characteristic length" determined by the shape of the EID function. Namely, ε is an upper limit of r at which $F(r)$ cannot be regarded to be approximately zero. In Fig.2.56, ε is about 3 μm. Define $A_i{}^{\varepsilon}$, which is an "extended" pattern of A_i, as shown in Fig.2.57. Then, we obtain intersections of $A_i{}^{\varepsilon}$ and other patterns and find A_{i_k}, $k = 1, \ldots, m$ for which $A_i{}^{\varepsilon} \cap A_{i_k} \neq \phi$. These patterns are the desired ones, they are adjacent to A_i.

Step 2: Representative points are chosen at the midpoints of the shortest path between A_i and adjacent patterns. Typical examples of representative points are shown in Fig.2.58.

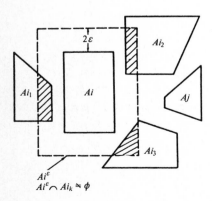

Ai^{ε}
$Ai^{\varepsilon} \cap Ai_k \neq \phi$

Fig. 2.57. Method of searching for patterns adjacent to A_i

Fig. 2.58. Typical representative points examples

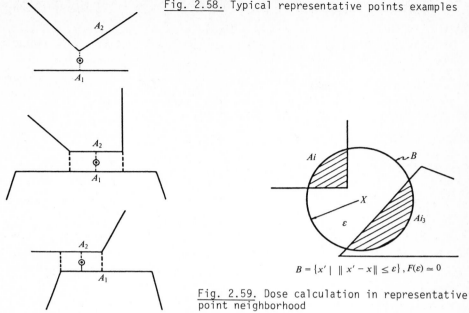

$B = \{x' \mid \|x' - x\| \le \varepsilon\}, F(\varepsilon) \simeq 0$

Fig. 2.59. Dose calculation in representative
point neighborhood

Step 3: Calculate the dose at the representative point, as shown in
Fig.2.59, i.e.,

$$E(x) = \iint I(y)Q(y)F(//x - y//)d^2y \quad , \qquad (2.21)$$

where $\theta(y) = 1$, if $y \in A$, $\theta(y) = 0$, if $y \notin A$. A denotes the set of points that
is included in one of the exposed patterns A_1, A_2, ..., A_n. Assume that the
exposure intensity $I(y)$ is constant over a separate pattern. Transforming the
point x to the origin, (2.21) can be integrated

$$E = \sum_j I_j \iint_{y \in A_j \cap B} \theta(y)F(//y//)d^2y \quad , \qquad (2.22)$$

where set B represent a small circular region with center x and radius ε.

Step 4: If the exposure at the representative point exceeds the threshold
value, add one to the smaller of $\ell(i)$ or $\ell(ik)$. If the exposure is less than
the threshold, repeat this procedure until correction is satisfactory for
$i = 1$, ..., n, and an appropriate incident beam is assigned to every pattern,
such that these patterns become separable.

c) Pattern-Shape Adjustment Techniques

The proximity-effect correction can be classified essentially into two methods: varying the exposure dose intensity and modifying the pattern shape.

This subsection deals with pattern-shape adjustment (PSA) [2.62] which is one variety of pattern shaping. Compared with varying the exposure dose intensity, the PSA technique has the following features:

(i) The exposure intensity distribution (EID) function decays exponentially with increasing distance. Thus, the dosage strongly depends upon the distance from the position of the exposed pattern, rather than upon the exposure dose intensity. Therefore, small shape reduction gives a good proximity-effect compensation.

(ii) Every pattern is exposed with the same electron-beam clock frequency in the PSA approach. This makes compensation simple from the view of software and hardware.

(iii) Patterns corrected by partial reduction of the original pattern shape have the tendency of becoming more complex than uncorrected patterns.

The correction procedure is as follows:

(1) Pick any pattern, and determine the representative among adjacent patterns.

(2) Calculate the exposure dose at the representative point by EID function integration, where the integration region is a circular area of radius ε, say the longest scattering distance, around the representative point.

(3) If at this stage the exposure dose does not exceed a given threshold value, it is not necessary to correct the pattern. Proceed to other representative points for compensation.

(4) Select the pattern to be modified, having the maximum contribution to the exposure dose at the representative point, say the highest integration value calculated in Step 2 above.

(5) If all the adjacent patterns have the size limitation like reduced patterns, proximity-effect compensation is impossible. Proceed to other representative points.

(6) Determine from the relative positions of the modified pattern and the representative point whether the modified pattern edge is one or two.

(7) Execute edge reduction.

(8) Obtain the new exposure dose at the representative point by integration of the exposure on the reduced patterns.

(9) If the calculated value exceeds the given threshold value, repeat the operation from Step 4 above.

(10) If the dose value is less than threshold, compensation is finished. Proceed to another representative point.

(11) After finishing the correction at all representative points, two attached patterns may have been detached by the above correction.

The effectiveness of the PSA technique in exposing LSI patterns using EB equipment has been discussed. The VLSI pattern fabrication will be described below. The EB equipment used was the Cambridge EBMF-2 system. The beam acceleration voltage was 20 kV, with beam diameter 0.05 μm and a 1 to 5 nA beam current. The resist material was PMMA of 500 nm thickness. Figures 2.60 and 61, respectively, show the uncorrected and corrected versions of part of the VLSI pattern. Figure 2.62 illustrates corrections due to computer simulation.

d) Dot-Beam Correction

Dot-Beam Correction is applied to the rounding of corners of a pattern. This rounding is caused by insufficient exposure doses at convex corners and by

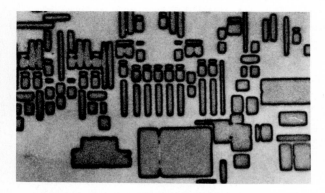

Fig.2.60

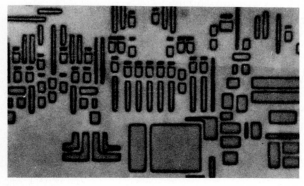

Fig.2.61

Fig. 2.60. Uncorrected EB exposure pattern
Fig. 2.61. Corrected EB exposure pattern. Minimum line width is 0.4 μm

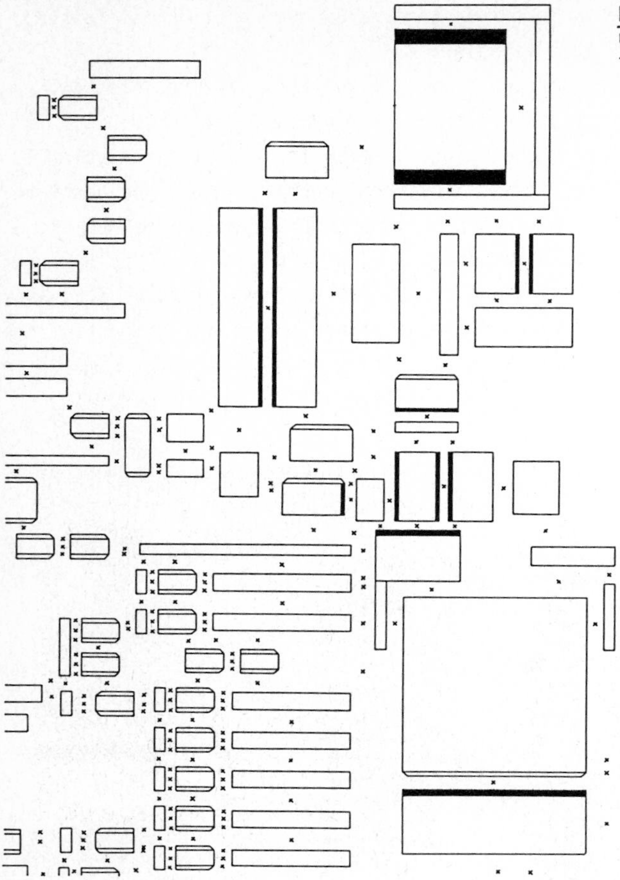

Fig. 2.62. Computer sim-
ulation plotting result
for Fig. 2.61

accumulation of exposure doses at concave corners, compared with the edge
center. To compensate for this rounding phenomenon, the following methods are
suggested.

For convex corners, a dot beam exposes simultaneously at, or immediately
outside the corner. For concave corners, the pattern is modified to decrease
the exposure dose. If the pattern contains only convex corners, e.g., a rectangular pattern, sufficiently good results are obtained by using the additional
dot beam only. However, patterns usually include both types of corners. In
these cases, the pattern is decomposed and modified to eliminate concave
corners, then the dot-beam correction method is applied.

Figures 2.63 and 64 show simulation results for the correction of a T bar
which includes both types of corners. Figure 2.63 is an uncorrected example
in which corner rounding appears. Figure 2.64 is a corrected example. First

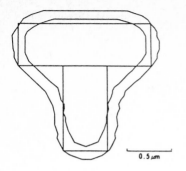 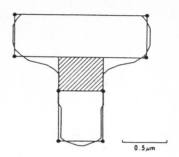

Fig. 2.63. Computer simulation of uncorrected pattern
Fig. 2.64. Computer simulation of corrected pattern using the dot-beam correction method. • denotes a dot beam. The diagonal lines represent the reduced area

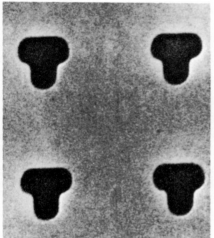 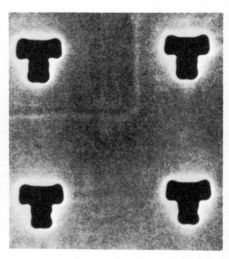

Fig. 2.65. Uncorrected EB exposed pattern

Fig. 2.66. Corrected EB exposed pattern

the T bar is decomposed and the shaded area is deleted, then additional dot beams are exposed at the corners of the two rectangles. Figures 2.65 and 66 show the uncorrected and corrected examples, respectively. For a pattern size of less than 0.5 μm, the dot beam must be immediately outside of each corner with size-depending doses.

e) Simultaneous Correction Method

This method is convenient for producing paired-pattern symmetry based on a careful consideration of the characteristics of the electric circuitry. The exposure dose E_i at any point (x_i, y_i) on a resist-coated wafer is obtained by

$$E_i = \sum_{j=1}^{n} I_j \int_{A_j} \int F(//r_i - r//) \, dx \cdot dy \qquad (2.23)$$

where A_j is the pattern unit with uniform exposure intensity, n the number of pattern units, I_j the exposure intensity on pattern unit, and F the EID function.

Let us consider points (x_i, y_i) $(i = 1, \ldots, m)$ as the representative points for exposure-intensity determination, then the exposure intensity I_j $(j = 1, \ldots, n)$ on pattern unit A_j is expressed by

$$\begin{cases} A_{11}T_1 + \cdots\cdots\cdots\cdots + A_{1n}I_n = E_1 \\ \phantom{A_{11}}\vdots \qquad\qquad\qquad\qquad \vdots \\ A_{i1}I_1 + \cdots + A_{ij}I_j + \cdots + A_{in}I_n = E_i \\ \phantom{A_{11}}\vdots \qquad\qquad\qquad\qquad \vdots \\ A_{m1}I_1 + \cdots\cdots\cdots\cdots + A_{mn}I_n = E_n \end{cases} \qquad \text{where} \qquad (2.24)$$

$$A_{ij} = \int_{A_j} \int F(//r_i - r//) \, dx \cdot dy$$

$E_i > E_{th}$: Pattern interior

$E_i \approx E_{th}$: Pattern boundary

$E_i \ll E_{th}$: Pattern exterior $\qquad . \qquad (2.25)$

For a detailed consideration of paired-pattern symmetry and circuitry characteristics, the following calculations are performed. The two expressions

$$\sum_j A_{i_1 j} I_j = E_{i_1} \qquad\qquad \sum_j A_{i_2 j} I_j = E_{i_2} = E_{i_1} \qquad (2.26)$$

for the two representative points corresponding to the paired patterns are replaced by

$$\sum_j (A_{i_1 j} + A_{i_2 j}) I_j = 2E_{i_1}$$

$$\omega \sum_j (A_{i_1 j} - A_{i_2 j}) I_j = 0 \qquad (\omega \gg 1) \qquad , \qquad (2.27)$$

and the least-square method is performed.

The techniques used for computation of (2.24) and (2.27) are: Replacing I_j and E_i by x_i and b_i, respectively, (2.24) and (2.27) are expressible as

$$A_x = b$$

$$A \in R^{m \times n} \quad , \quad x \in R^n \quad , \quad b \in R^m \quad . \qquad (2.28)$$

Generally, the number of pattern units does not coincide with the number of representative points, and m is not equal to n. When m < n, the solutions are underdetermined and nonuniform. When m > n, the solutions are overdetermined and generally there is no solution. Therefore, it may be assumed that the solutions of (2.24) and (2.27) can be regarded as the least-squared solution of the minimum norm. That is, if we let the solution be x*, then

$$x^* = \text{argmin}[\|x\|/\|Ax - b\| \to \text{min}] \quad . \tag{2.29}$$

This equation is solved by

$$(Ax - b)^T (Ax - b) \quad \to \quad \text{min} \quad . \tag{2.30}$$

It is given by

$$A^T A x = A^T b \quad \text{or} \quad x = A^+ b \tag{2.31}$$

where the inverse matrix is

$$A^+ = (A^T A)^{-1} A^T \quad . \tag{2.32}$$

As A is a large-size sparse matrix, the SOR technique is adopted.

By performing the following substitution

$$\tilde{A} = A^T A \quad , \quad \tilde{b} = \tilde{A}^T b \tag{2.33}$$

we obtain

$$\tilde{A}_x = \tilde{b}$$

$$A = \begin{bmatrix} & & -F \\ & D & \\ -E & & \end{bmatrix} \tag{2.34}$$

Using the Gauss-Seidel method

$$x^{k+1} = D^{-1}(\tilde{b} + Ex^{k+1} + Fx^k) \tag{2.35}$$

and over-relaxing the SOR method produces

$$\tilde{x}^{k+1} = D^{-1}(b + Ex^{k+1} + Fx^k) \tag{2.36}$$

$$x^{k+1} = x^k + w(x^{k+1} - x^k) \quad \text{or} \tag{2.37}$$

$$x^{k+1} = (D - wE)^{-1}[(1 - w)D + wF]x^k + w(D - wE)^{-1}b \tag{2.38}$$

where w is the over-relaxation factor (in this case, w = 1.8).

On the other hand, the exposure intensity has an upper and lower boundary. Therefore,

$$||Ax - b|| \rightarrow \min \quad . \quad x_1 \leq x \leq x_u \tag{2.39}$$

must be solved. This problem poses a quadratic program and the following revised formulation of the SOR method is used:

$$\tilde{x}^{k+1} = D^{-1}(b + Ex^k + Ex^{k+1} + Fx^k_0)$$

$$\tilde{\tilde{x}}^{k+1} = x^k = (\tilde{x}^{k+1} - x^k)$$

$$x^{k+1} = \begin{cases} x_u & , & \tilde{\tilde{x}}^{k+1} > x_u \\ \tilde{\tilde{x}}^{k+1} & , & x_1 \leq \tilde{\tilde{x}}^{k+1} \leq x_u \\ x_1 & , & x^{k+1} < x_1 \end{cases} \quad . \tag{2.40}$$

2.3.4 Warped-Wafer Correction

It is necessary during EB lithography to perform pattern alignment on each layer with high accuracy. The wafer warpage, which cannot be ignored with increasing wafer diameter, should be corrected so as to satisfy required precision specifications.

Warpage is produced in the wafer by heat-cycle processing, as shown in Fig.2.67. The pattern distortion due to warpage is considered to be three dimensional, but the EB exposure position correction can be thought of as two dimensional because of the depth of the beam focus.

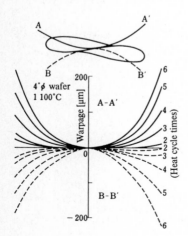

Fig. 2.67. The wafer warpage by heat-cycle processing

In order to compensate for pattern distortion, coordinates of sample points corresponding to alignment marks before warpage and after warpage are measured. Using the cubic interpolation to be described below, and based on the measured coordinates, the coordinates of every position on the wafer after warpage are estimated. Here, it is desirable that the number of sample points be as small as possible within the requirements of precision. The degree of warpage is considered to be virtually the same, provided the initial conditions and the wafer heat-cycle process are the same [2.62].

Therefore, if the distortion shift is measured at several model points on a typical wafer (called the model wafer), using the distortion shift of these model points and a few sample points, the distortion shift for any point on any wafer can be estimated.

a) Algorithm

This subsection deals with the algorithms used to determine the coordinates after warpage, using the sample and points.

First, the symbols employed in the algorithms are explained:

n_s: number of sample points,

n_m: number of model points,

n : number of lattice points, i.e., the sum of sample and model points,

(x_i^s, y_i^s), (x_j^m, y_j^m), (x_k, y_k): the coordinates before warpage, with the superscripts indicating here and below sample points, model points, and lattice points, respectively.

$i = 1, \ldots, ns$, $j = 1, \ldots, nm$, $k = 1, \ldots n$ with $(\tilde{x}_i^s, \tilde{y}_i^s)$, $(\tilde{x}_j^m, \tilde{y}_j^m)$, $(\tilde{x}_k, \tilde{y}_k)$ are the coordinates after warpage determined from model measurements (called the model experiment).

$(\hat{x}_i^s, \hat{y}_i^s)$, $(\hat{x}_j^m, \hat{y}_j^m)$, $(\hat{x}_k, \hat{y}_k)$ are the actual coordinates after warpage. The relations among the above coordinate definitions are shown in Fig.2.68.

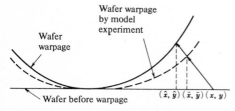

$(\tilde{x}-x, \tilde{y}-y)$ Distortion obtained from model
$(\hat{x}-x, \hat{y}-y)$ Distortion by wafer warpage

Fig. 2.68. The relations among the model and the actual warped

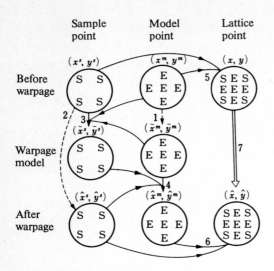

	Sample point	Model point	Lattice point

Fig. 2.69. The outline of the wafer-warpage correction algorithm

S : Sample point   E : Model point

-- ► Measurement   → Calculation

⟹ Final result

The outline of the algorithm is shown in Fig.2.69.

- Step 1: Estimate (x_i^S, y_i^S) from $(\tilde{x}_i^S, \tilde{y}_i^S)$ ($i = 1, \ldots, n_s$), and (x_j^m, y_j^m) and $(\tilde{x}_j^m, \tilde{y}_j^m)$ ($j = 1, \ldots, n_m$) using cubic interpolation (three degrees).

- Step 2: Estimate $(\hat{x}_j^m, \hat{y}_j^m)$ from (x_j^S, y_j^S) and $(\tilde{x}_j^S, \tilde{y}_j^S)$ ($j = 1, \ldots, n_s$) and $(\tilde{x}_i^m, \tilde{y}_i^m)$ ($i = 1, \ldots, n_m$) using cubic interpolation (one degree).

- Step 3: Consider the sum of (x_i^S, y_i^S) ($i = 1, \ldots, n_s$), and (x_j^m, y_j^m) ($j = 1, \ldots, n_m$) to be (x_k, y_k) ($k = 1, \ldots, n$). Also, consider the sum of $(\hat{x}_i^S, \hat{y}_i^S)$ and $(\hat{x}_j^m, \hat{y}_j^m)$ to be $(\hat{x}_k, \hat{y}_k)$.

The following operation is performed in Steps 1-7 in Fig.2.69.

  1) Measurement of model points after warpage by the model experiment;

  2) measurement of sample points after warpage;

  3) approximation of sample points after warpage by the model experiment;

  4) approximation of model point after warpage;

  5) definition of the lattice points by the sum of sample and model points before warpage;

  6) definition of the lattice points by the sum of sample and model points after warpage;

  7) transformation of the lattice points before warpage into the lattice points after warpage.

b) Cubic Interpolation Method

The function $Z = f(x,y)$ of a triangular surface, constructed with any three points from (x_i, y_i, z_i) $(i = 1, \ldots, n)$ on a curved surface, is approximated to the third degree of the function.

The procedures for the algorithm are:

Step 1: Construct a pattern of nonoverlapping triangles using points (x_i, y_i) $(i = 1, \ldots, n)$ on a cubic surface (Fig.2.70).

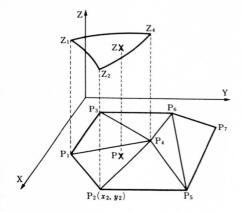

Fig. 2.70. Nonoverlapping triangles for a cubic interpolation

Step 2: Approximate the derivate function $(\partial Z_i/\partial x_i, \partial Z_i/\partial y_i)$ of the given points (x_i, y_i) $(i = 1, \ldots, n)$. $(\partial Z_i/\partial x_i, \partial Z_i/\partial y_i)$ is obtained using $Z_i = g(x_i, y_i)$ which satisfies the second degree of function $g(x_i, y_i)$, whose general formulation is $g(x_i, y_i) = a_1 + a_2 x + a_3 x^2 + a_4 xy + a_5 y + a_6 y^2$.

Step 3: Examine the triangle in which the selected point is located. Calculate the value at any point using the approximation in Step 2 obtained by this triangle.

The method for calculating the approximate value is as follows. The first degree of function Z on point (x, y) is given so as to become continuous on the edges of the triangle, i.e.,

$$Z = a_1 + a_2 x + a_3 y = \sum_{j=1}^{3} Z_j P_j(x,y) \quad , \tag{2.41}$$

$$P_j(x,y) = \frac{1}{C_{jkl}} (Z_{kl} + \zeta_{kl} x + \xi_{kl} y) = \frac{D_{kl}}{C_{jkl}} \tag{2.42}$$

where

$$Z_{kl} = x_k y_1 - x_1 y_k \quad , \tag{2.43}$$

$$\zeta_{k1} = y_k - y_1 \quad , \tag{2.44}$$

$$\xi_{k1} = x_k - x_1 \quad , \tag{2.45}$$

$$D_{k1} = \det \begin{bmatrix} 1 & x & y \\ 1 & x_k & y_k \\ 1 & x_1 & y_1 \end{bmatrix} = \det \begin{bmatrix} 1 & x_j & y_j \\ 1 & x_k & y_k \\ 1 & x_1 & y_1 \end{bmatrix} \tag{2.46}$$

For a function of the third degree, we have

$$Z = \sum_{j=1}^{3} \left[Z_j q_j(x,y) + \left(\frac{\partial z}{\partial x}\right)_j \gamma_j(x,y) + \left(\frac{\partial z}{\partial y}\right)_j S_j(x,y) \right] \quad \text{where} \tag{2.47}$$

$$q_j(x,y) = P_j(3P_j - 2P_j^2 + 2P_k P_1) + \delta q_j \quad , \tag{2.48}$$

$$\gamma_j(x,y) = P_j^2(P_1 \xi_{1j} + P_k \xi_{kj}) + \frac{1}{2} P_j P_k P_1 (\xi_{1j} + \xi_{kj}) + \delta \gamma_j \quad , \tag{2.49}$$

$$S_j(x,y) = P_j^2(P_1 \zeta_{1j} + P_k \zeta_{kj}) + \frac{1}{2} P_j P_1 (\zeta_{1j} + \zeta_{kj}) + \delta S_j \quad , \tag{2.50}$$

$$\delta q_j = 2\left[-A_j + (2 - 3\frac{L_1}{L_k} \cos \theta_j) A_k + (2 - 3\frac{L_k}{L_1} \cos \theta_j) A_1 \right] \quad , \tag{2.51}$$

$$\delta \gamma_j = \frac{1}{2}\Big[(\xi_{jk} + \xi_{j1}) A_j + (3\xi_{1j} + 5\xi_{jk} + 6\zeta_{1j} \frac{L_1}{L_k} \sin \theta_j) A_k$$
$$+ (3\xi_{kj} + 5\xi_{jk} + 6\zeta_{jk} \frac{L_1}{L_k} \sin \theta_j) A_k \Big] \quad , \tag{2.52}$$

$$\delta S_j = \frac{1}{2}\Big[(\xi_{jk} + \xi_{j1}) A_j + (3\xi_{1j} + 5\xi_{jk} - 6\zeta_{1j} \frac{L_1}{L_k} \sin \theta_j) A_k$$
$$+ (3\xi_{kj} + 5\xi_{jk} - 6\zeta_{jk} \frac{L_k}{L_1} \sin \theta_j) A_1 \Big] \quad , \tag{2.53}$$

$$A_j = \begin{cases} P_1 P_2 P_3 - \frac{1}{6} P_j^2 (3 - 5P_j) & (\text{in } \Delta T_j) \\[2mm] \frac{1}{6} P_k^2 (3P_1 - P_k) & (\text{in } \Delta T_k) \\[2mm] \frac{1}{6} P_1^2 (3P_k - P_1) & (\text{in } \Delta T_1) \end{cases} \tag{2.54}$$

where θ_j, L_j are the angle of P_i and the length of the edge opposite P_j, respectively. ΔT_j, ΔT_k, ΔT_1 are defined in Fig.2.71.

The relation between maximum error and number of sample points (SP's) and model points (MP's) is calculated by computer simulation, as shown in Table 2.6. With nine sample points, the maximum alignment error is 0.355 μm, but becomes 0.021 μm by addition of 72 model points (Fig.2.72).

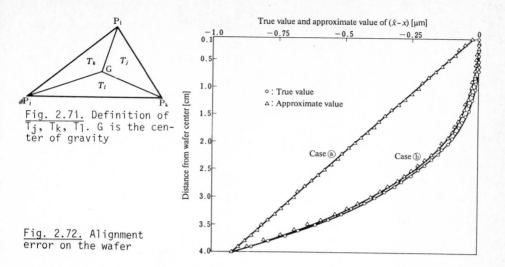

Fig. 2.71. Definition of $\overline{T}_j$, $\overline{T}_k$, $\overline{T}_l$. G is the center of gravity

Fig. 2.72. Alignment error on the wafer

Table 2.6. Maximum error based on the number of SPs and MPs

Case	SP	MP	Maximum error
a	9	0	0.355 µm
b	9	72	0.021 µm
c	25	0	0.044 µm
d	25	56	0.010 µm
e	81	0	0.005 µm

2.4 Wafer and Writing Systems

2.4.1 EB System with High Current FE Gun (VL-FI) [2.63]

The field-emission electron-gun and vector-scan delineator is a direct wafer writing system using an electron beam for the fabrication of the VLSI patterns down to submicrometer dimensions. This delineator was realized by using a newly developed field-emission electron-gun, as described in Sect.2.2.1c. The delineator has a circular electron beam with a Gaussian distribution profile, which can be focused to the smallest spot. This beam shape is superior to other beam shapes for fine-pattern writing with high precision. The small spot, however, causes that the writing time generally becomes longer than that in delineators using the other beam shapes (Sects.2.4.2,3).

To prevent that disadvantage, a vector scan method with steered beam is utilized, providing a delineator for fast fine-pattern writing. For example, it is suitable for proximity-effect correction, especially needed in the submicrometer pattern region. With this delineator one can change the scan speed,

Table 2.7. Major features of the FE electron gun and vector scan delineator (VL-F1)

Delineation method	Vector scan and step-and-repeat writing
Main use	EB direct wafer writing
Delineation area	Up to 3 inches
Scan field	2×2 mm^2 to 5×5 mm^2
Scan control method	Use of compensation type high speed and high precision 16 bit DAC
Address resolution	0.03 μm on 2×2 mm^2 field
Scan speed	1 m/s to 1 mm/s
Controlable figure	Rectangle, trapezoid, triangle, line, dot, etc.
Registration resolution	0.01 μm with reflected electrons
Registration accuracy	less than 0.1 μm
Stage positioning resolution	0.01 μm with laser interferometer
Input data format	PG 3000 or design data
Editorial function	Field partitioning, stretching, shrinking, posi/nega conversion, mirror-type reversal, etc.

the line pitch length, the scan direction, etc. in response to the pattern dimension and the pattern density, by a data pattern generator and ramp generators. These circuits belong to the function generators described in Sect.2.4.1c. The main features of such a delineator are represented in Table 2.7.

a) Electron Gun and the Electron-Optical System [2,64,65]

A field-emission (FE) electron gun provides high brightness and high emission current due to its fine emitter (Sect.2.2.1b). However, it is necessary to obtain a field-emission current for a delineator that is higher and stabler than for an electron microscope (at a probe diameter of about 1000 Å). It is more effiecient to use the total FE electron current as a probe current on the wafer. Then, the electro-optical lens system must be designed to transfer more electron current from the FE tip to the lens system with wide angle and lower aberrations.

A cross section of the electro-optical system is illustrated in Fig.2.73. This system consists of 4 vacuum chambers, an electron gun chamber, a first intermediate chamber, a second intermediate chamber and a work chamber, respectively, from the top of the figure. The lens system is built up of one electro-static lens, two electro-magnetic lenses (condensor lens and objec-

49

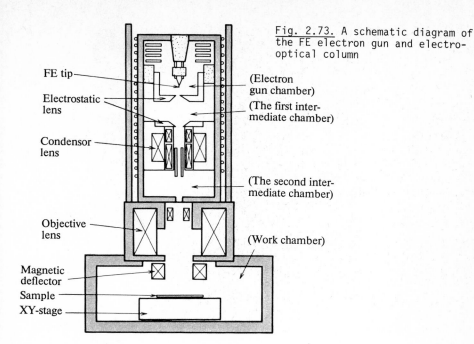

FE tip

Electrostatic lens

Condensor lens

Objective lens

Magnetic deflector

Sample

XY-stage

(Electron gun chamber)

(The first intermediate chamber)

(The second intermediate chamber)

(Work chamber)

Fig. 2.73. A schematic diagram of the FE electron gun and electro-optical column

tive lens), a post electro-magnetic deflector, an electro-static blanker, stigmators and aligners. The electro-static lens uses a Butler's lens (Sect.2.2.1b) of which the second electrode is combined with the condensor lens and connected to the earth potential like the column. On the other hand, the first electrode of the electro-static lens and the FE tip are supplied with negative potential against the second electrode to field-emit ion electrons from the tip and to accelerate the electrons. In this delineator, the separation of the tip and the first electrode is chosen so short that the electrostatic lens can have a wide aperture angle, in order to conduct much emission electron current into the electro-magnetic lenses. Furthermore, the magnetic lenses must be operated to enlarge the FE point of less than 0.01 μm to a size of around 0.1 μm, as described in Sect.2.2.2d.

As mentioned above, the unity of the second electrode with the condensor lens allows the operation of the condensor lens under a strong-magnetic-field condition. It is suitable to suppress aberrations of the condensor lens. Furthermore, as described in Sect.2.2.2b, the electrostatic lens can have a small magnification, thus giving small lens aberrations. Consequently, this electro-optical system can be operated with small aberrations in spite of the enlarged FE-source image.

If a high FE electron current is drawn for a long period with little fluctuation, there are the following problems. One must prevent (i) fluctuation of the electric field concentration around the tip due to changes of the tip surface (from smooth to rough) by residual-gas ion bombardment, and (ii) fluctuation of the work function on the tip surface due to residual gas absorption and migration. It is necessary to keep the vacuum pressure in the gun chamber so low and the chamber so clean that molecules absorbed on the chamber's wall cannot leave it during field-emitted electron bombardment. In order to realize these requirements, the gun chamber, the first intermediate chamber and the second intermediate chamber are evacuated by ion pumps, and the work chamber by a turbo-molecule pump. In addition, the differential pumping is achieved by locating orifices at interfaces of these chambers to keep an ultra-low gun-chamber pressure. Consequently, vacuum pressures in the gun chamber, the first intermediate chamber and the second intermediate chamber should be about $3 \cdot 10^{-10}$, $1 \cdot 10^{-9}$ and $1 \cdot 10^{-8}$ Torr, respectively, when the work chamber pressure is about $2 \cdot 10^{-6}$ Torr.

The gun chamber must be kept at the ultra-high vacuum even when the gun is operated to provide a small electron current. This is a possible reason why the absorbed molecules leave the surface of the gun-chamber wall in electron bombardment, which occurs around the FE tip during gun operation. The number of molecules increases more remarkably as the gun provides more FE electron current. It is also necessary to remove the molecules absorbed on the wall around the tip, in order to keep the gun chamber in ultra-high vacuum during the operation. Therefore, this column is designed to be baked out by a tape heater winding around the column, as indicated by circles in Fig.2.73.

Using such a design, this FE electron gun can provide a stable total emission current of 100 µA for a long time at room-temperature gun operation, i.e. operation of the gun without heating the tip. The tip temperature is not a room temperature during operation because of the ion bombardment and the field emission, but it is close to room temperature in comparison with the thermal field emission (TFE) electron gun (Sect.2.2.2b). Since the electro-optical system is designed to enlarge the FE-source image, the column becomes smaller than half of that used in the thermal emission (TE) electron gun. In this delineator, the distance between the tip and the sample is about 45 cm.

It has been said before [2.66] that an FE electron gun can provide more probe current than a TE electron gun in a region of less than 0.1 to 0.2 µm probe diameter, when the total FE current is about 10 µA (Fig.2.4). However, the FE-electron gun can stably provide a total FE current up to 100 µA as described above. Then, the region of the probe diameter, for which such an

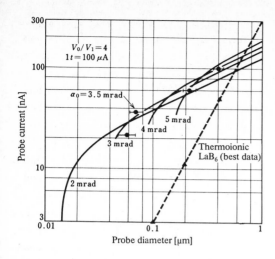

Fig. 2.74. Relationship between probe current and probe diameter, where V_0 is acceleration voltage, V_1 is emission voltage, and α_0 is a half aperture angle in objective lens. Solid lines are calculated results in this delineator, and broken line is in that using a LaB$_6$ electron gun

electro-optical system is more advantageous, can be expanded by improvements of both the vacuum system and the electro-optical system. Figure 2.74 gives relations between an electron-beam spot (probe) diameter and an electron-beam (probe) current, using the FE-electron gun. Solid lines are calculated results with electro-optical parameters of the system, ● denote experimental values for this delineator, and ▲ represent maximum experimental values obtained by another delineator using an LaB$_6$ electron gun.

The figure shows that the experimental values agree well with the calculated results, and this delineator can provide a beam current of more than 10 times that using a TE-electron gun of 0.1 µm probe diameter. In addition, the spot diameter, where the beam current in this delineator is equal to one in another delineator, is 0.5 to 0.6 µm. This value is 5 times larger than the previous result [2.66].

b) Mechanical System

An XY drive for EB lithography demands high accuracy, high speed and short settling time to write fine patterns by step and repeat. On the other hand, a laser-interferometric unit makes it possible to measure the stage position with high accuracy of better than 1 µm. Large-scale XY stages with high speed and high accuracy have been developed. However, the stage must be used in high vacuum and with electrons. This means that it must be shielded against electric and/or magnetic fields so that the delineator remains at high precision. Therefore, the stage must be satisfied with the following conditions: (i) the XY stage must not be magnetized, (ii) it drives smoothly without generation of undesired gas for the ultra high vacuum, (iii) it must do so without generation of dust particles from lubricated parts, and (iv) generation of heat in the driving motor must be suppressed. These conditions mean that magnetic materials such as steel cannot be used due to item (i), usual lubrication oil cannot be used due to item (ii), and the use of a slider as a driving guide must be avoided due to items (ii)-(iv).

Following these requirements and limitations, the XY stage uses non-magnetic material balls as rolling balls between the stage and the driving guide, the surface of which is hardened by a coating technique to obtain high position accuracy. The driving uses pulsed motors, one of which is placed outside the vacuum system and the other is directly linked to the Y stage in the vacuum system. The latter has an O-ring shielding and water cooling in order to prevent the generation of undesired gas and heat. Furthermore, to improve pattern delineation accuracy the motors must be shielded from the magnetic field; therefore one uses a bipolar driving technique [2.67] in a pulsed-motor driving mode. The use of the bipolar driving technique can decrease the heat generation in the motors.

The laser interferometer is made by combining a mirror bar for the X axis and another mirror bar for the Y axis, the shape being that of an L. A mirror is lapped within a rectangular error of 0.5 s and a flat within 0.1 λ; it is fixed near a position where the sample is held on the stage. In this delineator, the step-and-repeat drive can be carried out within a positioning error of 2 μm, which is measured by the laser interferometer and is compensated for by translation via the electron-beam deflection system. Consequently, the step-and-repeat drive can contribute to the delineator's speed-up. For example, it takes only 0.4 s from the maximum driving speed of 10 mm/s down to complete stopping.

It is necessary for fine writing to isolate a work chamber from outside vibration. The column must be supported by a stage with air dumpers. These can weaken the mechanical vibration from outside to less than 5% and 0.4% in the frequency range of 10 Hz and 100 Hz, respectively, when the load is 60 kg.

c) Electronic Control System

A block diagram of this control system is shown in Fig.2.75. Data flows are explained as follows. The delineation data, stored for each figure element

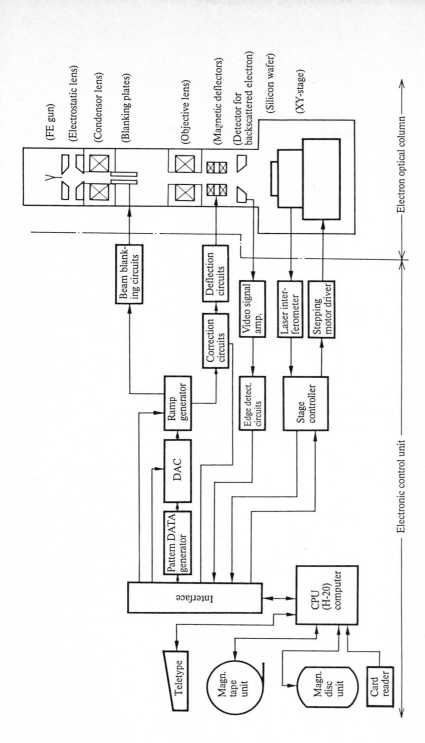

Fig. 2.75. A block diagram of the FE electron gun and vector scan delineator (VL-FL)

in a magnetic disc, are transferred from a mini-computer (H-20) to a pattern data generator through an interface. The pattern data generator is a micro-computer and consists of an ALU (arithmetic logic unit), memories, registers, etc. Here, the delineation data are decoded into commands for driving DACs (digital-analogue converters), ramp generators and blanking circuits, and destination addresses for the steered electron beam. These signals are trans-ferred into each electronic circuit in time sequence.

In this system, the delineation data consist of 5 words (a word has 16 bit). The pattern data generator determines the electron-beam starting address in X and Y, makes the electron beam jump there without writing, selects the scan direction and the separation between lines (line pitch), and appoints the destination addresses for the vector scan together with selecting either a beam writing mode or a beam jumping mode based on the time sequence. Con-sequently, vector-scan writing can be achieved. Thus, the pattern-data genera-tor can decode many commands and many electron-beam destination addresses from one figure data. This decreases the amount of data transmitted from the com-puter. Therefore, the pattern-data generator makes fast EB writing possible with only minicomputers.

The high speed and high precision DAC operates as follows. In general, a DAC is a circuit which converts the digital information to the analogue one and demands high speed and high precision with a lot of bits. The fact that the DAC has a lot of bits means that it can control an output signal up to high voltage with high accuracy, and write the resist pattern down to sub-micrometer dimensions over a wide area with high accuracy in the EB delineator. On the other hand, the writing time depends on the DAC settling time. If the bit number of the DAC is increased, the DAC settling time becomes longer than that in the DAC with a smaller bit number. This relation represents a trade off. The higher the speed used, the lower the precision obtained. In order to achieve a highly precise DAC with high speed, the DAC utilized consists of a highly precise standard ADC (analogue-digital converter) with 18 bit (mother ADC) and high-speed DACs with 16 bit (high DACs). A digital quantity is con-verted to an analogue voltage by high DACs, and the analogue voltage is ac-curately calibrated and compensated for by the mother ADC. A block diagram of this new compensation-type DAC [2.68] is shown in Fig.2.76. Before DAC oper-ation, one must generate data for the compensation of the high DACs. The data are stored in a digital memory as an error digit in response to the upper 8 bits of an input data. The error digit is determined by comparing the input data with an analogue/digital converted digit from the high DAC by the mother DAC. In operation, compensation is carried out by addition of the error digit in

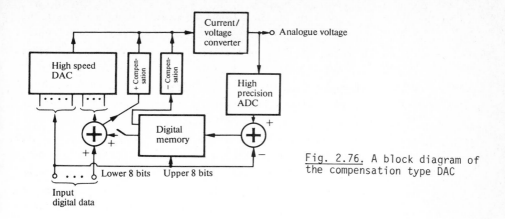

Fig. 2.76. A block diagram of the compensation type DAC

response to the upper 8 bits of the input data into the lower 8 bits of the input data based on the digital memory data. In the figure, a plus-compensation and a minus-compensation mean direct compensation of the analogue value from the high DAC in the special case that the compensation cannot be executed, as described above.

The ramp generator [2.69] is shown in Fig.2.77 and consists of circuits based on the Miller integrator. The input voltage v_i is a variable DC voltage for control of the EB writing speed. When v_i is applied to the integrator at time $t = 0$ by the switch SW [which turns to position (1) as shown in the figure], an increment Δv_0

$$\Delta v_0 = - \frac{v_i}{CR_0} t \qquad (2.55)$$

is obtained after time t due to the operational amplifier. C is the capacity of the condenser and R_0 is the input resistance of the amplifier. This delineator can select the writing speed by controlling the input voltage v_i. By turning the SW from (1) to (2) (earth potential), the electron-beam deflection can be completely stopped. An initial voltage v_0 determines the starting point in one vector scan, $-v_i/CR_0$ does the writing speed, and the terminal voltage represents an end point which is determined by DAC.

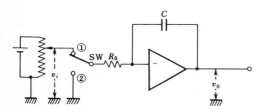

Fig. 2.77. Basic circuit of the ramp generator

89

The operation of the vector scan is explained as follows. In Fig.2.75, the output signal from the pattern-data generator is translated to the ramp generator through DAC, the output voltage of which is a ramp-generator destination voltage in response to the electron-beam destination address. The voltage controls the output voltage v_0 of the ramp generator. On the other hand, v_i has been determined by the interface and the blanking voltage is changed from 50 to 0 V to pass the electron beam through the blanking plates and lead it onto the sample. During driving the ramp generator, v_0 is observed to control the end point in the vector scan. When v_0 is approaching the destination voltage of DAC, the ramp generator assumes that one vector scan is just finished, and SW is turned from (1) to (2) as soon as possible. At the same time, it translates a control signal to the pattern data generator. Then, the pattern generator turns the blanking voltage back to 50 V to prevent electron incidence onto the sample, and proceeds to the next vector-scan step. The pattern-data generator can write fine patterns with accurately divided lines in itself for the vector scan based on data translated from the computer by controlling DACs, the ramp generators and the blanking circuits.

d) Computer System and Software

The delineator under consideration has a Hitachi mini-computer (H-20) as the main computer with 16 bit word length. It has built-in magnetic-core memories with 64 k word memory capacity and a cycle time of 0.065 μs/word. In addition, there are a magnetic-disk storage and a magnetic-tape storage as outer memory devices. The former consists of a fixed disk and an exchangeable disk cartridge with 4.9 M words storage capacity, and it has a rotational delay of 12.5 ms in the average, a total average access time of 35 ms, and a data transfer rate of 156.2 k words/s. The latter has a storage capacity of 10 M words/2400 ft, a storage density of 800 BPI (bit per inch); the data transfer rate is about 10 k words/s. Furthermore, there exists a PTR (photo tape reader) with a reading speed of 400 letters/s and a card reader with a reading speed of 310 cards/s as input peripheral devices. There is a line printer with a printing speed of 132 columns × 200 lines/min as an output peripheral device, and a data typewriter with 42 keys and two-shift operation, a printing speed of 600 letters/min and a tape punching speed of 600 letters/min.

Software for this delineator consists of a delineation main scheduler operated on a supervisor of the H-20 system software and an ISR (interrupt service routine), as shown in Fig.2.78. The ISR receives orders from each task through the supervisor, and executes an interrupt control of the XY stage, the DACs, the ramp generator, a DMA (direct memory address), and the pattern data

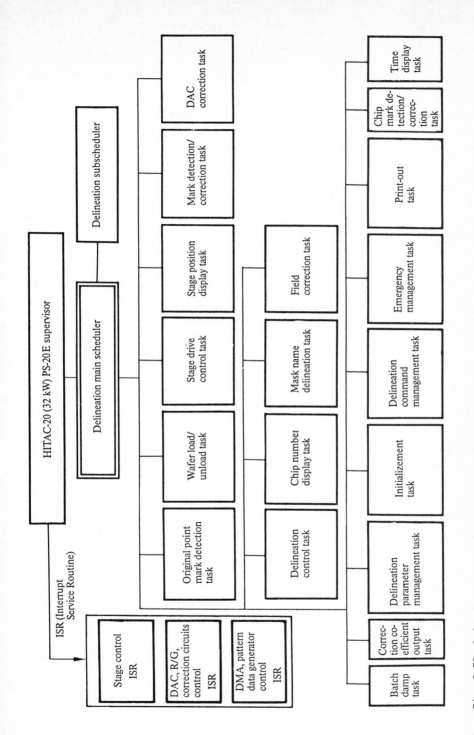

Fig. 2.78. Software construction of the delineator

91

generator. The delineation main scheduler carries out various task controls such as an original point-mark detection, a sample load/unload, a stage drive, a mark detection and correction, a DAC calibration, a delineation control, a deflection-distortion correction, an emergency stop, etc., together with a delineation subscheduler.

The data for delineation are constructed on the basis of the following two routes: One is to use the data which are directly entered into the computer (H-20) from the C/R (card reader), the DTY (data type writer) and the PTR. Another is to use the data which are converted to from a design file made by a system developed by APPLICON. In the direct data entrance, usage of the DTY is available for small-scale data. On the other hand, usage of data cards and data tapes with C/R and PTR is for large-scale data. In general, C/R is more useful than PTR because in C/R data are easily corrected.

Input data for this delineator consist of basic patterns such as a dot, a line, a rectangle, a trapezoid, a parallelogram and a triangle, which can be inclined at $0°$, $45°$, $90°$ and $135°$ against the X axis and are represented with 5 words. The other figure data can be formed on the basis of these patterns. These 5 words are constructed by 2 words for a starting address (X, Y) of the vector scan in a figure, one word for the length of the figure on the X axis, one word for another length in the Y axis and one word for the selection of the figure type, the inclination angle against the Y axis, etc. In addition, a proximity-effect correction can be carried out with the remaining bits due to changing of the scan speed, the line pitch length or the pattern dimension.

For conversion of design data to EB-delineation data by the APPLICON system, two softwares for data conversion are used (Fig.2.79). The design data are displayed as figures on a CRT of a graphic-design system. A designer makes additions and corrections to the design data and makes up a P/G (pattern generator) file or a design file. The former is for an optical pattern generator, and the latter is for an EB delineator. An EB delineation data processing system (Sect.2.3) makes a master EB delineation data file from the latter. Finally, an EB delineation data file is formed from that file with a large-scale pattern data-converting program (LPP) in H-20. The LPP has the function of not only data converting, but also line element partitioning for the vector scan, mirror revising against the X or Y axis, EB delineation data compressing and pattern dimension stretching/shrinking. In particular, the compressing function is significant for increasing the translation speed of the EB delineation data from the computer to the pattern data generator. In case of a series of the same pattern type and/or the same pattern dimension, input data after the first one are represented with reduced form (no 5 words). The

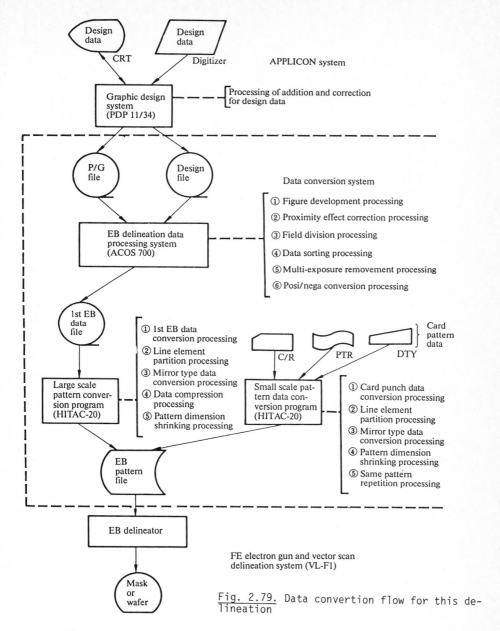

Fig. 2.79. Data convertion flow for this de-lineation

stretching/shrinking function is for one of the proximity effect corrections. In this delineator, magnetic tapes of the P/G file, the design file and the master EB delineation data file are used for device compatibility, and magnetic discs of the EB delineation data file are used for high translational speed.

e) Experimental Results [2.70,71]

The EB delineator using an FE electron gun is developed for fabrication of patterns including submicrometer dimensions. It is very important to investigate the proximity effect in this delineator. A variation of exposure dosage over a range of line widths and packing densities are shown in Fig.2.80 [2.72]. This exposure dosage is defined as the one which gives the same resist pattern as the input data. The pattern dimension was determined by definition of the middle points on edge slopes in the pattern profile as pattern edges with scanning-electron-microscope images. The ordinate of the figure represents the normalized exposure dosage with exposure dosage required for 1 μm lines at an infinitely large gap spacing. The figure gives the experimental result for the delineator under discussion (solid lines) together with CHANG's result [2.61] in the other delineator using an LaB$_6$ electron gun (broken lines). It shows that, for our delineator, the exposure dosage which is needed to obtain submicrometer wide lines decreases as the gap spacing decreases down to lower submicrometer dimensions. This tendency becomes strong as the gap spacing decreases. However, variations of the exposure dosage in the gap range larger than 1.5 μm become small. However, for Chang's result (broken lines), a deviation from exposure dosages which are needed to obtain 1.0 μm and 0.5 μm wide lines is about 50%, even in a gap range larger than 2.0 μm. Comparing these results shows that this delineator has a smaller proximity effect than the other one.

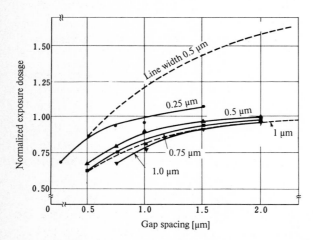

Fig. 2.80. Variation of exposure dosage with line and gap width; the solid line represents results using an FE electron gun, the broken line represents results using an LaB$_6$ electron gun

Submicrometer pattern writings with our delineator are: a 16 k bit MOS Al wiring pattern including 0.25 μm minimum line width written on a 0.5 μm thick PMMA resist on an Si wafer at constant exposure dosage (Fig.2.81). Figure 2.81a is an optical microscope image of the PMMA resist patterns, and Fig.2.81b is a narrow-angle image from a scanning electron microscope (SEM). As shown in Fig.2.81b, both the 2 μm squares and the 0.25 μm lines are achieved at a constant exposure dosage.

Figure 2.82 shows narrow-angle SEM images of overlaid patterns obtained by the EB direct wafer writing in our delineator. In the figure, the Al wiring pattern overlaid on the contact hole pattern for a reduced-size 16 k bit MOS RAM in VLSI is used in this experiment. A practical fabrication process for a semiconductor device needs a number of overlaying levels, putting etching and forming of the pattern between these levels, respectively. This 2-level overlaying experiment shows that it is possible to fabricate the VLSI pattern in each lithography step with the EB direct wafer writing.

The overlay experiment was carried out with four registration marks formed at four corners of a chip. The detailed procedure of the experiment is as follows.

(i) Application of PMMA resist (refer to Sect.3.6 for details) on an SiO_2 layer of an Si wafer including the registration marks.

(ii) Load of the wafer into this delineator.

(iii) EB direct writing of the contact hole-level pattern by destination of the EB writing condition.

(iv) Development, and chemical and thermal treatment of the resist after unload of the wafer from the delineator.

(v) Dry-etching of the SiO_2 layer with the resist pattern as a mask to form the SiO_2 contact holes.

(vi) After removing the resist by a plasma-asher, application of the new PMMA resist onto the SiO_2 contact hole layer.

(vii) Load of the wafer into the delineator again.

(viii) EB direct wafer writing of the Al wiring level pattern on the new resist.

(ix) Development of the resist after unload of the wafer from the delineator.

The 2-level overlaid pattern obtained by this process is shown in Fig.2.82 as an SEM image. Figure 2.82a is a top view, and Fig.2.82b is an oblique view. In the figure, the bottom layer of the square contact holes is the silicon substrate, the intermediate layer is SiO_2, and the top layer is the resist with the Al wiring pattern for a lift-off technique. Actually, the Al wiring

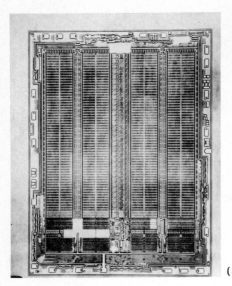 

(a) (b)

Fig. 2.81. PMMA resist patterns of a reduced 16k bits MOS Al wiring level on an Si substrate with 0.25 µm minimum line width; (a) optical microscope overall top view image and (b) SEM nallow angle image (0.5 µm PMMA resist thickness)

(a) (b)

Fig. 2.82. SEM images of 2 level overlaid pattern with a contact level and an Al wiring level in 16 kbits NMOS RAM chip reduced to about 0.83×0.42 mm^2 chip ($\sim 1/8$ of chip) containing about 0.75 µm minimum line width; (a) top view of overlaid pattern, and (b) oblique view of the pattern

pattern will be completely formed by Al deposition on this sample and removing the resist (the lift-off technique). The figure shows that the minimum line width of the Al wiring pattern is about 0.75 μm, and the overlay error between the contact hole level and the Al wiring level is smaller than 0.1 μm.

2.4.2 Variable-Shaped-Beam Lithography (VL-S2)

A new technique for electron beam exposure using a variable-shaped beam has been developed in Japan [2.73,74]. The technique enables both high-speed exposure and high resolution. The variable-shaped-beam lithography system called VL-S2 is a prototype which utilizes this new technique. Here, we describe the development of VL-S2 and give a brief description of the system's design and performance.

It is well known that electron-beam exposure yields a significantly better system than photo-pattern generation and photo-mask making. Electron-beam exposure offers high resolution, but it is not yet practical because its full potential for high-speed exposure and pattern generation has not been developed. Many experimental results for high-resolution pattern generation have been reported. The most significant problem for practical application of high-speed exposure is cost. In relation to this, as very large-scale integrated circuits (VLSIs) are being developed, the number and variety of exposure patterns are increasing rapidly. Consequently, the amount of pattern data the electron-beam exposure system must handle is increasing, too. This means high-electron beam shoot speeds on wafers and high-speed transfer of large quantities of pattern data to the exposure system are required for the VLSI system.

The shaped-beam method has been developed as an alternative to the technique of generating patterns which are printed in one step. In the shaped-beam method, projection exposure is done after dividing the required patterns into segments. In this manner, the time required for electron-beam control in time-series signal transfer can be significantly reduced, thereby reducing the overall exposure period. In the variable-shaped beam method, the segment size is variable and thus high-resolution pattern exposure is achieved. The double exposure of the fractional pattern width sections produced by fixed-size shaped beams is eliminated in this method. The amount of data required to do the variable-shaped-beam method is less than for generating patterns with spot beams under time-series control. This is because the data for the patterns of individual segments can be compressed. The variable-shaped-beam exposure method thus meets the requirements of high-speed bulk control data transfer.

Table 2.8.   Performance characteristics of VL-S2

Beam size	0.5 to 5 μm
Beam current density	>1 A/cm^2
Field size	2.5 mm × 2.5 mm
Stage S and R speed	<0.5 s/2.5 mm
Exposure area	110 mm × 110 mm
Pattern accuracy	0.2 μm - including butting and overlaying accuracies
Throughput*	<18 min/4-in diameter wafer, direct exposure including overhead

* Pattern density: 5×10^6 shots/cm^2 including 1 μm width patterns
Total pattern number: 2.5×10^8 shots/4-in diameter wafer
Resist: PGMA (sensitivity, 1.2×10^{-6} C/cm^2)

The VL-S2 system uses this variable-shaped-beam method with a data processing system which can handle enough pattern data to control exposure of 1 Mbit VLSI chips. VL-S2 can be used for direct or mask exposure of wafers for devices having patterns whose minimum line width is approximately 1 μm. The time for direct exposure of a 5,000.000 shots/cm^2 chip (equivalent to a 1 Mbit LSI chip) over a 10 cm wafer is less than 18 minutes including overhead time. The pattern exposure time for mask exposure is less than 10 minutes excluding overhead time (Table 2.8).

Figure 2.83 gives a picture of the VL-S2 system. The interface system is shown on the right, and the electron-optics system and stage-control system are in the middle and the left parts of the figure, respectively.

a) VL-S2 System Configuration

The variable-shaped-beam lithography system (VL-S2) consists of the following three main subsystems (Fig.2.84):

- Computer subsystem
Including computer, magnetic disk, and interface unit.
- Controller subsystem
Controls various functions including the data transfer function. D-A converters are included.
- Electron-optics subsystem
Includes the electron-beam column and the X-Y stage.

In this system configuration, the flow of exposure data is as follows.

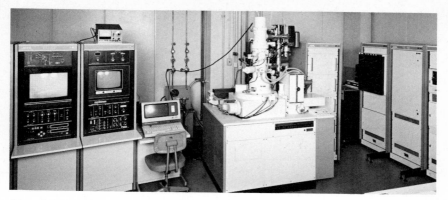

Fig. 2.83. Variable-shaped-beam lithography system (VL-S2)

The exposure chip data which have been converted to the system format are transferred from a magnetic tape to a magnetic disk as data files. The exposure data on the magnetic disk are transferred as chip data in DMA mode through the high-speed data transfer controller, and stored temporarily in the high-capacity buffer memory, where each rectangle data corresponds to one word. Then, word transfer is performed in a high-speed transfer cycle initiated by the exposure start signal. Each rectangle data is corrected for distortion according to the field deflection distortion data measured in advance. Then the data are transferred to the high-speed data controller and input to the register corresponding to the D-A converter. There are two pairs of D-A converters, one pair of X and Y converters (10 bit) for beam variation and one pair of X and Y (16 bit) for beam positioning. After the voltage is applied to the deflection plate through each D-A converter, a blanking signal is added, the electron beam is passed through the column, and then the sample substrate is exposed. The maximum rectangle size which can be processed in this system is 5 μm × 5 μm. Figures larger than this are exposed by simply butting these rectangles together. Data must therefore be converted beforehand by using special data conversion software for rectangle division and exposure sequence. The field size in the VL-S2 system is 2.5 mm × 2.5 mm. Chip data larger than this are exposed by field butting after the chip data are divided into portions of field size or smaller. Chip data are divided into fields beforehand by software exclusively used for this purpose, as in rectangle division. At the same time, data must be sorted so that pattern exposure can be performed according to a fixed sequence to obtain the desired pattern in the field or chip.

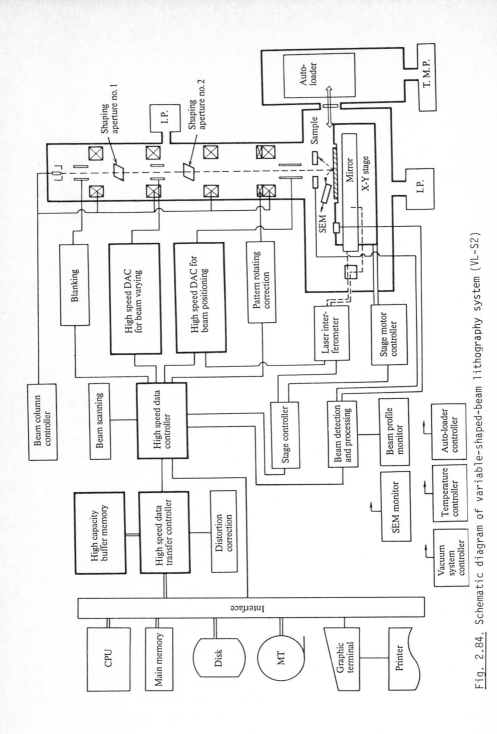

Fig. 2.84. Schematic diagram of variable-shaped-beam lithography system (VL–S2)

The exposure sequence is performed for the entire chip, and chip patterns are completed by butting several fields together. The exposure of the field is handled by the vector-scan system, so the stage does not move during exposure and moves intermittently to position for the next field. The stage movement is continuously and precisely supervised by a laser. Any shift between the specified and the actual stop position is fed back to a servo loop in the beam-position control system, which then corrects the position to provide exact field butting. In mask exposure, the entire mask substrate area is exposed, but in direct exposure, the exposure must be overlaid to match the pattern shaped in the previous process. For alignment in direct exposure, it is possible to scan the preset chip alignment mark for each chip on the wafer with the electron beam to obtain the correct exposure position by processing the reflected electron signals for the exposure of the next pattern layer. When the chip is four fields or larger, three-chip alignment marks are placed on the circumference of the first field for high-precision overlaying exposure, without setting any marks inside the chip.

b) VL-S2 Electron-Optics System

The VL-S2 electron-optics system employs a fixed-area aperture image created inside the electron column, which is projected onto the sample substrate. The ray diagram of the electron-optics system is shown in Fig.2.85. The lens includes four beam condensers and two apertures for variable shaping in the intermediate section. The electron beam emitted from the electron gun (using an LaB_6 single crystal) is projected through the first shaping aperture by the first lens (called the illumination lens). The rectangle beam is then projected into the second shaping aperture by the second lens (called the projection lens). In this case, the beam-projection position with respect to the second shaping aperture is varied by applying voltage to the deflection plate (shaping deflector) between the two shaping apertures. The size of the rectangle beam passing through the second shaping aperture is thereby changed. The rectangle beam is demagnified by the third lens (called the demagnifying lens), and the demagnified rectangle beam is finally projected onto the sample substrate by the fourth lens (called the projection lens). The beam-projection position on the sample substrate is controlled by the beam-positioning deflector after the fourth lens. The electrostatic post-lens deflector used in VL-S2 has a deflection distortion of ±0.4 µm (max) and the amount of deflection-distortion variation in a 100 µm $\times$ 100 µm mesh is less than 0.05 µm.

The characteristics of the basic rectangle beam have a significant influence on the resulting exposure patterns in the variable-shaped beam method.

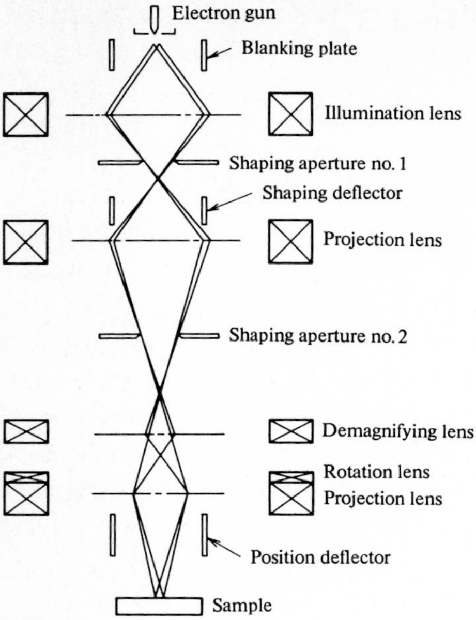

Fig. 2.85. Electron optics system of VL-S2

If the beam sides are not orthogonal with respect to the deflection direction (this need not be considered for a spot beam), the accuracy of consecutive exposure patterns will degrade. At the same time, the linearity of variable characteristics derived from the variable-beam method and the orthogonal characteristics in various X and Y directions must be assured [2.75]. If the current density of the rectangle beam is not uniform and a sharp current rise is not obtained on the edge of the beam, exposure will be uneven and the resolution may be reduced. Accordingly, VL-S2 has adopted various measures for ensuring the electron-optical characteristics unique to the variable-shaped beam method, and these characteristics are automatically adjusted by inputting a check program before exposure. For example, a rotation lens near the fourth lens is used to rotate the rectangle-beam image so that the rectangle-beam side is oriented in the field-deflection direction. The current density distribution (beam profile) characteristics such as beam size, density distribution, the sharpness of current rise, etc., are measured by differentiating the current obtained by scanning the wired Faraday's cage in the X and Y directions. The results are:

- The 5 μm × 5 μm beam size can be set with an accuracy of 0.05 μm.
- The width of the beam foot is about 0.15μm (10 to 90%) less than the 5 ± 0.05 μm square beam size (half-value width) with 1.3 A/cm^2 current density.

- Beam current density distribution in the flat area is within 3% peak to peak.
- The difference of current density distribution in the X and Y direction is less than 0.08%.

c) VL-S2 Stage System

VL-S2 uses a step-and-repeat and a screw-drive system driven by a reliable small-scale high-speed pulse motor. The sample stage is attached on guide rails and supported by a precision ball on the hyperplane stool to minimize pitching and yawing. The stage speed is within 450 ms/2.5 mm over the entire movement range of 110 mm × 135 mm.

The VL-S2 stage system has an autoloader attached to the exposure chamber, which holds up to twelve sheets of 100 mm diameter wafers and 125 mm square mask substrates. The wafers and substrates are loaded into the exposure chamber in the proper sequence automatically. Wafers and mask substrates are set in their respective holders, which are set in magazines and loaded in the autoleader chamber.

d) VL-S2 Data Transfer Control System

Exposure is performed according to the flow of exposure data explained in the above Subsect.2.4.2a. The high-capacity buffer memory is made of 16 Kbit MOSRAMs, with a total capacity of 10.5 MBytes, including the Ecc. Pairs of 16-bit and 10-bit D-A converters and converter amplifiers are used, and the 16-bit D-A converter amplifier subsystem has a 400 ns settling time, which is a new level of performance. Under this system, data are transferred sequentially from the buffer memory and when the D-A converter settles, the electron beam is shot. This shot cycle determines the actual exposure time. The shot cycle measured with blanking signals is 1.65 μs when the blanking-off (electron beam is being shot on the sample substrate) time period is 1 μs. This value measures the practical resist sensitivity, and is much higher than shot cycle values of other existing exposure devices.

The high speed data transfer controller organizes specific pattern data in the exposure block into groups, stores the pattern groups in the high-capacity buffer memory, and transfers pattern groups either as required or repeatedly to configure the chip patterns. In this manner, the buffer memory is used effectively, and high-density bulk chip patterns can be efficiently exposed.

e) VL-S2 Software System

The VL-S2 software system consists of the following main programs:

- Conversion program

Converts CAD data (e.g., PG 3000 data) into exposure data format.
- Exposure control program

Controls exposure processing.
- Editor and utility program

Used for support of exposure processing software.

Using input modification parameter values, the conversion program converts exposure data, from PG 3000/3600 or Calcomp data format, for example, into exposure data format by biasing (shifting an entire pattern area), inch/milli-meter conversion, scaling (demagnification/magnification of entire patterns), mirror conversion, resizing (specific pattern size modification), and positive negative conversion.

The exposure control program is used for mask number input, exposure sched-ule specification, selection of exposure-mode direct exposure (including re-gistration method) or mask exposure, chip configuration, and specification of exposure patterns for each chip. The exposure control program is also used for an operator to communicate with the display on a character display termina Once the pattern file for the chip and the parameters are specified, the ex-posure sequence is performed automatically.

The stage system, beam current, the shaped-beam control system, and the de-flection system can be adjusted by the software of the adjustment program.

f) VL-S2 Pattern Exposure

An experimental example of field butting and pattern overlaying is shown in Fig.2.86 [2.76,77]. The experimental procedure was as follows:

(i) Fifty or more chip areas, each 10 mm square, are butted together in 2.5 mm × 2.5 mm on a 100 mm diameter wafer, and the first layer patterns are exposed.

(ii) The wafer is extracted from the device, reset in the wafer holder, and then the second layer patterns are exposed using the alignment marks.

Figure 2.86 shows test patterns on butting parts at the four corners of the field where errors are most likely to occur. Figure 2.86a shows a wafer on which the first-layer patterns have been developed; the boundary of the four adjacent fields is on the diagonal line on the central grid-shaped patterns. The scale-shaped patterns on the boundary line are error-detection scale pat-terns exposed in each field area. Adjacent error-detection scale patterns are

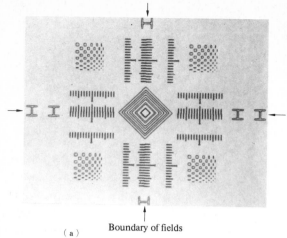

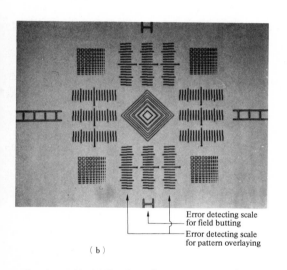

Fig. 2.86. Photomicrographs showing the results of butting and overlaying (a) first layer patterns for field butting (b) overlaid patterns of two layers

(a)　Boundary of fields

(b)　Error detecting scale for field butting
Error detecting scale for pattern overlaying

paired so that the butting error of the two fields can be determined. This error-detection scale includes a main scale and a vernier scale made up of 1 µm width line patterns. It has 0.1 µm error-detection resolution. Thus, the center lines of the main scale and the vernier scale coincide, and the fields are butted with an accuracy better than 0.1 µm. The maximum field butting error in this figure is approximately 0.1 µm. The patterns on both sides of the butting scale are the main scales for overlay error detection (Fig.2.86a). Figure 2.86b shows second-layer patterns that have been exposed by the above-described procedure. The format of the overlay error-detection main scale and vernier scale can be more clearly understood than in Fig.2.86a.

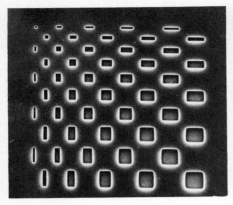

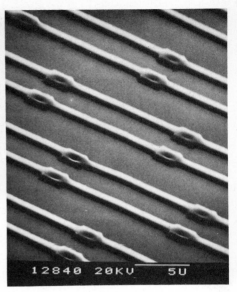

Fig. 2.87. SEM image of rectangle test patterns (5μm x 5μm - 0.5μm x 0.5μm)

Fig. 2.88. SEM image showing the result of overlaying of a device model patterns

The overlay accuracy is within 0.1 μm in this exposure example. The rectangle pattern groups at the four corners shown in Fig.2.86a are exposed patterns varying in size from 5 μm × 5 μm to 0.5 μm × 0.5 μm by a factor of 0.5 μm, and a scanning electron micrograph of these patterns is shown in Fig.2.87. The oblique exposure patterns in Figs.2.86a and b are used to judge the field butting conditions with patterns varying from 5 μm width to 1 μm width by a factor of 1 μm. An oblique line can then be exposed with high accuracy in this exposure condition, under which errors are most likely to be produced. Repetition of these tests has revealed that it is sufficient to expose 0.5 μm width lines, and the accuracy of field butting and pattern overlay over the 10 mm × 10 mm chip area is within ±0.2 μm anywhere on 100 mm diameter wafer.

An example of overlay exposures for a model device pattern is shown in Fig.2.88. This MOSRAM model pattern contains more than 1 Mill. contact hole patterns, and its pattern density is about 5 Mill. rectangle patterns/cm^2 whose throughputs are tested simultaneously. First, the 1 μm × 1 μm contact hole part was exposed and dry-process Si etching was performed. Then, resist (PGMA) was coated and the metal layer (1 μm width in line parts and 2 μm × 2 μm in contact parts) was exposed. The micrograph indicates that the overlay error is less than 0.2 μm.

A test was conducted to check the exposure speed of the VL-S2 system MOSRAM metal conductor model patterns were used, chip sizes were 9.6 mm square, each chip contained about 1600 line patterns of 1 μm width. About 4.6 Mill. rect-

Fig. 2.89. Photograph of VLSI
chip patterns on a 4-in wafer
by direct exposure method
(52 1-Mbit equivalent chip
patterns)

Fig. 2.90. SEM image of VLSI
chip patterns (minimum line
width; 1μm

angle patterns were exposed. The equivalent number of memory cells was more
than 1 Mbits, chip layout on the wafer was in 10 mm × 10 mm pitch, and up to
52 chips could be made on a 100-mm-diameter wafer. The total number of shots
was about $2.5 \cdot 10^{8}$, and the direct exposure method was used. An external view
of the test wafer is shown in Fig.2.89 and a partially-magnified electron
micrograph of the chip is given in Fig.2.90. These patterns were exposed with
PGMA resist, and all 52 chips on the wafer were exposed.

The experimental results were:

- The overhead time was about 165 seconds including the time for wafer
cassette insertion/ejection, column checking, wafer alignment, etc.
- The exposure time was about 885 seconds including the time for the various
processing required by the direct exposure method such as stage movement, mark
detection, etc.
- Thus, the total processing time for all 52 chips was 17 minutes and
30 seconds.

107

2.4.3 Raster-Scan-Type Electron-Beam Delineator (VL-R1, VL-R2)

The electron-beam delineators VL-R1 and VL-R2 are of a raster-scan type. The VL-R1 system was developed for making master masks and reticles of VLSI patterns with a minimum line width of 1 μm. The VL-R2 system was developed for making masks and direct-writing of VLSI patterns down to 0.5 μm.

a) VL-R1 Writing Scheme

The writing scheme of the VL-R1 system is shown in Fig.2.91. A circular electron beam scans repetitiously the substrate with a width W in one direction, while the work-stage moves continuously in the other direction perpendicular to the beam-scan direction. The scanning pitch P is equal to the minimum address unit and is also equal to the beam diameter. The narrow stripe area drawn by a single continuous motion of the stage is termed "frame". Completing the first frame drawing, the work-stage moves towards the beam scanning direction at a distance of the scanning width W. The second frame is drawn with backward motion of the work stage. Thus, the circuit pattern is delineated frame by frame in sequence from one end of the substrate. The writing scheme of the VL-R1 system enables one to delineate such a large chip pattern as a reticle or composite chip pattern in the same delineating time as a single chip pattern.

The VL-R1 system has three modes of operation, A, B, C, listed in Table 2.9. The A and B modes are for master-mask making and/or direct writing, and the C mode is for the 10× reticle making. The minimum line width of A, B and C

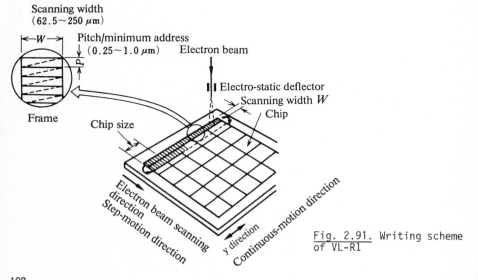

Fig. 2.91. Writing scheme of VL-R1

Table 2.9. Writing modes of VL-R1

Mode	A		B		C
Minimum line width [μm]	1.0		2.0		4.0
Beam diameter [μm]	0.25		0.5		1.0
P [μm]	0.25		0.5		1.0
W [μm]	62.5	125*	125	250*	250
Address [bit]	250	500*	250	500*	250

* Address frequency = 20 MHz

modes is 1, 2, 4 μm, respectively, and is furnished by four addresses. The scanning width of the three modes is 62.5, 125 and 250 μm and consists of 250 addresses. A and B modes are used for making master masks for 1× projection aligners, X-ray aligners or electron-beam aligners. The C mode is for making reticles for reduction-type projection aligners or photo repeaters.

In this writing scheme, a net writing time[3] T_E is defined as

$$T_E = A/f \cdot a^2 \quad [s] \quad , \tag{2.56}$$

where A is the writing area [mm^2], f is the address frequency [MHz] and a is the address unit [μm]. In order to decrease the writing time, it is necessary to increase the address frequency f. The maximum frequency, however, is limited to $f \leq J/S$, where S is the resist sensitivity [μC/cm^2] and J is the beam current density [A/cm^2]. The standard address frequency of the VL-R1 system is 10 MHz, which is able to expose high-resolution but low-sensitivity resist PMMA. To increase the writing speed for higher-sensitive resist the A and B modes can be operated at an address frequency of 20 MHz. In this case, the scanning width becomes double and consists of 500 addresses.

In principle, the raster-scan-type system has to process a much larger amount of writing data than the vector-scan-type system, because the raster-scan-type system scans the whole area of the substrate. In the case of delineating 100 mm^2 of writing area at the 1 μm address unit, the total amount of the data of 10^{10} bit should be processed at a data-transfer rate of 10 Mbit/s. The VL-R1 system provides a high-speed data processing system with a 160 MByte magnetic disk incorporated with a data compaction technique applying the regularity of LSI patterns and the raster-scan method. Consequently, current high-density LSI patterns can be delineated at high writing speed.

---

[3] The total writing time is the sum of the net writing time, the beam flyback time, the stage-stepping time and the substrate load-unload time.

Pattern accuracy is also an important performance of an electron-beam delineator. The system was designed to achieve a pattern accuracy of ± 0.1 μm in the A mode, ± 0.2 μm in the B mode and ± 0.4 μm in the C mode by allocating allowable error for electron-optical, mechanical and electrical subsystems.

b) VL-R1 System Configuration

The system's block diagram of VL-R1 is shown in Fig.2.92. The system consists of four subsystems: the electron-optical, the mechanical, the electrical and the software subsystem.

The electron-optical subsystem illustrated by Fig.2.93 is of the cross-over type and consists of three magnetic lenses. The electro-static deflectors are provided for beam deflection and high-speed blanking. The maximum scan width is 350 μm. A single-crystal LaB_6 cathode is used to obtain high beam current, higher than 1000 nA at a beam diameter of 0.5 μm. The cross-over image formed at the electron gun is focused into the deflection center of the beam-blanking deflector by the first condensor lens. The second condensor lens is used to change the beam diameter which ranges from 0.25 to 1.0 μm. The objective lens is employed to adjust the focus condition of the beam. Accelerating voltages of 20 and 10 kV are available. The electron-optical column is covered with a magnetic shield to prevent an external magnetic field from affecting the beam trajectory.

The work stage installed in the housing chamber can be moved at a maximum speed of 32 mm/s in continuous motion by a DC servo-motor. The stage is driven in the beam-scanning direction at a speed of 5 mm/s by a pulsed motor. The rise time from stationary state to steady-motion state is about 0.2 s. The work stage is supported by mechanical bearings. Pitching and rolling of the stage motion is less than 5 seconds. In order to isolate the housing chamber and the electron-optical column from floor vibration, the chamber is supported by air cushions. The chamber and the electron-optical column is evacuated by ion pumps to obtain clean vacuum. The pre-chamber which can contain 10 mask cassettes is attached to the housing chamber and is evacuated through turbo pumping. The mask cassette is transferred from the pre-chamber to the work stage by the auto-feeder system. In order to minimize the pattern error caused by temperature change a constant-temperature water flow system is provided. Consequently, the temperature of the housing chamber and the electron-optical column is controlled to less than 0.1°C.

The electrical subsystem consists of a high-speed data processing unit, a mini-computer TOSBAC-40D and control interfaces. Pattern data of MANN-3000 format are converted to an intermediate format; the R1 format is stored in a

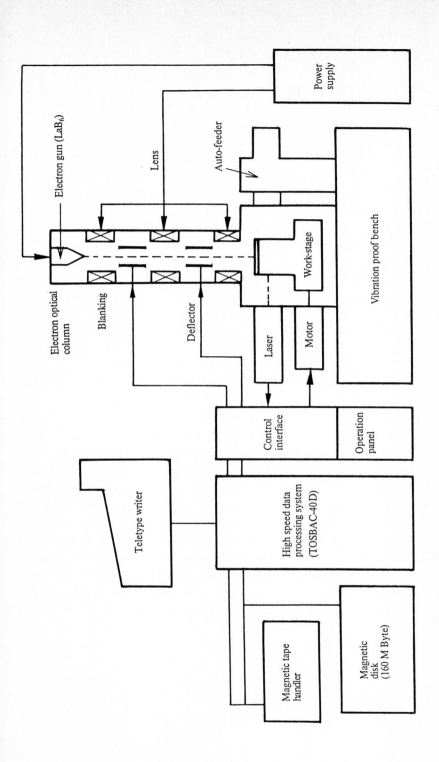

Fig. 2.92. System block diagram of VL–R1

111

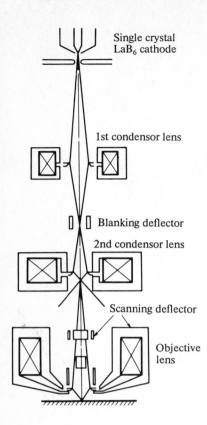

Fig. 2.93. Electron optical system of VL-R1

Single crystal
LaB$_6$ cathode

1st condensor lens

Blanking deflector

2nd condensor lens

Scanning deflector

Objective
lens

160 MByte magnetic disk. Before delineation, R1-format data are converted to bit data and stored in the magnetic disk. At the delineation the bit data are transferred to a part of the bulk memory of the minicomputer, which acts as a buffer memory, and then transferred to the writing circuit through DMA. The stage position is measured by the laser interferometer, and the position data are transferred to the stage-position circuit and timing circuit. Each beam scanning is started by the pulse generated in the timing circuit.

The software of the VL-R1 system consists of a data control program and a system control program. The data control program has functions of pattern data input, edition, conversion and data transfer. Scaling and rotation of circuit patterns and different chip-pattern compositions are possible in the data control program. The system control program manages delineation sequence, stage motion, autofeeder and system error detection.

Figure 2.94 shows a photograph of the VL-R1 system. The typical performances of VL-R1 are listed in Table 2.10.

Fig. 2.94. Overview of VL-R1

Table 2.10. Typical performances of VL-R1

Substrate	Hard mask (125 mm square or less) Wafer (100 mm diameter or less)				
Maximum writing area [μm]	100 mm ×100 mm				
Minimum line width [μm]	1		2		4
Minimum address P [μm] (beam diameter)	0.25		0.5		1
Scan width W [μm] (frame width)	62.5	125	125	250	250
Address bit	250	500	250	500	250
Writing speed [cm^2/min]	0.27	0.54	1.05	2.00	3.62
Address frequency [MHz]	10	20	10	20	10
Writing time (125 mm mask)	–	180	–	50	30
Maximum dose [C/cm^2]	$3 \cdot 10^{5}$	$1.5 \cdot 10^{-5}$	$3 \cdot 10^{-5}$	$1.5 \cdot 10^{-5}$	$3 \cdot 10^{-5}$
Accelerating voltage [kV]	20 (10)	800 nA			
Writing pattern accuracy	<0.1 μm (σ)				
Input data format	MANN #3000				
Data storage	Magnetic tape, 160 Byte magnetic disk pack				
Main functions	Black and white reversal, scaling 32-words ID, registration				

c) VL-R1 Delineation Results

The VL-R1 system can be used for high-resolution positive electron resist PMMA, because an electron dose of 50 μC/cm^2 is available at the address frequency of 10 MHz by employing single-crystal LaB$_6$ cathodes. Figure 2.95 shows a 1 μm

Fig. 2.95. PMMA resist pattern
delineated by VL-R1

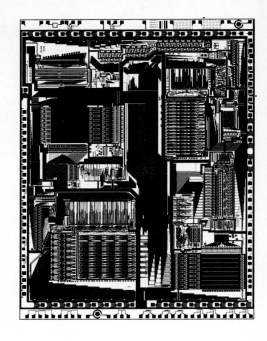

Fig. 2.96. A photograph of a
reticle of high density LSI

thick PMMA resist pattern of an LSI circuit whose minimum feature size is 1 μm.
The development was done in isoamylacetate; a resist pattern with steep profile was obtained. Figure 2.96 illustrates an example of a high-density LSI
reticle of 67×88.5 mm^2. The writing time of this reticle was less than 30
minutes.

d) VL-R2 Writing Scheme

The VL-R2 system provides a new writing scheme with direct writing capability
of high density VLSI patterns composed of 5 Mill. rectangles/cm^2. This is
equivalent to the pattern complexity of 1 Mb RAMs and has a throughput capability of ten times higher than that of the VL-R1 system. Figure 2.97 is a
schematic illustration of the VL-R2 writing scheme, which utilizes the variable-size beam concept combined with the electrical scanning of the beam in
one direction, and the mechanical continuous movement of the work stage in the
perpendicular direction. The beam size in the scan direction (H direction) is
preset at either of two operation modes of the system, which differ in the
address units: 0.125 μm and 0.25 μm. The beam size in the stage-movement direction (V direction) is varied up to 4 μm according to patterns to be drawn.
At the delineation of circuit patterns composed of rectangles parallel and
orthogonal to the scanning direction, the beam size is fixed for one scanning
duration. At the delineation of circuit patterns with slanted rectangles,

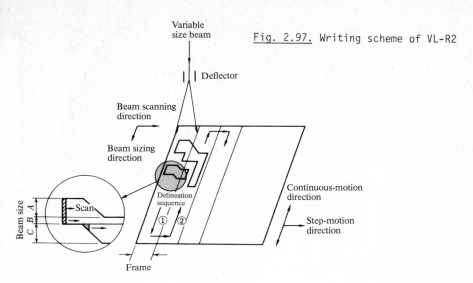

Fig. 2.97. Writing scheme of VL-R2

either the beam size, or both the beam size and the beam position are varied
at an adequate rate according to the direction and the angle of slanted rectangles. The variable-size beam scans 250 µm in width on a substrate in H direction (frame width), but is capable of scanning 350 µm so as to register each
layer in direct writing on the wafer. The beam is also capable of scanning
250 µm in the V direction for skipping the blank area of the pattern to increase the writing speed by reducing the number of beam scannings. The work
stage moves at an optimum speed predetermined by calculating the number and
the size of subcells in each frame. The maximum stage speed is 100 mm/s. In
the raster-scan-type system, the net writing speed may be increased by increasing the address frequency, but it is limited by resist sensitivity and
beam-current density; it is also limited by the pattern-data transmission
rate. In the VL-R2 system, a data compaction technique is also effectively
incorporated to increase the data-transmission rate. VL-R2 can delineate
VLSI patterns with slanted lines without loss of writing time.

e) VL-R2 System Composition

Figure 2.98 gives a schematic diagram of VL-R2. The electron-optical column,
shown in Fig.2.99, consists of an electron gun with a single-crystal LaB_6
cathode, 5 magnetic lenses, 2 beam-sizing apertures and electro-static deflectors for beam blanking, sizing and scanning. Beam sizing is carried out
by shifting the first aperture image onto the second aperture by deflecting
the beam trajectory. An LaB_6 chip with apex radius 480 µm is used to illuminate uniformly the first aperture with enough brightness. Two condenser lenses

115

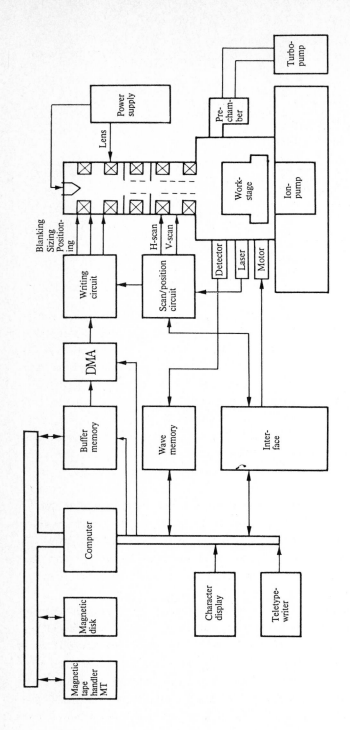

Fig. 2.98. Schematic diagram of VL-R2

116

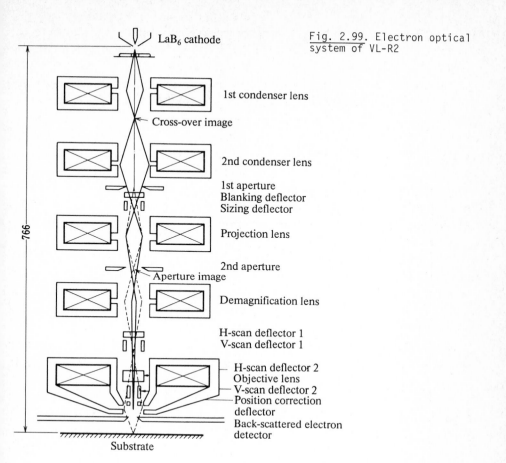

Fig. 2.99. Electron optical system of VL-R2

are used to adjust the illumination condition and the position of the cross-over image to satisfy the Koeller illumination condition. The demagnification lens and the object lens demagnify the aperture image by 40 to the substrate. In addition to the beam-scanning deflectors, a set of deflectors is provided to shift the beam position in the V direction for delineating slant lines. A current density of more than 50 A/cm^2 is attained.

The mechanical design of VL-R2 is similar to that of VL-R1, except for the stage-support mechanism. Air bearing is implemented for the V direction stage support to drive accurately and smoothly the work stage at the speed of up to 100 mm/s. The cassette-loading mechanism is improved to have a good registration accuracy by loading accurately the cassette on the stage.

The system's computer is a TOSBAC-40D with 512 kByte memory. A magnetic-tape handler is used for input of pattern data with PG-3000 and Calma format.

Converted and compact data, system-control data and programs are stored in a 160 MByte magnetic disk. At the delineation, converted pattern data in the magnetic disk are transferred to a high-speed buffer memory with a capacity of 2 MByte. The data in the buffer memory are transferred to a writing circuit through DMA. The writing circuit decodes the compacted data to blanking data, sizing data and position data, which are supplied to deflectors in the electron optical column. The scan-position circuit receives the stage-position data measured by a laser interferometer, and génerates the beam-scanning wave form to be supplied to the scanning deflectors. This circuit makes skip-scan and variable stage speed writing possible, a special feature of VL-R2.

A special registration technique provides for direct writing. Before loading the wafer cassette to the pre-chamber, an optical pre-aligner is used to adjust the wafer setting position to the cassette. Before delineation, registration marks on a wafer are scanned. The mark signal detected from back-scattered electrons is transferred to a wave memory where the signal is processed and converted to digital data. The data are stored in the magnetic disk as registration data. Registration data are used to correct the start position and the angle of the beam scanning.

The software of the VL-R2 system is shown in Fig.2.100. The operating system (R2-OS) was specially developed for this system so as to be able to process a large amount of data and to have functions required in the system such as a test of hardware. The VL-R2 software consists of two main programs: a data con-

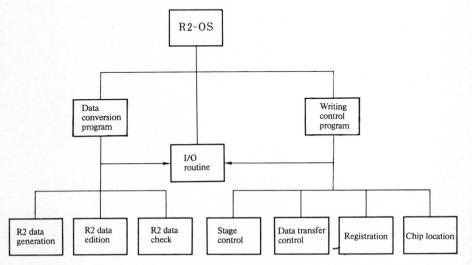

Fig. 2.100. Software of VL-R2

Fig. 2.101. Overview of VL-R2

version program and a writing control program. The data conversion program consists of three sub-programs: data generation, data edition and data check. The
tasks of the data-conversion program are data format conversion from MANN
PG-3000 or Calma to R2 format, scaling, soating, resize, rotation, chip location including different chip composition and data check by displaying patterns
on a graphic display terminal. The writing control program consists of four
sub-programs: stage-motion control, pattern-data control, registration and
chip location. The tasks of the writing control program are sequence control
for delineation including block-writing and circular-chip location for direct
writing, calibration and adjustment of the system.

Figure 2.101 shows an overview of the VL-R2 system. Typical performance
characteristics of VL-R2 are listed in Table 2.11. A maximum writing speed of
8 cm^2/min is obtained. Overlay accuracy measured by the double-exposure method
for 125 mm square mask was 0.05 μm (σ). The overlayer registration accuracy
in direct writing for a 100-mm-diameter wafer is a shift of 0.17 μm r.m.s.
with 0.06 μm (σ).

f) VL-R2 Delineation Results

Figure 2.102 shows a SEM photograph of a resist pattern of a high-density
submicrometer circuit model which is made by using the scaling function of
the data conversion program. The electron resist was EBR-9 (poly-tri-fluoro-
ethyl-α-chloroacrylate) developed in the Cooperative Laboratories.

g) Conclusion on VL-R1 and VL-R2

The VL-R1 system efficiently delineates master-mask patterns for LSIs of 1 μm
minimum feature size and 10× reticle patterns. The writing time of the reticle

119

Table 2.11. Typical performances of VL-R2

Substrate	100 mm diameter wafer, 125 mm square mark
Maximum writing area	105 mm ×124 mm
Minimum line width	0.5 μm 1.0 μm
Address unit	0.125 μm 0.25 μm
Scanning width	250 μm
Address frequency	30 MHz
Maximum beam size	4 μm
Beam current density	50 A/cm^2
Accelerating voltage	20 kV
Pattern accuracy	0.1 μm (σ)
Registration accuracy	0.2 μm
Maximum writing speed	100 mm□/12 min
Input data format	MANN #3000, PG 3000, Calma
Main functions	Black and white reversal, scaling, resize, mirror

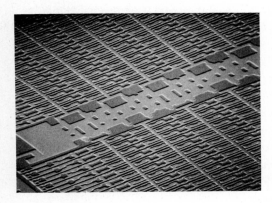

Fig. 2.102. EBR-9 resist pattern delineated by VL-R2

is less than 30 minutes so that an electron-beam delineator of the VL-R1 tpye is currently in practical use in the production of LSI masks. The VL-R2 system successfully shows its possible capability for high-speed delineation of future VLSIs, essentially in direct writing on wafers. The VL-R2 system will become an effective delineator by optimizing the total system taking advantage of advances in related technologies.

3. Pattern Replication Technology

In semiconductor device fabrication, optical lithography has widely been employed as the pattern replication method. Today, three methods are used: the shadow casting method, the imaging method, and the holographic method.

3.1 UV Replication Technologies

The method originally introduced in the wafer lithography process was the shadow casting technology which is still most popular. When the necessary conditions are optimized, patterns as small as 0.5 µm can easily be replicated even with visible light. Consequently, shadow casting has been frequently used as the most powerful method for manufacturing small chips with fine patterns. However, for the fabrication of LSI, this method becomes inadequate for the following reasons: (i) the method is liable to cause serious damage to both photomask and wafer while they are brought into contact, and (ii) it shows quite different resolution characteristics for a slight change in the gap between photomask and wafer.

In the imaging method, the photomask and the wafer are separated by a certain distance thus avoiding any damage caused by pressure contact between photomask and wafer during replication. Then the method is suitable for the fabrication of LSI, when its resolution capability is sufficiently improved. Table 3.1 lists imaging lithography systems that have been developed for semiconductor device fabrication.

Table 3.1. Exposure schemes for full wafer surface patterning

Exposure scheme	Optics
one shot	refraction, catadioptric
scanning	reflection, catadioptric
step and repeat	refraction, catadioptric

With regard to refractive optics, almost diffraction-limited lenses have become fabricable by virtue of considerable progress in glass-material technology and computer-aided design techniques. The numerical aperture of imaging optics seems to be limited to about 0.4 in practical use, because large numerical aperture leads not only to high resolution but also to a shallow focal depth. Meanwhile it is obvious that resolution improves with shorter wavelength. Much effort has been put into the development of UV optics, but a practically useful UV lens has not yet come out. A lens with wider image field is quite beneficial from the viewpoint of throughput; however, image field size and resolution capability are the largest trade-offs for high performance optics. Figure 3.1 illustrates such high performance lenses that have been developed for semiconductor-device lithography.

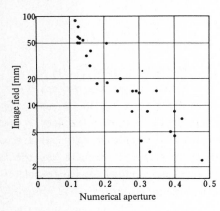

Fig. 3.1. High-performance lenses developed for semiconductor device lithography

The greatest advantage of reflection optics is its freedom from wavelength effects: high throughput is expected because of the adaptability to the full spectral-sensitivity range of the photoresist. Moreover, a higher resolution capability is also expected by exploiting shorter wavelength regions.

Typical reflection optics, employed in semiconductor-device fabrication, is composed of a pair of concave and convex mirrors which are arranged to be nearly concentric. Such an arrangement produces almost an aberration-free ring zone at a certain distance from the optical axis. The usable zone width is so narrow that it is quite necessary for the purpose of full wafer surface patterning to scan the photomask and the wafer simultaneously. Therefore, the resolution characteristics of a mirror projection system depends not only on its optical performance but also on its mechanical performance.

Catadioptric optics has a much wider zone than reflection optics. By introducing certain refractive elements, it is applicable not only to the scanning exposure system but also to the step-and-repeat exposure system.

Table 3.1 describes three sorts of exposure schemes for full wafer surface patterning. The one-shot exposure scheme has an advantage in constructing a simple and low-cost pattern replication system, but, for an increased wafer size, a larger image field is required and the resolution capability decreases. The scanning exposure scheme fits all kinds of imaging optics, especially reflection optics which has a long arc-shaped image area. The step-and-repeat exposure scheme is quite useful for refraction optics with high-resolution capability, but has a small image field. In addition, it has another advantage in alignment accuracy because of its capability to correct each exposure position. Thus, step-and-repeat systems along with some scanning exposure systems are expected to be the most powerful lithographic methods in the production of VLSI, until the problems of E-beam and X-ray systems are successfully solved.

In the following, we shall give an introduction to two kinds of step-and-repeat exposure systems (stepper), experimentally developed in 1977 in the Cooperative Laboratories, VLSI Technology Research Association.

3.1.1 Reduction Stepper (VL-SR 2)

The system developed so far is illustrated by a photograph (Fig.3.2) and the schematic drawing in Fig.3.3; its specifications are listed in Table 3.2. The imaging optics is mounted vertically to stabilize its performance as well as the photomask setting to keep air-borne dusts off the mask surface. The entire operation is controlled by a set of microcomputers. In the following subsections the imaging optics and the alignment mechanism are discussed in more detail.

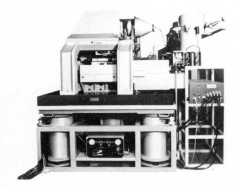

Fig. 3.2. Overview of VL-SR 2

a) Imaging Optics

The imaging performance of an optical system improves with an increase in reduction. However, larger photomasks are then necessary which cause some prob-

123

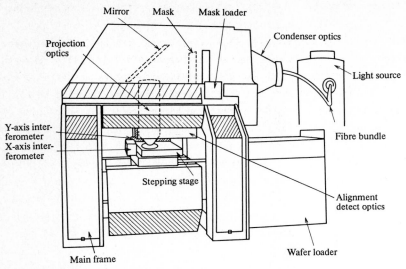

Fig. 3.3. Schematic drawing of VL-SR 2

Table 3.2. Specifications of VL-SR 2

	Specification
projection optics	1/10 reduction, N.A. = 0.28
wavelength	g-line (436 nm)
unit image field	14.5 mm diameter (10 mm × 10 mm)
resolution	1 μm
mask size	12.5×12.5 cm^2
mask setting	automatic
wafer size	75 mm, 100 mm, 125 mm
wafer feed	automatic
exposure scheme	step and repeat
alignment	automatic
alignment accuracy	0.3 μm (groval); 0.1 μm (step by step)

lems. Therefore, the reduction ratio of the optics employed in this system has been chosen to be 10:1.

In order to realize a uniform imaging performance and a minimum image distortion over the image field, a telecentric arrangement was employed on the image side of the optics.

The source of illumination is a short-arc high-pressure mercury lamp. Its light is collected by collecting elements. After the removal of heat radiation,

124

the light is monochromatized and fed into a condenser optics through a bundle
of optical fibres.

The coherence of the illumination critically influences the image contrast,
and the system is thus designed to choose the preferable coherence by changing
the diameter of the fibre bundle. The contrast transfer characteristics of the
imaging optics is exhibited in Fig.3.4; it was obtained by measuring spatial
frequency patterns transferred to a positive photoresist.

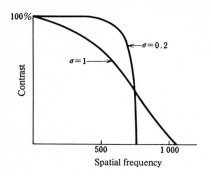

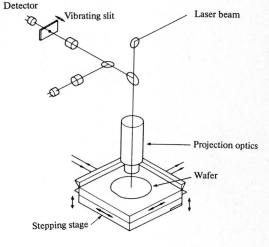

Fig. 3.4. Contrast characteristics
of developed lens

Fig. 3.5. Focusing optical arrangement of VL-SR 2

Figure 3.5 illustrates the focussing optics. A laser beam passes through
the imaging optics and is focussed into a spot on the wafer surface. The beam
reflected from the surface comes back through the optics, is folded by a half
mirror and forms a conjugate image on an axial position sensor. According to
the information from the sensor, the wafer stage is driven vertically to keep
the correct focus. As the sensing accuracy is better than 0.2 µm, automatic
focussing is satisfactorily performed. ,

Figure 3.6. shows some scanning electron micrographs of resist patterns
in 1-µm-thick positive photoresist, exposed under $\sigma = 0.25$ illumination condition. Although a slight contrast degradation is seen in 0.8 µm lines and
spacings, patterns larger than 1 µm have sufficient contrast and steep profiles.

As another example, Fig.3.7 shows 1.0-µm-wide resist patterns formed across
0.5 µm polysilicon steps. The dimensional change in resist patterns beneath
the steps seems to be negligibly small.

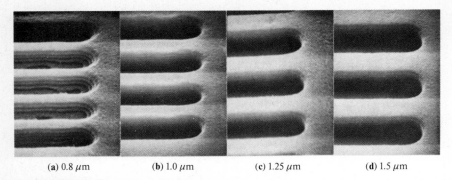

(a) 0.8 μm   (b) 1.0 μm   (c) 1.25 μm   (d) 1.5 μm

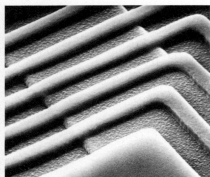

Fig. 3.6. Scanning electron micrograph of developed resist patterns

Fig. 3.7. 1.0 μm wide resist patterns formed across 0.5 μm polysilicon steps

b) Alignment Mechanism

Two modes of alignment are utilized in the system under consideration: off-axis groval alignment and TTL step-by-step alignment.

The off-axis groval alignment performs through the optical arrangement shown in Fig.3.8. A pair of cross marks are prepared near the periphery of the wafer for the position measurement. The cross marks are registered with a pair of biaxial electrooptical microscopes, and so the information on the position is obtained. Exposure positions are then calculated.

In an ordinary off-axis alignment system, the problem has been observed that alignment errors sometimes occur due to the run-out and the drift of relative positions of the imaging optics and the measuring microscopes. In our system, the problem was successfully solved by applying an automatic position correcting function which is operated by measuring the positions of both imaging optics and microscopes prior to the exposure. The alignment accuracy obtained with this system is shown in Fig.3.9.

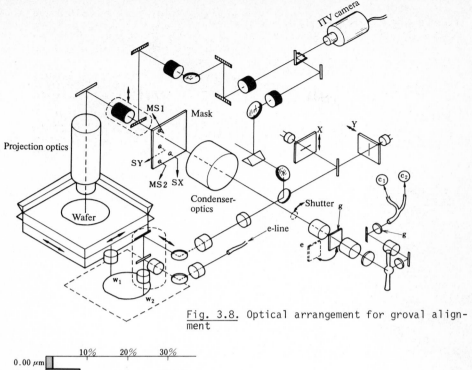

Fig. 3.8. Optical arrangement for groval alignment

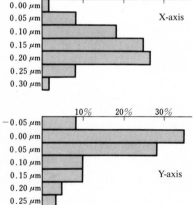

Fig. 3.9. Alignment accuracy obtained with groval alignment

The optical arrangement for the step-by-step alignment is shown in Fig.3.10. Each step to be aligned is previously prepared with certain alignment marks, for example in scribing lines. The marks of each step are detected through the imaging optics and then the exposure position is corrected. In the usual case, it is unnecessary to realign all the steps and, therefore, the system provides a zonal realignment function.

127

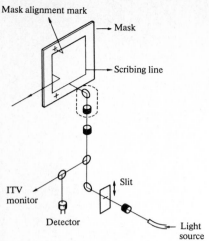

Mask alignment mark

Mask

Scribing line

Slit

ITV monitor

Detector

Light source

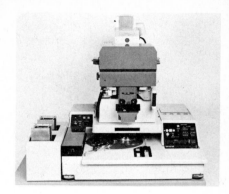

Fig. 3.10. Optical arrangement for step by step alignment

Fig. 3.11. Overview of VL-SR 1

Table 3.3. Specifications of VL-SR 1

	Specification
projection optics	1/1, telecentric, N.A. = 0.16
wavelength	g (436 nm), h (405 nm)
unit image field	42 mm diameter (30 mm × 30 mm)
resolution	2 μm
mask size	5×5 cm^2, 6.75×6.75 cm^2
wafer size	50 mm, 75 mm, 100 mm
wafer feed	automatic
exposure scheme	step and repeat
alignment	automatic
alignment accuracy	0.3 μm

3.1.2 1:1 Stepper (VL-SR 1)

The system developed is illustrated by a photograph (Fig.3.11) and schematically shown in Fig.3.12; its specifications are given in Table 3.3.

a) Imaging Optics

In order to obtain a large image field and a long focal depth, a 1:1 image optics was employed. This optics is designed in a symmetrical configuration: every nonsymmetrical aberration such as coma, distortion, etc. is intrinsically neglected. Consequently, by correcting other aberrations exactly, high resolution and wide field optics with telecentric configuration was developed

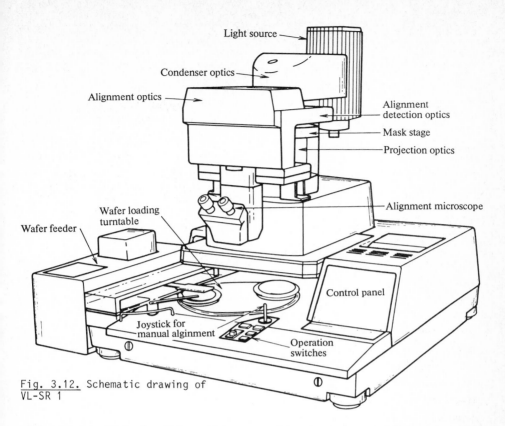

Fig. 3.12. Schematic drawing of
VL-SR 1

with success. Employing this optics, we could reduce the stepping motion
remarkably and achieve a very high throughput.

The light source for this optical system is a 500 W short-arc high-pressure
mercury lamp. The light is collected, monochromatized after removal of heat
radiation and then fed into condenser optics so as to save the amount of
usable radiation. Relatively high coherence is employed to improve the image
contrast. Figure 3.13 shows a scanning electron micrograph of some resist
pattern with 2 μm lines and spacings, obtained with this system.

Fig. 3.13. Scanning electron micrograph of developed resist patterns

b) Alignment Mechanism

The automatic wafer alignment function operates on the step-by-step principle utilizing optical scattering at the edges of alignment marks previously formed on the wafer (Fig.3.14). The light beam from a He-Ne laser is reflected by a polygon mirror, passed through the imaging optics, and scans the alignment marks on both the photomask and the wafer. The shape of the marks are determined on two orthogonal lines rotated by 45° in order to obtain the positional information for both the X and Y axes simultaneously in one scan. Only the scattered light from the wafer marks can be received by the detector. Reflected light coming directly from the wafer surface is blocked by a stopper located on the optical axis of the detection optics. Using this optical arrangement, positional information sufficient for alignment can easily be obtained even with marks which have a step height lower than 1000 Å.

The throughput of this system was measured to be about 60 pieces of 7.5 cm wafers per hour in fully automatic alignment and exposure operation.

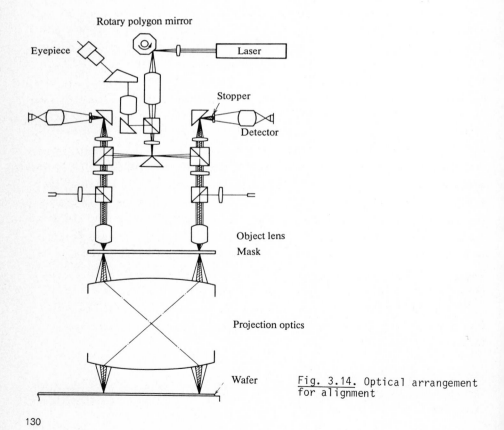

Fig. 3.14. Optical arrangement for alignment

3.2 Deep-UV Projection System

The technology of pattern writing on silicon wafers utilizes a mirror projection and reduction step-and-repeat systems, departing from contact printing and proximity printing systems. For the future, the most important fine patterning technology will be the soft X-ray lithography system using a very small scale synchrotron as an ideal mass production system. Although it is possible to build up such a machine now, its application in a production line is very difficult at present for different reasons. Therefore, the deep-UV light projection system is proposed to span the gap between the last and future technologies. The physical basis for this is that the patterning resolution is proportional to the wavelength, so the UV projection system will find its place between violet-light and X-ray systems.

3.2.1 Light Source

There are various deep-UV-light soureces such as D_2, Hg, Xe-Hg lamps and lamps containing heavy metals, etc. The D_2 lamp radiates low-intensity deep-UV light, and needs a water cooling unit. It is difficult to produce a powerful D_2 lamp, and its lifetime is short. The low-pressure Hg lamp provides high intensity, 184.9 and 253.7 nm radiation. However, it is impossible to manufacture a powerful low-pressure Hg lamp. The super-high-pressure Hg lamp is emitting very little deep-UV light, but in a remarkable degree thermal light. The lamp containing Cd radiates little energy in the region of 230 to 240 nm because of self-absorption of Cd. In addition, it is difficult to control the cooling in its arc-shaped lamp. Because of the drawbacks of the above-mentioned lamps, a point source high pressure 2 kW Xe-Hg lamp was used. Such a light source easily provides increased brightness in the range of 200-260 nm. The spectral energy distribution of this lamp is shown in Fig.3.15. A high-pressure Xe-Hg lamp can easily have a higher brightness by shortening the arc length. However, the longer the arc length is, the samller is the output. On the other hand, the lifetime of this lamp decreases with the shortening of its arc length. An arc length of 3.5 mm seems to be a compromise.

As expected, this lamp is emitting a reasonable deep-UV radiation from 200 - 260 nm. Lamps for a wavelength below 200 nm will be developed in the future.

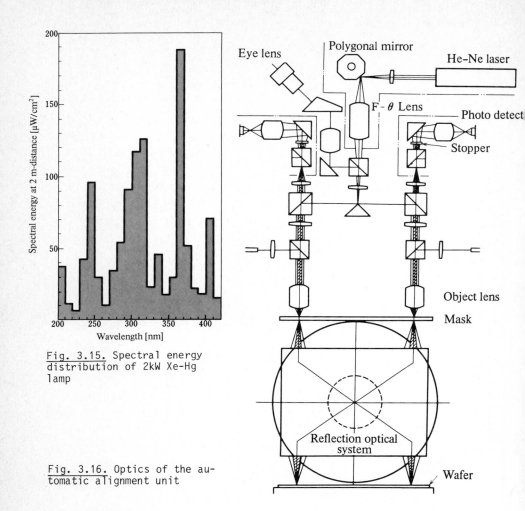

Fig. 3.15. Spectral energy distribution of 2kW Xe-Hg lamp

Fig. 3.16. Optics of the automatic alignment unit

3.2.2 Alignment

High alignment accuracy is the key to patterning VLSI devices. Figure 3.16 demonstrates the optics for the automatic alignment unit. A He-Ne laser is used as the light source. Its beam is scanned across the photomask and the wafer by a polygonal mirror. The F-θ lens keeps the scanning speed of the beam constant. The scattered light from the edges of each alignment mark on the photomask and on the wafer is detected photoelectrically. Each direct reflection from the photomask is blocked by a stopper. The relative deviation between photomask and wafer can be calculated. Then the alignment mark on the wafer is aligned to the alignment mark on the photomask by a pulsed motor drive. By using this automatic alignment system, the alignment accuracy is about ±0.3μm.

3.2.3 Overall System

Today there are two deep-UV systems available, the proximity system and the projection system. The former has already been announced by Canon, Inc. The latter system will be described in this section.

The optics of the deep-UV projection system (VL-MR1) is shown in Fig.3.17. It is a reflection-mirror projection system with a deep-UV source. The equipment is able to expose wafers with a diameter of 12.5 cm in a single scan. An advantage of the mirror projection system is its efficiency for deep-UV light because it minimizes the chromatic aberration and does not absorb deep-UV light. The light source is of the 2 kW Xe-Hg type mentioned in Sect.3.2.1 with cooled metallic terminals. The point light source is changed to an arc-shaped beam with three concave mirrors to achieve an exact focus of the last mirror projection optics. A cold mirror is used to reflect the deep-UV light and to suppress the temperature rise in the photomask and in the wafer by

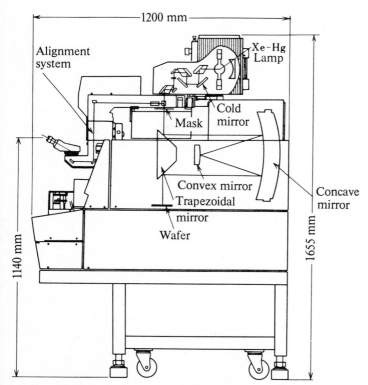

Fig. 3.17. Optical system of deep-UV reflection mirror projection

133

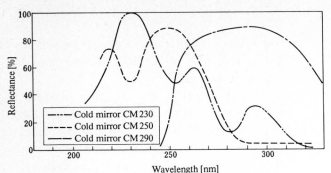

Fig. 3.18. Spectral reflectance of cold mirrors

infrared light. The mirrors CM 230, CM 250 and CM 290 are devised for maximum reflectance at each 230, 250 and 290 nm wavelength. The spectral reflection of the cold mirror is shown in Fig.3.18. The expanded arc-shaped beam is projected onto a mask through a 1 - 1.5 mm wide slit.

The reflection-mirror projection system consists of a trapezoidal mirror and a pair of spherical mirrors. In this system, the curvature-radius ratio of the concave mirror and the convex mirror is 2 : 1. Their focal points are offsetted about several mm to minimize the astigmatism of the sagittal and the meridional fields. An F number of 3.5 was chosen. In the case of a glass-lens system, the smaller the F number, the higher is the resolution. But in the case of a reflection-mirror projection system, the larger the F number, the higher is the resolution. In addition, an increase of the F number increases the depth of the focus tolerance, especially at shorter wavelengths. This optics has a maximum resolution of 1 μm with a focus tolerance of ±4 μm using 230 nm wavelength light; it projects mask patterns 1 : 1 to the wafer.

The photomask is fabricated on a synthetic quartz substrate which is transparent to deep-UV light. The wafer is vacuum chucked from the back surface to minimize the wafer distortion. The groove width and spacing of the vacuum chucking jig is designed to minimize the wafer distortion. The vacuum-chucking jig is then pulled up by vacuum so that the wafer surface touches nails at several points which keep the wafer surface within the focus tolerance. The photomask and the wafer are mounted on a moving frame supported by air-bearings on linear rails. The air-bearings are used for smooth scanning of the frame by a linear motor. The entire frame is moved in the y direction so that the arc-shaped patterns scan and expose across the wafer. The linear scanning by air-bearings minimizes the distortion between wafer and mask to less than 0.1 μm at scanning operation. In this system, there is no roof mirror and the number of mirrors is minimized. The patterning compatibility for mask and wafer and between different machines is good.

Table 3.4. Specifications of a deep-UV mirror projection system

Item	Specification
mask size	4"□, 5"□ and 6"□
wafer size	75mm, 100 mm and 125 mm diameter
optics	1 : 1 reflection mirror projection
wave length	200 - 260 nm
resolution	1 μm
automatic wafer feed system	option
exposure system	1.5 mm width slit scanning system
scanning speed	12.5 - 1.25 mm/s
scanning irregularity	<1%
alignment system	automatic alignment system
alignment accuracy	0.3 μm
automatic exposure control	error <2%

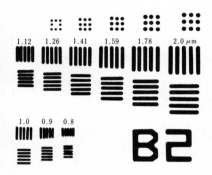

Fig. 3.19. Resist pattern by deep-UV lithography. (Resist: White Resist, 0.5 μm)

The major characteristics of the system are listed in Table 3.4 and some resulting resist patterns are demonstrated in Fig.3.19. Without question, the deep-UV mirror projection system can be improved for higher resolution by developing shorter-wavelength lamps and a better control system of the atmosphere around the optical path. Further improvements point towards better materials for resist and mask, design of optics for a 0.5 μm resolution and the application of the scan and step-and-repeat system.

3.3 X-Ray Lithography

X-ray lithography is a technique for mask-pattern replication using soft X-rays in the range of 0.4 - 5 nm. Figure 3.20 illustrates a simplified

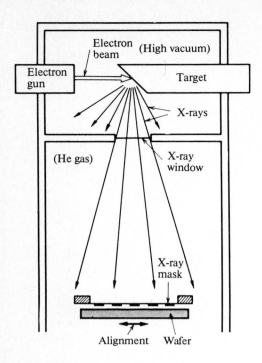

Fig. 3.20. Schematic diagram of an X-ray lithography system

X-ray lithographic system. X-rays that are produced by bombarding a target with an electron beam in high vacuum are led into a helium-filled exposure chamber through an X-ray window (beryllium foil), irradiating the X-ray resist on a wafer through an X-ray mask. A small gap is usually kept between the mask and the wafer (what is called "proximity exposure method"). In the case of pattern overlaying, the mask and the wafer should be aligned before exposure.

X-ray lithography was first reported by SPEARS and SMITH in 1972 [3.1,2], and this technique attracted attention because of its capability of very-high-resolution replication. Since then it has been actively studied to adopt it for practical use. Numerous papers on experimental equipment and masks have meanwhile appeared. Because of problems both in equipment and masks, X-ray lithography has not yet been introduced in production lines. It is, however, expected as an important technique for manufacturing VLSI because it may be easy to replicate submicrometer patterns.

3.3.1 Problems of X-Ray Lithography

It seems to be suitable to sketch advantages and disadvantages of X-ray lithography.

a) Features

The advantages of X-ray lithography are as follows:

(i) High resolution: The wavelength of soft X-rays is much shorter than the dimension of the device patterns, so that diffraction effects can be neglected. Even proximity exposure hardly reduces resolution.

(ii) High aspect ratio: Resist patterns can easily be generated with high aspect ratios (height-to-width ratios) because X-rays propagate along straight lines through even thick layers of resist, and backscattering is small.

(iii) Insensitivity to dust: Dust particles little generate defects in replicated patterns, because X-rays have a high transmissivity to ordinary dust.

On the other hand, there are some disadvantages:

(i) Difficulty of fabricating masks: It is not easy to obtain masks having sufficient strength and dimensional stability because the absence of solid material with sufficient transparency to soft X-rays requires the mask substrate to be extremely thin.

(ii) Difficulty of precise alignment: It is not easy to obtain an overlay accuracy suitable for fine patterns across the wafer, because it is difficult to deflect X-rays and therefore either the mask or the wafer has to be moved for alignment.

(iii) Long exposure time: The low efficiency of conventional X-ray sources and the difficulty of collimating X-rays requires resists which can be exposed with low-power X-rays. At present the existing X-ray resists have too low a sensitivity.

b) Geometrical Distortions

Normally, two types of geometrical distortions affect the efficient utilization of X-rays. At first, because of the difficulty in collimating X-rays, a diffused X-ray beam is used for exposure. As shown in Fig.3.21, the existence of a mask-to-wafer gap s causes the penumbral blurring δ due to the finite size of the X-ray source and the geometrical distortion Δ (often called runout or run-off), caused by nonnormal incidence of X-rays at the edge of the wafer. δ reduces the resolution and the accuracy of the pattern width, Δ reduces the pitch accuracy between the patterns. In the case of pattern overlaying, Δ' (fluctuation of Δ, due to fluctuation of s) is more annoying than Δ, because Δ' causes registration errors of the patterns. δ and Δ' are given by

$$\delta = -\frac{sd}{D} \ ,$$

(3.1)

$$\Delta' = -\frac{s'R}{D} \quad , \tag{3.2}$$

where d is the diameter of the X-ray source, D is the distance between the X-ray source and the mask, s' is the gap fluctuation, and R is the radius of the field to be exposed.

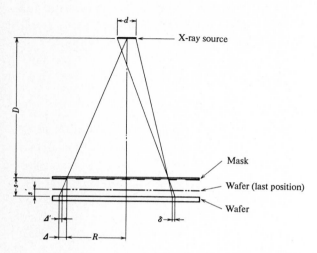

Fig. 3.21. Geometrical relation in X-ray proximity printing

Among these parameters, d is limited by the cooling capacity of the target, R is decided by the wafer size, and s is limited by the mechanical tolerance. Therefore, the finer the pattern and the larger the wafer, the longer is D. The required exposure time becomes longer proportional to D^2.

c) Wavelength Selection

Both exposure time and resolution are influenced by the X-ray wavelength. As shown in Fig.3.22, the absorption coefficient, except for the discontinuities at absorption edges, decreases with decreasing wavelengths. Consequently, X-ray attenuation by mask substrate and the X-ray window decrease, and unfortunately the resist sensitivity decreases, too. Therefore, there is an optimum wavelength for minimum exposure time under a definite condition. It is obvious that the exposure time can be shortened by using a mask substrate, an X-ray window and an environment, whose materials have the lower absorption coefficients. Furthermore, its thickness should be as thin as possible. Under the condition mentioned, there is an optimum wavelength in the lower wavelength region.

138

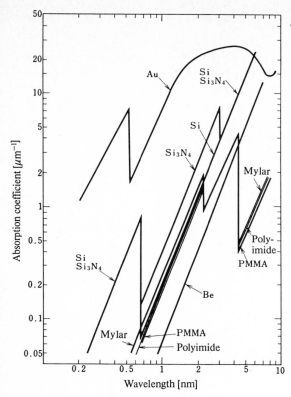

Fig. 3.22. Absorption coefficients for some of materials in the soft X-ray region

Regarding the resolution, the shorter wavelength is usually disadvantageous, because in order to obtain the required contrast of the mask, the shorter wavelength demands thicker absorber patterns that are more difficult to form finely. In addition, the shorter-wavelength radiation increases secondary effects by X-rays within the resist such as the Auger effect. For practical use, the wavelength should be 0.4 - 5 nm with regard to resolution, and 0.4 - 1.4 nm with regard to exposure time.

3.3.2 X-Ray Source

The principle of electron-beam excitation in which an accelerated electron beam is focussed onto the target is widely used for the generation of X-rays. High-power sources being on the market for industrial use produce hard continuous X-rays with wavelengths less than 0.1 nm. Therefore, this type of source cannot be used as a soft X-ray source. In an X-ray lithography source, the electron gun has to operate at relatively low voltages (<25 kV), and then it is possible to get characteristic soft X-rays from the target. The physical process is based on the fact that an accelerated electron can expel one electron from the K shell of the atom, and therefore an electron of the L shell can fall into the vacancy generated in the K shell. Then, characteristic X-ray (K_α line) are emitted, having the wavelength corresponding to the energy difference between the K shell and the L shell. Table 3.5 lists the wavelength of the characteristic X-rays and the radiated power in µW/sr for 1 W power input for several target materials [3.3]. Only a small fraction ($\sim 10^{-4}$) of the power input is converted into the radiated power. Stationary targets were commonly used in an early stage of X-ray exposure by an electron-beam excitation. However, sufficient X-ray intensities could not be obtained, because only low electron-beam power (~ 1 kW) was permissible due to the low cooling efficiency of the stationary target. Consequently, rotating targets are applied to increase the cooling efficiency. By this method, 10 to 25 kW power input can be achieved. When using low-sensitivity X-ray resists, reasonably short exposure times cannot be obtained even by this high-power source.

Table 3.5. X-ray target efficiencies

Target	Wavelength [nm]	Efficiency [µW/W/sr] v = 20 kV
C ($K\alpha$)	4.48	11.6 (10 kV)
Cu ($L\alpha$)	1.334	19 (8 kV)
Al ($K\alpha$)	0.834	55
Si ($K\alpha$)	0.713	60
Mo ($L\alpha$)	0.541	53
Rh ($L\alpha$)	0.460	65
Pd ($L\alpha$)	0.437	68

High-power targets are usually installed in demountable tubes. The use of such tubes requires vacuum pumping systems, but it is easy to exchange targets and filaments in the tubes. Fairly high power input is allowed by using rotating targets, but the mechanical vibration and the difficulty of vacuum seal are unavoidable for such targets. Therefore, Bell Laboratories developed a stationary conical-shaped target cooled by a high-velocity water flow. 4 - 8 kW power can be obtained using this target [3.4]. A new transmission target has also been developed [3.5,6]. This target is a large stationary plane-source, consisting of many component targets (constituent parts of the plane source). Each target has both a slit structure (a narrow aperture for the transmitted

X-rays) to control penumbral blurring and geometric distortion, and a foil target in the form of either a circle or a square. Reduction of the gap between mask and wafer allows a close assembly of the sample (mask and wafer) to the source, and leads to shorter exposure times. In other words, the target can easily be designed according to pertinent requirements, such as exposure time, penumbral blurring and geometric distortion.

High conversion efficiencies can be obtained by using soft X-ray emission from laser-produced plasmas. When high-power laser beams ($\sim 10^{14}$ W/cm^2) are irradiated onto solid targets, such as iron and aluminum, high-intensity soft X-rays are emitted isotropically at the focus. It was experimentally confirmed that the laser energy is converted into X-rays with an efficiency of about 10 - 25%. The high conversion efficiency from laser radiation to X-rays is due to a favorable electron energy distribution which produces both bound-bound and free-bound transitions in the highly ionized atoms. In addition, the high ion density ($\gtrsim 10^{21}$ cm^{-3}) and the high radiative transition probability of the collisionally excited states combine to yield a highly emissive radiator. As an example, 100 J, 1 ns laser pulses focussed onto a solid iron target were converted into X-rays with an efficiency greater than 25% [3.7]. Most of the observed X-rays consist of spectral lines and are concentrated in the soft X-ray region between 0.3 and 1.5 keV. However, the laser plasma source has some basic problems, such as the increase of the laser-pulse repetition rate and the optimization of laser energy, pulse width and target element, when using this source for X-ray exposure.

IBM has developed a new X-ray tube [3.8]. It operates by creating a hot, dense plasma of carbon ions and electrons through which an electron beam passes, causing X-rays to be emitted by the carbon ions. The X-ray tube has a relatively simple structure, produces short bursts of intense X-rays with a 10 - 100 Å wavelength and can produce a resist exposure in 100 ns.

The synchrotron radiation source for X-ray lithography has such character- istics as high intensity, wavelength tunability and small beam divergence. Due to the good synchrotron radiation collimation, penumbral blurring is elimi- nated and the gap between mask and wafer is limited mainly by diffraction. It was reported that, for gaps up to 1 mm, these diffraction effects would not interfere with subsequent device fabrication steps for 1 μm linewidth de- vices [3.9]. The wavelength of synchrotron radiation is in the $\sim$0.1 to $\sim$10 nm range and soft X-rays of a few nm in wavelength are obtained with high in- tensity. However, the synchrotron radiation source is huge in size and very expensive.

3.3.3 X-Ray Masks

An X-ray mask consists of absorber patterns on a soft X-ray transmitting substrate and a framework supporting the substrate. As mentioned in the preceding
section, there is no solid material sufficiently transparent to soft X-rays
(Fig.3.22). Consequently, the mask substrate should be extremely thin in order
to keep the absorption loss low. The material for the X-ray mask substrate
has to fulfill the following conditions:

(i) a small value of the absorption coefficient in the soft X-ray region;

(ii) the possibility to make a membrane, and sufficient strength in membrane
form;

(iii) resistance to high temperatures and to strong chemicals during the
mask fabrication processes;

(iv) a smooth and flat surface, and maintenance of flatness;

(v) dimensional stability;

(vi) transparency to visible light if an optical mask alignment is used.

It is difficult to satisfy all these conditions, a fact that is a disadvantage of X-ray lithography. On the other hand, as a material for absorber
patterns, gold is generally used because of its large absorption coefficient
in the soft X-ray region and its easy processing. However, it is not completely
opaque to soft X-rays so that absorber patterns require about 0.4 - 0.8 μm in
thickness.

In the following we shall discuss some attempts using different materials
for the mask substrate.

a) Inorganic Materials

Silicon membranes were first reported as X-ray masks [3.10]. Starting with
absorber patterns fabricated on a heavily boron-doped layer of the front surface in a silicon wafer, the silicon is etched with a solution of ethylen-
diamine, pyrocatechol and water from the back side except for the part to
form the framework and ribs. By etching the heavily boron-doped silicon very
scarcely with the etching solution, the boron-doped layer remains as a membrane

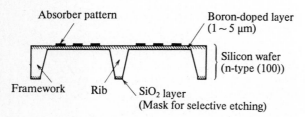

Fig. 3.23. Schematic cross
section of a silicon-mem-
brane X-ray mask

142

Such a silicon membrane mask is shown in Fig.3.23. The spacing of crystal lattice planes of silicon is shortened by boron doping and the membrane that is supported mechanically by the silicon framework is under tension and very flat. The disadvantages of this silicon membrane mask are: it is brittle and fragile, its fabrication processes are very complex, and it is opaque to visible light.

Si_3N_4 membranes can also be fabricated on silicon wafers. In this case, the doping of the surface of the silicon wafer is replaced with a chemical vapor deposition of Si_3N_4. Afterwards the same processes are used as for silicon membranes. Although Si_3N_4 membranes are transparent to visible light, they are more fragile than silicon membranes. $Si_3N_4/SiO_2/Si_3N_4$ sandwich-structure membranes that were reported by N.T.T. (Musashino Electrical Communication Laboratory) are much better than Si_3N_4 membranes with regard to strength [3.11].

Membranes of other materials such as Si_3N_4/SiO_2, Al_2O_3, BN, SiC, Ti have been investigated by several authors. But at present, there is no solution which may predominate in the near future.

b) Organic Materials

Commercially available polyester films, such as Mylar, can be used for X-ray mask substrates. Fabrication steps for these masks are bonding the film to the framework and forming the absorber patterns. Their advantages are: substrates are completely transparent to visible light, large masks are available without ribs, and the fabrication procedures are very simple. They have, however, some disadvantages: the surface of the films are rough and have many dust particles, and the dimensional stability is not sufficient.

Polyimide is becoming very popular as an organic substrate material [3.12, 13]. Polyimide membrane masks are fabricated by the following processes. Polyamic acid which is spin-coated on a glass or silicon substrate is polymerized by heating. After bonding a framework on the coated surface of the substrate, it is etched away. The framework can be formed from this silicon substrate by selective etching. Owing to the difference in thermal expansion coefficients between polyimide and the substrate material, the polyimide membrane is supported by the framework due to tension. Polyimide membranes have very flat surfaces and excellent heat resistance.

Using this method, the VLSI Cooperative Laboratories have fabricated polyimide conformable masks having a 3 µm thickness and a 100 mm diameter. Its structure has the following features: gold absorber patterns were enbedded within the membrane in order to reduce damages of the mask and the wafer by contact, the surface of the membrane was grooved lengthwise and crosswise in

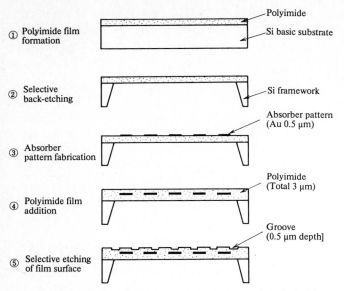

① Polyimide film formation — Polyimide / Si basic substrate

② Selective back-etching — Si framework

③ Absorber pattern fabrication — Absorber pattern (Au 0.5 μm)

④ Polyimide film addition — Polyimide (Total 3 μm)

⑤ Selective etching of film surface — Groove (0.5 μm depth]

Fig. 3.24. Fabrication steps for a grooved polyimide-membrane X-ray mask

order to obtain complete contact to the wafer and easy separation from the wafer. Good results have been obtained by the application of this mask in contact printing [3.14]. Figure 3.24 shows the fabrication steps for this polyimide conformable mask.

There are reports on polyimide reinforced inorganic membranes, too.

3.3.4 Alignment

For the alignment in X-ray lithography the mask or the wafer has to be moved mechanically since there is no possibility to deflect the exposure beam as in the case of EB replication. So the alignment aparatus resembles that of UV lithography. The only differences are the mask and wafer gap to be controlled more carefully and higher accuracy to be required because of dealing with finer IC patterns in X-ray replication.

One of the methods to detect the relative positions of mask and wafer is to use the X-ray itself as detection means. The alignment mark made of the X-ray absorbing material is put both on the mask and the wafer. They are irradiated by X-ray, and the relative position is measured by detecting the transmitted X-ray.

The optical detection method uses a laser beam. Through-holes are opened both on the mask and the wafer, and the transmitted beam is detected. When the

144

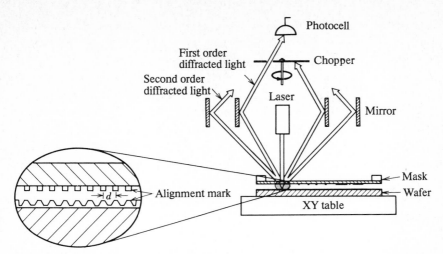

Fig. 3.25. Principles of interference-type alignment method (compare the right-hand and left-hand diffracted light intensities from two grid mark)

wafer is driven by a bimorph-type piezoelectric transducer, the laser beam passing through both mask and wafer hole marks varies, and the transmitted light intensity has many frequency components. Among them the intensity of the light at the basic frequency is proportional to the mask and the wafer displacements. Filtering out this basic frequency, a high S/N ratio detection is attained in detection [3.15].

Another advantageous method is to use the diffraction of the laser beam by a grid [3.16]. The grid mark is made both on mask and wafer. When the grid mark is irradiated with the laser beam, the latter is diffracted to the plus and minus sides. Comparing the intensity of both diffracted parts, the relative position of the mask and the wafer can be determined with an accuracy of 0.02 μm (Fig.3.25).

To increase the detection accuracy still further, as described above, the development of a mask or wafer stage driving system is necessary.

Conventionally, a pulsed motor drives such a stage. In this case, the rotational motion has to be converted into a linear movement, and usually this requires a displacement demagnifying mechanism. Backlash and friction resistance cause a deterioration of the translational resolution. To overcome this problem an alignment apparatus has been developed in the Communication Technology Laboratory (NTT), whose stage is suspended by an elastic plate and driven by an electrodynamic transducer [3.17]. The translation resolution of the order of 0.01 μm with a 20 μm stroke can be achieved.

An alignment apparatus which uses a piezoelectric linear motor has been developed in the VLSI Cooperative Laboratories. It can perform a coarse alignment whose initial misalignment is of the order of 200 µm and a fine alignment down to 0.1 µm with only one driving unit [3.18]. The piezoelectric linear motor has one piezoelectric element which expands and contracts along its length, and two piezoelectric elements which expand and contract along its radius to clump the spindle. These three elements are controlled to do a worm-like motion, thus the spindle is translated linearly and continuously along its length. This piezoelectric motor has a high translation resolution (6 nm/2 V) and a long stroke (25 mm) which is enough to do the fine alignment and the coarse alignment with the same motor. Its speed is widely variable (0 - 20 mm/min).

With this alignment apparatus the result shown in Fig.3.26 was obtained. The picture demonstrates the distribution of the residual mask and wafer misalignment after completion of the automatic alignment, whose mask and wafer gap is in the range of several ten µm's. Figure 3.27 illustrates the result of residual misalignment after mask and wafer are brought in contact, and the gap between mask and wafer is evacuated. Both figures give the data of the repeated alignment with the same mask and wafer pair. White circles in Fig.3.27 show the distribution of the misalignment when the wafer is brought in contact with the mask. It is obvious that by raising the wafer after alignment to touch the mask, a fairly large lateral shift shows up, but its direction and its magnitude vary little.

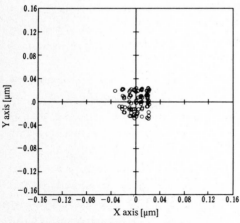

Fig. 3.26. Distribution of alignment accuracy (before contact)

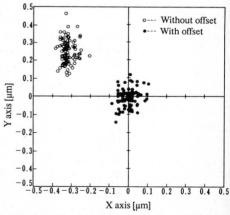

Fig. 3.27. Distribution of alignment accuracy (after contact)

After the magnitude of misalignment is fed into the microcomputer, the wafer is separated from the mask and is realigned to the point whose absolute misalignment is equal but of opposite sign with regard to that read in.

Then the wafer is raised again to touch the mask. After the offset alignment, as described above, is performed, the residual misalignment becomes very small, as shown by the black circle in Fig.3.27.

3.3.5 Examples of Experimental Systems

Table 3.6 compares some typical experimental X-ray lithography systems reported on in the literature.

Table 3.6. Specifications and performances of several experimental X-ray lithography systems

	Hughes Research Lab. (USA)	Bell Telephone Lab. (USA)	Musashino Lab., N.T.T. (Japan)	VLSI Cooperative Lab. (Japan)
input power [kW]	10	4	20 - 25	10
max. accelerating v. [kV]	15	25	20 - 25	20
target material	Al	Pd	Si	Al
target type	rotary	stationary	rotary	rotary
source-mask distance [mm]	240	500	350	300
exposure environment	He	He	Air	He
wafer number	1	1	1	6
wafer size [mm diam.]	76	76	76	100
position detecting method	optical microscope (visual)	optical microscope (visual)	vibration method (automatic)	light intensity comparison (automatic)
alignment-stage actuator	pulse-motor	manual	electrodynamic, etc.	piezoelectric linear motor

The system of Hughes Research Laboratories (USA) consists of a contact-type mask aligner being on the market for UV lithography, and an X-ray source having a water-cooled aluminum rotary target. After aligning, the X-ray source is moved to the position right above the mask. LSI chips of CMOS/SOS were fabricated with this system and their characteristics were evaluated [3.19].

In the system of Bell Telephone Laboratories (USA), the mask-wafer combination that is aligned at the alignment station is moved under the X-ray source for exposure. This X-ray source has a water-cooled palladium stationary target for the production of shorter X-rays (0.44 nm) [3.4].

The system of Musashino Laboratories of N.T.T (Japan) consists of an automatic mask aligner using visible light and a high-power X-ray source with a water-cooled silicon rotary target. After aligning, the position-detecting unit shifts aside, and then the X-ray source moves downward the mask and subsequently irradiates the mask in air atmosphere [3.20]. In order to reduce the X-ray attenuation by air, the X-ray window is placed in close proximity (3 mm) during exposure. The mask aligning device has a fine-adjusting mechanism for the X, Y and θ directions and, furthermore, for the Z and Z_θ directions because of the non-contact method.

VLSI Cooperative Laboratories have constructed a system for very high throughput and high alignment precision suitable for 1 μm patterns [3.14]. The mask alignment unit is completely separated from the exposure unit. After the alignment is finished, the mask and the wafer are combined in vacuum, and this mask-wafer assembly is carried to the exposure unit, and then is exposed. The application of piezoelectric worm-type transducers for moving alignment stages has improved the alignment accuracy (Sect.3.3.4). The output power of the X-ray source is not very high, but the throughput of this system is very high owing to the following means for increasing the X-ray use efficiency:

(i) Six wafers are exposed in parallel.

(ii) To shorten the distance between X-ray source and mask, contact printing with a conformable mask is used (Fig.3.21; in spite of short D, Δ' is minute owing to minute s').

(iii) Aligning and exposing are done in parallel simultaneously.

(iv) The aligned mask and wafer set can be put on the exposure unit in air and yet be exposed in helium environment to avoid X-ray decay in air.

Figure 3.28 shows cross-sectional views of the exposure unit, and Fig.3.29 gives and overall view of the X-ray exposure system. This system has the following performances: the alignment precision is better than 0.25 μm, a throughput of 50 wafers (100 mm diameter) per hour using negative electron-beam resist being on the market (PGMA). This throughput is comparable to the throughput on the UV exposure system. Figures 3.30 and 31 represent photomicrographs of patterns replicated by this system. They demonstrate the capability of this system for overlaying replication of 1 μm patterns, and if overlaying is not necessary, the replication of submicrometer patterns.

The alignment precision is determined not only by the performance of the aligner but also by the precision of the masks, the geometrical distortions of replicated patterns caused by nonnormal incidence of X-rays, and by the wraps and the expansion/contraction of wafers. The wafer size will be enlarged still further. Therefore, it is not easy to perform X-ray lithography with

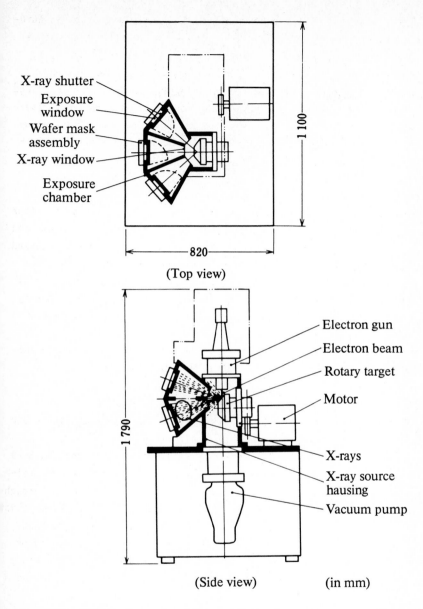

X-ray shutter

Exposure window

Wafer mask assembly

X-ray window

Exposure chamber

1 100

820

(Top view)

Electron gun

Electron beam

Rotary target

Motor

X-rays

X-ray source hausing

Vacuum pump

1 790

(Side view)   (in mm)

Fig. 3.28. Schematic cross sections of the X-ray exposure unit constructed by the VLSI Cooperative Lab.

Fig. 3.29. Overall view of the X-ray lithography system constructed by the VLSI Cooperative Lab.

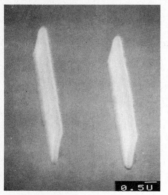

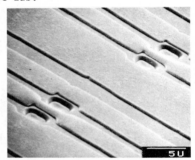

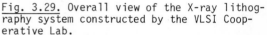

Fig. 3.30. Photomicrograph of resist patterns in PMMA. Resist patterns (0.2-0.3 μm width, 2 μm height) stand on the wafer

Fig. 3.31. Photomicrograph of the result of the overlay replication. Resist (EBR-1) patterns are overlaying on the SiO_2 patterns of the wafer

overlaying fit for practical use in the region of submicrometer. This is one of the serious problems to be solved in the near future. The dividing exposure method (step-and-repeat method) with step alignment may be a solution to this problem under the condition that far more powerful X-ray sources and far more sensitive X-ray resists are developed.

3.4 Electron-Beam Projection

Conventional photolithographic techniques for microcircuit production are limited in resolution to minimum line widths of around 1 - 2 μm. X-ray lithography systems and electron projection systems have been studied instead to

150

achieve high-throughput capabilities with a resolution in the submicrometer range. This section describes a demagnifying electron projection system.

3.4.1 Demagnifying Electron Projection Method

A self-supporting metal-foil mask is illuminated by electrons and imaged onto a wafer coated with an electron-sensitive resist. An example of the system is shown in Fig.3.32 [3.21]; its performance is listed in Table 3.7. The equipment has been developed by the Cooperative Laboratories, VLSI Technology Research Association, and will be described in the following.

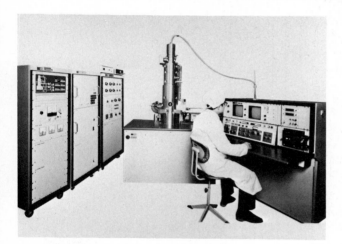

Fig. 3.32. Demagnifying electron projection system

The system consists of an electron-optical column including a wafer-moving stage, a vacuum unit, a scanning electron microscope display unit for alignment and control panels for lenses and deflectors. Figure 3.33 shows the basic design of the electron optical column and the ray diagram for both projection and alignment modes. The electron-optical column consists of an electron gun, three magnetic condenser lenses, two magnetic projection lenses and electromagnetic deflectors.

The system has two operation modes, the projection mode and the alignment mode. The projection mode is used to project a mask pattern onto a wafer and the alignment mode for mask registration. In the projection mode shown by the solid line in Fig.3.33 the demagnified image of the electron source (LaB_6 cathode) is formed at the deflection center of two sets of deflection coils mounted between the second and the third condenser lens. The third condenser lens then acts to provide parallel illumination at the mask plane.

Fig. 3.33. Schematic of the system

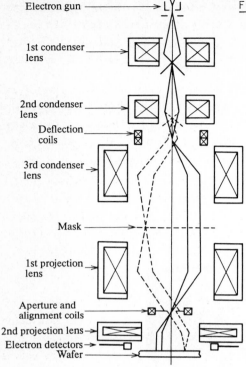

Electron gun

1st condenser lens

2nd condenser lens

Deflection coils

3rd condenser lens

Mask

1st projection lens

Aperture and alignment coils

2nd projection lens

Electron detectors

Wafer

(Solid line)   projection mode
(Dotted line) alignment mode

Table 3.7.   Performance of demagnifying electron projection system

electron gun	LaB_6 cathode
accelerating voltage	20 kV
illumination system	three magnetic lenses
mask size	20 mm square
projection system	two magnetic lenses
reduction ratio	1/4
projection method	scanning method
alignment	scanning electron microscope image
field size	3 mm square
resolution	0.2 μm
alignment accuracy	±0.2 μm
exposure time	3 s (PMMA)
wafer size	7.5 cm, 10 cm, 12.5 cm
column size	2.1 m high, 250 mm in diameter

The illumination beam has about 1 mm in diameter on the mask. In the system the scanning projection method is applied and the beam is magnetically scanned across the entire mask pattern (20 mm square) with deflection coils. The entire mask pattern is projected onto the wafer. There are about 300 - 500 scanning lines.

Since this method makes it possible to dynamically adjust the illumination angle, depending upon the position of the illumination beam on the mask, aberrations of projection lenses can be optimized and the distortion of projected images can be minimized. The illumination angle at the mask plane is adjusted by changing the power of each set of the above deflection coils and by shifting the effective deflection center in synchronism with beam deflection. Improvements on image resolution and distortion can be also achieved because some adjustment of the third condenser and projection lens is possible depending on the position of the beam. The scanning projection method has the features that a large image field can be obtained using a relatively small electron-optical column.

Electrons passing through the open parts of the mask are imaged by two magnetic projection lenses (symmetric magnetic doublets) onto the wafer [3.22]. The reduction ratio of the projection lenses is 1 : 4 and an image resolution of 0.2 μm can be obtained. Besides the scanning projection method, another projection method can be carried out [3.22,23]: the projection of the entire mask pattern at a time onto the wafer using a flood beam.

The alignment mode shown by the dotted line is used for registration. High alignment accuracy (±0.2 μm) can be achieved with the system because scanning electron microscope images of alignment marks at the mask and wafer planes are used, and condenser lenses, projection lenses and deflection coils operate in nearly the same mode as in projection. This is illustrated in more detail in alignment procedures in Subsection b below.

In another system [3.24] electrons passing through fine alignment slits on the mask are scanned on alignment marks which are arranged parallel to the mask slits on the wafer for mask registration without using a scanning electron microscope image. A 7.5 - 12.5 cm wafer is mounted on a wafer-moving stage which is stepped and repeated by a pulse-driven system between exposures. The main problems of the system are as follows: (i) the design of an electromagnetic lens with a large pole-piece bore diameter; (ii) the alignment procedures, and (iii) the preparation of the self-supporting metal foil mask.

These items will be discussed in detail below. The systems reported so far are at present only equipments for experimental use. We are expecting further improvements in the preparation of a self-supporting foil mask and in the system itself.

One of the problems realizating such a system is the design of the electron optics with high resolution and low distortion over a large image field. Magnetic lenses are, in general, used for the projection, and the third condenser lenses which are the core of the electron optics. An important condition for the system is the demand of high resolution in the submicrometer range over an image field of a few mm square on the wafer. Therefore, a magnetic lens with a large pole-piece bore diameter and a large gap and with negligible aberrations of higher than third order is required for the lenses mentioned. It is advisable to choose the lens with a bore diameter and gap of about 5 - 10 times larger than the dimension of the mask, and with the sufficiently large lens excitation from the view of the aberrations. In order to determine these values, an empirical equation for the conventional lens with a small bore diameter may be utilized as the first-order approximation.

$$H_1/D_1 = H_2/D_2, \quad N_1I_1 = -N_2I_2, \quad k_1^2 = k_2^2$$

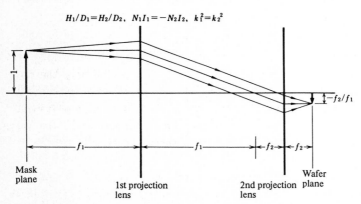

Fig. 3.34. Schematic of projection lenses composed of magnetic symmetric doublet

An example of projection lenses arranged by this approach is shown in Fig.3.34. The lenses are composed of a symmetric magnetic doublet. The two lenses are separated by the distance $(f_1 + f_2)$, the focal length f_1 of the first projection lens and that of the second projection lens, f_2. The mask plane is located at the focal point of the first lens and its image is projected at the focal point of the second lens. This is demagnified by a factor of $-f_2/f_1$. The two lenses are similar in geometry.

The pole-piece-gap to bore-diameter ratio of the first lens (H_1/D_1) is equal to that of the second (H_2/D_2); $(H_{1(2)}$ gap of the first (second) lens, $D_{1(2)}$ bore diameter of the first (second) lens). The excitation parameters k_1^2, k_2^2 are set equal, too, and their magnetic fields are acting in opposite directions. The factor k^2 is defined by $k^2 = (e/8 \text{ mV}) B_{max}^2 R^2$ (V being the accelerating voltage). Therefore, the Ampere turns of the two lenses $N_1 I_1$ and $N_2 I_2$ are chosen by $N_1 I_1 = -N_2 I_2$. Under these conditions, Seidel's aberrations and the rotation and magnification chromatic aberrations except five aberrations (spherical aberration, isotropic astigmatism, isotropic coma, field curvature and axial chromatic aberrations) vanish. Therefore, the values of H/D, f and k^2 may be chosen by the designers to optimize the above remaining five aberrations.

The following values are selected for the sample shown in Fig.3.34: $D_1 = 140$ mm, $D_2 = 35$ mm, $H_1 = 200$ mm, $H_2 = 50$ mm, NI = 1000 AT, $f_1 = 200$ mm, $f_2 = 50$ mm. The demagnification factor of the mask image is 0.25.

Thus, the bore diameter and the gap of the second projection lens must be fixed sufficiently large that aberrations higher than third order have no influence on the performance.

b) Alignment Procedures

The alignment mode shown by the dotted line in Fig.3.33 is used for mask registration and focussing. Here, the condenser lenses are converted into a large-field scanning electron microscope. A change-over to the alignment mode from the projection mode, and vice versa, can be done by adjusting only the excitation of the second condenser lens. The operations of other condenser and projection lenses remain constant in both modes. Therefore, high alignment accuracy can be achieved because the deviation of an optical axis, the image shift due to variations in the strength of the second condenser lens is extremely small.

In the alignment mode the probe focussed onto the mask plane is scanned across the alignment marks of the mask by the same deflection coils used in the projection mode. The probe at the mask plane is simultaneously imaged to form a second probe in the wafer plane, thus this probe is scanned across the alignment marks on the wafer.

The superimposed scanning-electron-microscope images of the alignment marks on the mask and on the wafer are displayed on a CRT using backscattered-electron detection at the wafer plane. Here the standard display technique of a scanning electron microscope is used. The resolution of the alignment mark images is better than 0.1 μm. The superimposed image displayed on a CRT is shown in Fig.3.35 where a transmission image of the mask mark and a reflection image of the wafer mark are superimposed. Figure 3.35 illustrates an example of the final mark images of the mask and the wafer after the alignment is completed. The rough alignment is carried out by moving a wafer-stage and the fine alignment by using the alignment coils located in the projection

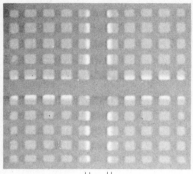

Fig. 3.35. Superimposed scanning electron microscope images of the alignment marks of mask and wafer

3 μm (Mask mark)

4 μm (Wafer mark)

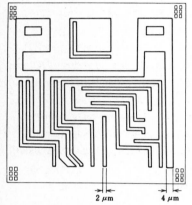

Fig. 3.36. Two complementary masks and resist pattern exposed using them

(a) Resist pattern exposed using masks A and B

2 μm   4 μm

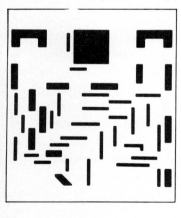

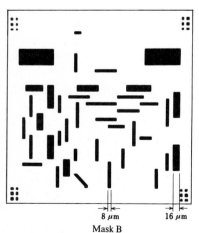

Mask A

Mask B

8 μm   16 μm

(b) Complementary masks

lenses. High alignment accuracy can be achieved because four superimposed images of the alignment marks at four corners on the mask plane and the wafer plane are simultaneously displayed on a CRT and aligned correctly.

Figure 3.36 demonstrates an example of composed patterns exposed on a wafer by using two masks. Here a set of complementary masks described in Subsection c is used. Mask A is first exposed to the resist on a silicon wafer after it is aligned. Then mask B is exposed onto the first exposed pattern (mask A) after the alignment. The resist pattern composed by the two exposures should appear without any offset in the case of exact alignment, and yields the overlay accuracy of better than ±0.2 μm.

c) Self-Supporting Metal Foil Mask

The mask consists of a self-supporting metal foil without substrate. It must be thin enough to minimize electron scattering, but large enough in size. A mask of 20 mm square in size and about 2 - 3 μm in thickness is used in the system. The foil mask is prepared by combining UV photolithography with the electroplating technique. Figure 3.37 shows a sketch of the process. An original master of the foil mask pattern is prepared on the glass substrate by using conventional photolithography.

The resist pattern is produced on a silicon wafer covered with a thin-film metal using the original glass mask. The foil mask pattern (electroplated foil) is finally produced by means of an electroplating technique. Nickel is usually

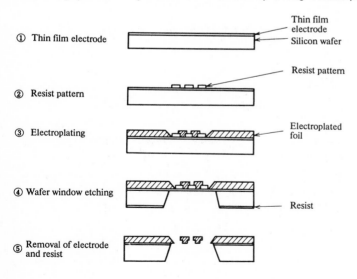

Fig. 3.37. Preparation of the self-supporting metal foil mask

taken as material for the foil because of mask strength. It is 2 - 3 μm thick. Besides the electroplating technique there is the vacuum evaporating technique. However, the electroplating technique is superior to the vacuum evaporating technique because the stress of the metal foil is smaller and the growing speed is larger. The foil of which the stress is large is easy to break. The thicker it is the more desirable it is from the viewpoint of foil strength; however, it must not be thicker than 2 - 3 μm when mask features have a line-width of less than 2 μm. (The linewidth on the wafer is less than 0.5 μm for a demagnification factor of 1 to 4.) Therefore, the outer region of the mask may be electroplated sufficiently thick to strengthen the mask.

More accurate fine patterns may be obtained, and the foil strength will become larger, if some improvement in the electroplating solutions and some adjustment of electroplating speed make the metal particles in the electro-plated layer finer. Unnecessary parts are finally removed from the mask by etching, and the silicon substrate is used as support of the foil. The "stencil problem" occurs because the mask consists of a self-supporting foil. It is im-possible to prepare the mask features which will provide an electron-opaque area, completely surrounded by an electron-transparent area. Some methods have been proposed to overcome this problem.

The first one is a method called four-step deflection/exposure technique [3.25] in which four continuous exposures are carried out using alignment coils and a mask with the elementary apertures of, for example, 5 μm square spaced on a pitch of 10 μm. In this technique, the dimensions of exposed patterns are restricted to even integer multiples of the elementary square aperture, and the minimum linewidth to twice of that.

Another one is the fine supporting grid method [3.26]. A mask with the pattern features supported by a very fine grid is used here. For example, a fine grid (1 μm thick with a linewidth of 1 μm spaced on a pitch of 5 μm) is fabricated by combining UV photolithography with electrochemical plating and etching techniques. The principle of this method is that a fine grid in a stencil should disappear in the resist pattern as a result of the edge shift due to electron scattering in the resist. An edge shift of 0.125 μm will be required for a demagnification of 1 to 4 so that a grid composed of 1 μm wide lines does not appear. Since the edge shift depends on the electron dose which illuminates the resist, a relation between edge shift and optimum dose must be established.

Some problems on the fabrication techniques for a fine grid need to be solved.

The third technique is the complementary mask method [3.27]. As shown in Fig.3.36 one original pattern is divided into two complementary masks and a pattern composed by the two exposures is recorded on a wafer. It is impossible or extremely difficult to prepare a self-supporting metal foil mask which has rectangular islands and a lot of long, peninsular patterns, as shown in the resist pattern of Fig.3.36. Two registrations and two exposures may be carried out to obtain a composed pattern using two complementary masks A and B into which an original pattern is divided.

It is then necessary to add a lap of about 0.5 μm to each mask, for example, for a demagnification of 1 to 4. This is, at present, the best method to solve the "stencil problem" because the preparation of a mask is easier compared with that of other methods, if sufficiently high alignment accuracy is achieved. However, damage, shrinkage, and so on, of the foil in mask preparation due to the self-supporting foil mask are the problems to be still solved. It is not exaggerated to say that the improvement of production techniques for the mask holds the key to the future of a system of this kind.

3.4.2 Photocathode Pattern Transfer System

A photocathode pattern transfer system transfers circuit images onto a silicon wafer covered with an electron sensitive resist using photo-electrons emitted from a photocathode mask. This type of systems can be classified in the following way: a 1 : 1 transfer system [3.28] and a demagnified transfer system [3.29]. This subsection describes the 1 : 1 transfer system.

Figure 3.38 illustrates the principle of the 1 : 1 photocathode pattern transfer technique. A photocathode mask is held parallel to a silicon wafer. When the mask is illuminated from behind with an ultraviolet lamp, electrons are emitted from the clear area of the photocathode mask and are focussed

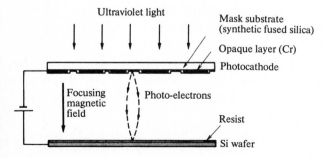

Fig. 3.38. Principle of 1:1 photocathode pattern transfer system

onto the silicon wafer by electric and magnetic fields between mask and wafer, so that the mask patterns are transferred to the wafer at unit magnification. The main advantages offered by this system are: (i) low-cost equipment, (ii) high throughput, (iii) high resolution in the range of submicrometers, (iv) large depth of focus, (v) mask fabrication similar to conventional, (vi) magnification correction for wafer expansion.

a) Imaging System

An imaging system in this transfer technique is formed of uniform and co-axial electric and magnetic fields. Photoelectrons will converge in focus onto the wafer provided that the following condition is respected:

$$B = \pi \sqrt{\frac{2m}{e}} \frac{\sqrt{V}}{d} (1 - \sqrt{\frac{V_0}{V}} \cos \theta)^{-1} \quad , \tag{3.3}$$

where B denotes the magnetic-field intensity, V is the accelerating voltage, d is the mask-wafer spacing, V_0 is the initial kinetic energy and θ the emission angle of the photo-electron, respectively. Chromatic aberration caused by the spread of the initial kinetic energy dominantly determines the ultimate resolution of this transfer system. Edge definition (δ) due to chromatic aberration is described below $\delta \approx 0.5$ d V_h/V [3.30], where V_h is the half width of the initial kinetic energy of the photo-electrons, which is determined by the photo-emitter and the illumination source. To realize high resolution, narrow distribution of the initial kinetic energy of the photo-electrons is required. High electric and magnetic fields between mask and wafer are also necessary. The edge definition can be smaller than 0.1 μm by suitable selection of V_h, V and d.

The depth of focus can be derived from (3.3) and is about several tenths of micrometers. This is very favorable to transfer patterns over steps on the wafer.

The electric and magnetic fields of the practical system are not perfectly uniform, resulting in image distortion. Magnetic fields have to be uniform to 10^{-5} in order to keep distortions smaller than 0.1 μm. Electric field in-homogeneity arises from the mask and wafer holder configuration and the wafer bowing. The distortion due to mask and wafer holders is reproducible, so that it is not serious and can be reduced by the careful design of the holders. On the other hand, the distortion due to wafer bowing is unreproducible because the wafer bowing varies during the processing. This type of distortion causes degradation of the registration accuracy. For a registration accuracy of better than 0.1 μm, it is necessary to control the wafer bowing severely and to flatten the wafer using an electrostatic holddown chuck [3.31].

b) Photocathode Mask and Illumination Source

The mask used in this transfer system is similar to a conventional optical
mask made of chromium on a glass substrate. This point is of great advantage
in manufacturing masks. The differences from optical masks are as follows:
A thin layer of photo-emitter is evaporated on the patterned surface. The
mask substrate is required to be transparent in the UV. A conducting layer
transparent to UV light is necessary between the photo-emitter layer and the
patterned surface in the case that the photo-emitter is an insulator.

High current density is one of the important properties for this system.
The exposure time t depends on the emission current density J and the sensitivity S of electron-sensitive resist, i.e., $t = S/J$. As the current density
is several $\mu A/cm^2$, an exposure time of less than 10 s is sufficient even for
a low sensitive resist. The combination of a CsI photo-emitter and a low-
pressure mercury discharge lamp [3.32], where the wavelength of 1849 Å is
used, is the best one that has ever been reported. In this combination an
emission current density of about 5 $\mu A/cm^2$ is easily achieved, and the half
width of the initial kinetic energy of the photo-electrons is about 0.3 eV.
It has been reported that the CsI photocathode deteriorates with use, and
the lifetime of a CsI photocathode is up to about 50 exposures [3.33]. From a
practical viewpoint, an increase of lifetime of the photocathode may be required.

c) Alignment System

In the photocathode pattern transfer system the detection of backscattered
electrons, which is used in an electron beam pattern delineator or a demagnified electron beam projector, is not possible because the backscattered electrons are trapped in the vicinity of the alignment marker by the electric
field between mask and wafer. Two alignment methods have been proposed. One
method uses an etched through-hole on the wafer as the alignment marker, and
electrons passed through the hole are detected [3.34]. The other method utilizes "Bremsstrahlung" X-rays generated when photo-electrons strike the alignment marker made of a heavy element on the wafer [3.35]. The translational
corrections are made by the deflection coils and the rotation is carried out
mechanically. Both methods give an alignment accuracy of $\pm 0.1 \sim \pm 0.2 \mu m$.

d) System Description

Several photocathode pattern transfer systems have been reported from Philips
[3.36,37], Electron Beam Microfabrication [3.38] and other manufacturers. In

this section, the experimental system developed in the Cooperative Laboratories
VLSI Technology Research Association is described [3.39].

Specifications of this system are listed in Table 3.8. The focussing magnet
is made of a Helmholtz-type superconducting magnet in order to realize high
resolution over the whole area of 10 cm^2 wafer. The superconducting magnet can
carry a large exciting current, thereby generating an intense and homogeneous
magnetic field with simple winding configuration. The superconducting magnet
can also be operated in the persistent current mode, thereby generating a
stable magnetic field without ripple or drift. Moreover, the superconducting
magnet can make temperature control of the whole system easy because it does
not generate Joule's heat. The maximum field intensity of this magnet is 3 kG.
The homogeneity is better than $6 \cdot 10^{-5}$ over the transfer area of 90 mm in
diameter. There is a slow decrease of $1.7 \cdot 10^{-6}$/min in the magnetic field intensity in the persistent current mode, which is caused by the residual electrical resistence in the coil. However, this order of decrease does not cause
any degradation in the resolution capability. Figure 3.39 shows a schematic
of this system.

Table 3.8. Specifications of experimental photocathode pattern transfer
system

mask size [mm]	100 $^\phi$
wafer size [mm]	75 $^\phi$, 100 $^\phi$
spacing between mask and wafer [mm]	5, 7.5, 10
accelerating voltage [kV]	≤20
magnetic field [kG]	≤3
photocathode	CsI
illumination	1849 Å from low-pressure mercury discharge lamp

Figure 3.40 demonstrates an example of the developed PMMA resist patterns
transferred at 20 kV, whose minimum line width is less than 1 μm. This photograph shows that submicrometer patterns with step profile in 1 μm thick resist
layers are successfully transferred by this system. The exposure time is
15 s for PMMA resist and it is a few seconds for high-speed electron resist.
The depth of focus was found to be about ±25 μm for submicrometer pattern
size. Figure 3.41 demonstrates an example of 1 μm line-and-space patterns
transferred in 1 μm thick PMMA resist layers over 0.5 μm Si steps.

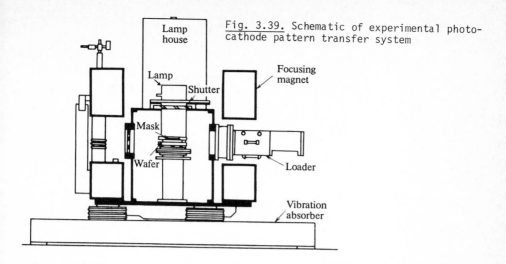

Fig. 3.39. Schematic of experimental photo-cathode pattern transfer system

Lamp house

Lamp

Shutter

Focusing magnet

Mask

Wafer

Loader

Vibration absorber

Fig. 3.40. SEM photograph of submicron patterns transferred in PMMA resist layer (pitch size 1 μm)

Fig. 3.41. SEM photograph of 1 μm line-and space patterns transferred over 0.5 μm Si steps

There remain some problems to be solved, but this pattern transfer system will be one of the most useful tools for submicrometer lithography when a high-performance imaging system suitable for high accelerating voltage operation has been developed.

3.5 Radiation-Sensitive Resist for Microfabrication

Radiation-sensitive resists are basically necessary in the microfabrication process of all kinds of semiconductor devices, especially IC's. Two types of resists are classified as positive and negative (Fig.3.42).

Photoresists being sensitive in 300 - 450 nm wavelength range are used widely in the today's IC fabrication on the LSI level. The practical resolution of photoresists is about 1 - 2 μm due to light diffraction and scattering phenomena.

Table 3.9. Microfabrication resist systems

Radiation Source	Resist	Remark
deep UV (200 - 300 nm)	deep UV resist	suitable for nearly 1 μm pattern transfer
electron beam (10 - 30 keV)	electron-beam resist	suitable for submicron pattern generation and transfer
X-ray (0.4 -4 nm)	X-ray resist	suitable for submicron pattern transfer

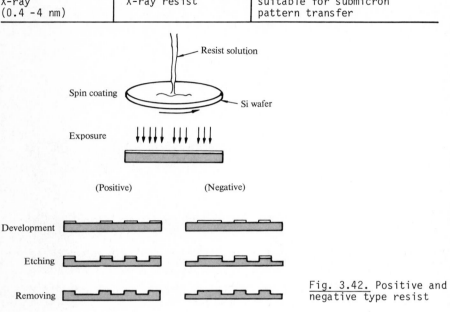

Spin coating — Resist solution — Si wafer

Exposure

(Positive)   (Negative)

Development

Etching

Removing

Fig. 3.42. Positive and negative type resist

Radiation-sensitive resists with a higher resolution than that of photo-resists are required in VLSI fabrication with submicrometer pattern (Table 3.9).

Up to the present, it is known that photoresists can be used as deep-UV resists, although their sensitivity and resolution are not always high enough.

Chloromethylated polystyrene (CMS) [3.40], white resist [3.41] are the best known resists representing the negative type deep-UV materials. On the other hand, a positive-type resist with a very high quality has not been developed yet. Some deep-UV resists are described in [3.42].

3.5.1 Electron-Beam Resist

In principle, an electron-beam resist must have high sensitivity and resolution for high throughput in the VLSI fabrication process. In addition to this, a very high etching resistance is sought for because of adaptation to the dry-etching process which is suitable for submicrometer etching and needs high chemical and thermal stability of the resist.

Up to now, a positive-type polymethylmethacrylate (PMMA) is used as high-resolution resist (<0.1 μm), but it is of low sensitivity, a disadvantage for high-speed delineation.

Between the sensitivity S [Coulomb/cm^2], the delineation speed v [cm^2/min], and the electron beam current I [A] there is the following relation for electron beam delineation:

$$S = 60 \ I/v \quad . \tag{3.4}$$

Thus, a high-speed delineation needs a highly sensitive resist.

a) Positive-Type Electron-Beam Resist

Because a positive-type resist is designed so that the main chain is scissored and the molecular weight is lowered by electron-beam irradiation, the radiated area becomes soluble. Between the initial molecular weight M_n and the scissored molecular weight $M_n{}^*$, the following equation exists [3.43,44]:

$$\frac{1}{M_n{}^*} - \frac{1}{M_n} = KQG_s \quad . \tag{3.5}$$

Here, K is a constant, Q is the electron-beam irradiation [Coulomb/cm^2], and G_s is the main-chain scission efficiency [1/100 eV].

From (3.5), large M_n resist is expected to be highly sensitive, but this type of resist is generally known for its difficulty to form uniform layers.

On the other hand, G_s can be enlarged by a modification of the molecular structure [3.45].

Equation (3.5) shows that sensitivity is dependent on the break of the main chain, but also on the solubility property of the scissored polymer.

The quantitative relation between sensitivity and solubility is not clear. But, empirically, polymethacrylate with a fluorinated alkyl ester side chain dissolves even by a slight decrease in molecular weight.

In EBR-9 listed in Table 3.10, the ester part is substituted by a trifluoroethyl group in order to increase the solubility difference owing to the above-mentioned empirical fact. Furthermore, the Cl substitution of the α-position is based on a quantum chemical prediction [3.45]. The sensitivity of EBR-9 is a hundred times higher than that of PMMA, and its resolution is attained to 0.1 μm. Its glass transition temperature is higher than that of PMMA. A submicrometer pattern of EBR-9 is shown in Fig.3.43.

Table 3.10. Electron beam and X-ray resists

a) Resists developed in Cooperative Labs

Resist	Type	Use	Sensitivity Electron beam [C/cm²]	Sensitivity X-ray [J/cm²]	Resolution [μm]	Thermal stability	Remarks	Molecular structure
EBR-1-U	pos.	electron X-ray	$8 \cdot 10^{-7}$	$34 \cdot 10^{-3}$ (AlKα)	0.3	good		$\left(CH_2-C{\small\begin{array}{c}CH_3\\ \|\\ COOCH_2CCl_3\end{array}}\right)_m \left(CH_2-C{\small\begin{array}{c}CH_3\\ \|\\ COOCH_3\end{array}}\right)_n$
EBR-9	pos.	electron	$8 \cdot 10^{-7}$		0.1	good	high reso-lution	$\left(CH_2-C{\small\begin{array}{c}Cl\\ \|\\ COOCH_2CF_3\end{array}}\right)_n$
CP-3	pos.	electron	$4 \cdot 10^{-7}$		0.3	excellent	excellent dry etching resistance	$\left(CH_2-C{\small\begin{array}{c}CH_3\\ \|\\ COOCH_3\end{array}}\right)_m \left(CH_2-C{\small\begin{array}{c}CH_3\\ \|\\ COOC(CH_3)_3\end{array}}\right)_n$
P(MMA-DMM)	pos.	X-ray		$7 \cdot 10^{-3}$ (AlKα)	0.5	good	high sensi-tivity	$\left(CH_2-C{\small\begin{array}{c}CH_3\\ \|\\ COOCH_3\end{array}}\right)_m \left(CH_2-C{\small\begin{array}{c}COOCH_3\\ \|\\ COOCH_3\end{array}}\right)_n$
CNR	neg.	electron	$0.9 \cdot 10^{-7}$		0.6	good	excellent dry etching resistance, slight post-crosslinking effect	$\left(CH_2-C{\small\begin{array}{c}H\\ \|\\ COOCH_2CH_3\end{array}}\right)_m \left(CH_2-C{\small\begin{array}{c}Cl\\ \|\\ CN\end{array}}\right)_n$

Table 3.10. (continued)

Resist	Type	Use	Sensitivity Electron beam [C/cm²]	Sensitivity X-ray [J/cm²]	Resolution [µm]	Thermal stability	Remarks	Molecular structure
DER	neg.	electron	$1\cdot10^{-7}$		0.2	excellent	excellent dry etching resistance, little post-crosslinking effect	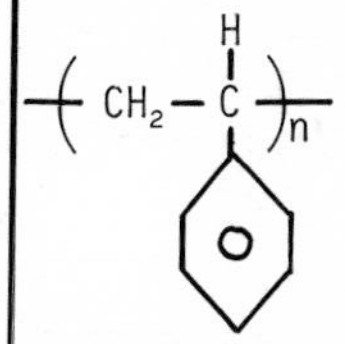 $\left(CH_2-\overset{H}{\underset{}{C}}\right)_n$ with phenyl-epoxide ring
PDBA	neg.	X-ray		$1.5\cdot10^{-3}$ (AlKα)	0.5		high sensitivity	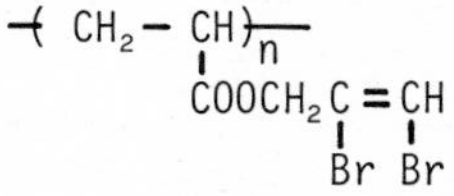 $\left(CH_2-CH\right)_n$ $COOCH_2C\!=\!CH$ with $Br\ Br$

b) Conventional resists

Resist	Type	Use	Sensitivity Electron beam [C/cm²]	Sensitivity X-ray [J/cm²]	Resolution [µm]	Thermal stability	Remarks	Molecular structure
PMMA (IBM)	pos.	electron X-ray	$500\cdot10^{-7}$	$500\cdot10^{-3}$ (AlKα)	0.1	fair	low sensitivity, high resolution	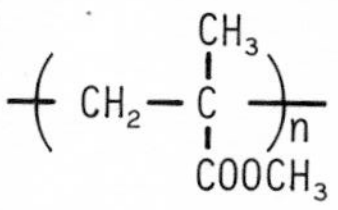 $\left(CH_2-\overset{CH_3}{\underset{COOCH_3}{C}}\right)_n$
COP (Bell Lab Labs.)	neg.	electron X-ray	$6\cdot10^{-7}$	$159\cdot10^{-3}$ (PdLα)	0.5	fair	fair post-crosslinking effect	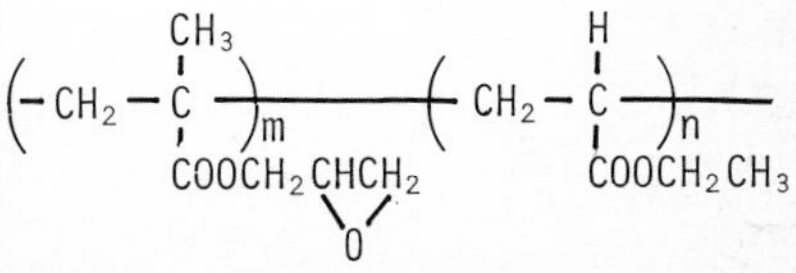 $\left(-CH_2-\overset{CH_3}{\underset{COOCH_2CHCH_2}{C}}\right)_m \left(CH_2-\overset{H}{\underset{COOCH_2CH_3}{C}}\right)_n$

Polymethacrylate desolubilized by thermal cross-linkage of its ester side chain is known to be easily developed by a low dose of electrons. EBR-1, EBR-1-U [3.46] and CP-3 [3.47] are resists of this type. They are shown also in Table 3.10. A microlithographic pattern with a minimum line width of 0.5 μm of CP-3 is shown in Fig.3.44.

 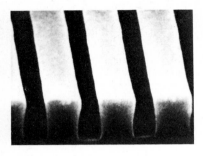

Fig. 3.43. Pattern formed with EBR-9. (Left) Vertical groove width 0.2 μm. Horizontal groove width 0.5 μm. (Right) Groove width; upper 0.3 μm, bottom 0.4 μm

Fig. 3.44. Pattern formed with CP-3. (Left) Minimum line width 0.5 μm (Right) 0.8 μm line width pattern on 1.5 μm step

b) Negative-Type Electron-Beam Resist

Negative-type electron beam resist is crosslinked in all three dimensions and desolubilized by electron-beam irradiation. In this case, the relation between the initial molecular weight M_w and the initial gel dose Q [Coulomb/cm^2] is given by

$$K'/M_w G_x = Q \quad , \qquad\qquad (3.6)$$

where K' is a constant, and G_x is the crosslinking efficiency [1/100 eV].

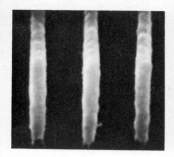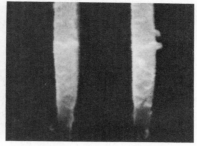

Fig. 3.45. Pattern formed with DER. (Left) Line width 0.2 μm. (Right) Slightly overhanged pattern; upper edge 0.5 μm, bottom 0.4 μm

The sensitivity of negative-type resists is mainly determined by M_w and G_x. Generally, the crosslink group in the resist having large G_x is not thermally stable and a crosslinkage occurs gradually for long periods after electron-beam irradiation (post crosslinking effect). In order to suppress this phenomenon, some additives were tried to add, which resulted in lowering the sensitivity.

As thermal crosslinkage results between crosslink groups, a block copolymer CNR is developed being composed of one monomer unit with a crosslink group (A), and the other one without it (B).

$$\left(A \right)_{n_1} \left(B \right)_{n_2} \left(A \right)_{n_3} \ldots \left(A \right)_{n_{i-1}} \left(B \right)_{n_i} \; ;$$

n_i is a positive integer.

CNR is a highly sensitive resist, the post crosslinking effect being suppressed. This resist has also a highly thermal stability for dry etching.

Although polystyrene with high resolution and without the post crosslinking effect was not considered to be useful for high-speed delineation because of its low sensitivity, polystyrene having high molecular weight with narrow weight distribution, which is called DER, is found to be of higher sensitivity than that estimated by (3.6) [3.48]. The high aspect-ratio pattern of DER is shown in Fig.3.45.

c) Problems of Electron-Beam Resist

In electron beam delineation, there is an important problem well known as the proximity effect. This phenomenon results from electron scattering in the resist layer, the silicon substrate, and at the resist-silicon interface. As electron scattering is independent of pattern size, the proximity effect markedly affects small-size patterns. Although the proximity effect can be

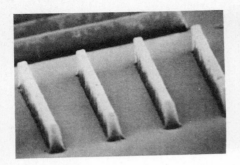

Fig. 3.46. Pattern formed with P(MMA-DMM). (Left) Line width 0.5 µm.
(Right) Line width 0.2 µm

decreased by correcting the delineation program, e.g., electron dose, delineation size, it cannot perfectly be overcome. In order to reduce this effect,
a high-contrast resist is required.

3.5.2 X-Ray Resist

X-ray patterning has a large merit in that diffraction and interference effects can be ignored because of their very short wavelength (0.4 - 4 nm). Therefore this techique offers very high resolution. On the other hand, a highly
sensitive resist is a necessary condition due to the low-energy-density radiation from X-ray tubes being in use today.

a) Positive-Type X-Ray Resist

X-ray resist composed of elements with a high mass absorption efficiency is
expected to be highly sensitive. In this direction, PMMA blended with ZnI_2
which has a higher mass absorption efficiency was developed and its efficiency achieved to 18 mJ/cm^2. However, its high sensitivity was found to result
from an insolubilizing effect by ZnI_2 addition to PMMA [3.49] as in the case
of EBR-1 and CP-3.

Another, highly sensitive resist is P(MMA-DMM). Its solubility strongly
changes by X-ray irradiation [3.50], giving a sensitivity of 7 mJ/cm^2. A
micropattern achieved with P(MMA-DMM) and a linewidth of less than 0.5 µm
is shown in Fig.3.46.

b) Negative-Type X-Ray Resist

PDBA containing the element Br shows a high sensitivity of 1.5 mJ/cm^2 to AlKα
line. The resolution is in the range of 0.5 µm with very little scan.

3.5.3 The Future of Electron-Beam and X-Ray Resists

Up to the present, microfabrication resists have been developed directed to a resolution of more than 0.1 μm and a sensitivity of more than 1 $\mu C/cm^2$ for electron beam and 10 mJ/cm^2 for X-rays, respectively. These requirements are now reached. In the near future, the development of resists will be seen to branch into two areas: On the one side, the practical performance has to be improved, especially the compatibility to conventional process steps. On the other side, the achieving of resolution exceeding the submicrometer range is one of the most important targets. In order to achieve this, the contrast ratio of resist needs to be improved as well as the delineator and the delineation methods.

4. Mask Inspection Technology

With the advent of masks, an inspection technology had to be developed. Initially it relied on the human eye, but in VLSI technology high-speed, high-precision inspection became necessary to handle large amounts of fine patterns

Section 4.1 summarizes principles of man inspection and discusses requirements on light and electron beam for mark inspection. In Sect.4.2 we introduce some examples of mask-inspection systems in existence and in conclusion we discuss trends of future technology.

4.1 Principles of Mask Inspection

In the advancement from LSI to VLSI, the patterns on masks become finer and more complex. This requires high-speed inspection of the fine patterns, thus promoting the automation of instruments in use. Distinctive features of mask-inspection systems developed for high-speed, fine-pattern inspection to be treated in Sect.4.2.1, are summarized on the basis of Table 4.1.

In Table 4.1, the spot diameter is assumed to be 0.5 μm for the photo system and 0.2 μm for the electron-beam (EB) system. The diameter in the former system is comparable to the utilized wavelength which determines the lower limit of the diameter, whereas the diameter in the EB system is larger than the wavelength by two orders of magnitude or more. The superiority of the EB system is therefore evident. From items 3 and 4 of the table we infer that the lenses for the electron beam must have a long focal length and a small semi-convergence angle. This explains the fact that the beam energy in the EB system is the same as that in the photo system (see item 6) despite the much higher brightness of five orders of magnitude or more as given in item 5. The EB system has a two orders of magnitude larger focal depth and a wider scanning field (see items 7,8). Moreover, the scanning speed of the EB system is 150 times faster, as illustrated by item 9. From these data it is clear that an electron beam provides the potential for a high-speed inspection system of fine patterns.

Table 4.1. Feature comparison of the photo and electron mask inspection systems

Item	Photo	Electron
1 Spot diameter [μm]	0.5	0.5
2 Distribution	Gaussian	Gaussian
3 Distance between objective lens and sample [mm]	3	134
4 Semiconvergent angle of objective lens [mrad]	424	2.7
5 Source brightness [W/cm^2 str]	10^5	$2 \cdot 10^{10}$
6 Beam energy [mW]	0.1~0.2	0.1
7 Focal depth [μm] (at conf. disc diam. of 0.1 μm)	±0.2	±30
8 Max. allowable scan field [mm^2]	0.1	5
9 Scan speed [μm/s]	40	5250
10 Wave length [Å]	6328	0.1
11 Normalized quantum energy	1	$6 \cdot 10^4$
12 Normalized shot noise	1	250
14 Source	He-Ne laser	LaB$_6$ thermal emission gun Acc. Volt. 20 kV

As will be mentioned in the following subsection, a high signal-to-noise (SN) ratio is necessary for accurate detection of the signal. Shot noise (quantum noise), being a limiting factor in a low-resistance noise detector, is proportional to $\sqrt{n}$, n being the number of photons in the light or the electron beam [4.1]. The energy of a photon, being inversely proportional to its wavelength, is small in comparison to that of the electron with a short wavelength. As can be seen from Table 4.1, shot noise of electrons is about 250 times larger than that of photons providing an equal flow of radiant beam energy. This should be taken into account when high-speed detection is required due to a small electron-beam current. At present, a beam energy greater than 0.1 mW is sufficient, depending on the contrast discussed in Sect.4.1.1a.

Comparing the two systems we may note that light inspection techniques have a long history, they are stable and handy systems. EB inspection systems have attractive features with regard to stability, convenience of operation and price.

4.1.1 Signal Generation and Processing

a) Detection of Signal [4.2]

For mask inspection, a signal must be detected which contains mask information on light or electrons emitted from the mask when the mask itself is irradiated by light or electrons. This can be done by using reflected or penetrating light and reflected or secondary electrons, respectively. The SN ratio should be large enough for accurate mask inspection. For the electron detection, the SN ratio can be written as

$$S/N = (-1 - 1/\sqrt{\theta})n/2B \quad , \tag{4.1}$$

where n is the maximum detected electron number, θ is the ratio of maximum n to minimum n (this is called *contrast*), and B is the detected frequency band width.

Here, the thermal noise of the detection system is omitted, which is possible in a modern system. For a large SN ratio, θ as well as n should be increased. In the case of light, the equivalent value of n is large enough and θ can be large when transmitted light is detected as well. Therefore, the SN ratio for a light system is usually larger than that for an electron system.

b) Transmission of Signal

A line (one-dimensional) information taken from a plane (two-dimensional) mask information must be transmitted sequentially to a computer for processing. Thus, an immense information must be processed within a short period of time through a transmission line with a wide frequency band coming from an inspection of a large mask area with fine patterns. This necessarily increases the band width B in (4.1) and decreases the SN ratio leading to a low-detection accuracy. For example, the inspection of a 1 mm^2 chip by a 0.1 μm minimum unit in one second needs a band width of at least 100 MHz, for which an amplifier is hardly available at present. So the inspection of a mask with fine patterns would take too much time in practical cases, if the parallel transmission of the signal or the compression of its information by a pretreatment would be carried out. For these technologies many problems are to be solved.

c) Miscellanous

High contrast is important for a large SN ratio according to (4.1). A high contrast will easily be achieved when the transmitted light is used for detection, while a proper arrangement of detectors or a selection of pattern material should be done to improve the contrast when electrons or reflected light are to be detected. For electrons, heavy metals such as tungsten or gold are useful for high contrast (instead of chromium which is usually used) [4.3].

Normally, one cannot pick up correct mask information by irradiating electrons on a usual glass mask because of its charge collection. This can be easily eliminated by a low-conductive mask [4.4]. Such a mask can be favourably used in a photo-process line by its eminent features that it is free from pattern break or dust adhesion due to charge up [4.5].

4.1.2 Mask Pattern Inspection Technology

a) Beam Width and Dimension Accuracy [4.6]

In order to measure a pattern-edge position or the pattern width, signal from patterns scanned by a narrow electron beam should be detected with a high SN ratio. It is theoretically possible to determine an accurate position by an infinitesimal beam width, but accurate measurement cannot be done in the end because the signal produced by the beam is too small for a sufficient SN ratio. Here, one might consider the dimension inspection accuracy obtained by a pattern which is scanned by a beam of finite width.

By scanning a rectangular pattern by an electron beam with Gaussian distribution, a strong reflected signal is detected when the beam hits the pattern, while only a weak signal is available when the beam does not hit it. Figure 4.1 shows the calculated amplitude variation of the signal from a rectangular pattern scanned by a constant current beam with differing half beam width. It can be seen that a correct measurement is possible on pattern widths wider than one half of the beam half-value width, provided that the width measurement is done at half the maximum signal amplitude by using a Gaussian beam. This is because the half-value width of the signal obtained by scanning a pattern with an infinitesimal half-value-width beam is equal to that of the signal obtained with the beam having twice the half-value width of the pattern width.

We now discuss the maximum width of a beam which can be used to measure the distance between two parallel lines with the line width a and the line spacing a, as shown in Fig.4.2. S_H and S_L are the maximum and the minimum values of the signal detected by scanning the lines with a Gaussian beam, respectively. The ratio S_H/S_L gives the contrast. The distance between the lines down to 1/5 times the half-value width can be determined if the SN

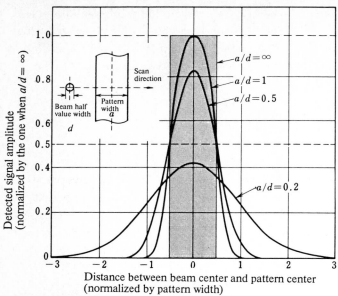

Fig. 4.1. Detected signal intensity when a pattern (with width a) is scanned by a beam (with half value width d)

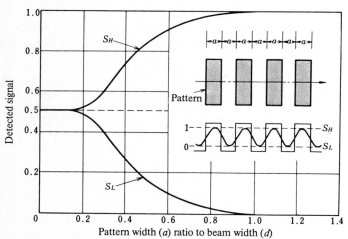

Fig. 4.2. Relations between detected signal and ratio of pattern width to beam width

ratio is large enough, even if the contrast decreases to 3/100. It is possible to measure a line separation down to 1/5 of the beam width.

As mentioned above, the long side of a rectangle under inspection must be long compared to the beam half-value width, as in the case of pattern-dimension inspection. If the pattern is a square or a circle, the maximum

176

detected signal is less than that shown in Fig.4.2 when the beam width exceeds the pattern width. Therefore, the beam half-value width should be small enough compared with the minimum detectable size when the patterns are squares or circles, as in the case of defect inspection.

b) Dimension Inspection and Its System

Mask-dimension inspection is divided into pattern-size inspection for measuring the pattern width, and pattern-position inspection for the measurement of the pattern position. The pattern-size inspection determines the difference of pattern-edge positions, while the pattern-position inspection concentrates on measuring the coordinate of the pattern from an original point. The pattern-size inspection, where the difference of the two edge positions is obtained, is easy to do correctly because an error of the measurement system can be excluded by a subtraction process, except for the system error. On the other hand, pattern-position inspection is difficult to carry out correctly even with a complex measurement system. Its accuracy depends upon the mechanical stability of the measurement system as well as the positional detection accuracy of the detector.

Table 4.2 shows a list of mask-dimension inspection systems so far reported [4.7]. As mentioned above, pattern-size inspection is so easy to do that a variety of its systems is reported. Only a few fully automated inspection systems are, however, capable of measuring pattern position both in and between chips. At present, almost all the systems use light (including laser light) for detection means; only one practical system utilizes an electron beam which has been tested by experiments. Fully automated systems capable of pattern-size and pattern-position measurement will be described in Sects.4.2.1,2 in more detail.

4.1.3 Mask Defect Inspection

a) Defect Decision

In order to inspect a mask defect, the defect should be recognized by any method. At the beginning, the mask defect inspection was mainly done by human eyes. A person decided on defects not resulting from design data of individual patterns but from specific features of patterns. Therefore, efficient defect detection can be achieved by the features of defects experimentally known from the process with which the mask is made. This is called *feature extraction*, which includes the following cases:

177

Table 4.2. List of dimension inspection systems [4.7]

System name	Measurement method edge detection/size detection		Measur. range [μm]	Min. pattern [μm]	Min. display [μm]	Repeat-ability [μm]	Detection means	Measure-ment
Screw micrometer	Magnified image is aligned with a hair line, displacement is read by a micrometer		2~100	2·5	0.1	>0.3	light	manual
Image clearing (Vickers)	Split images are aligned, displacement is read by an encoder		12~200	12	0.01	0.075	light	manual
MPA (transmission type) (Nikon)	Detection by a slit-type photo-electric micrometer and a linear encoder		1~100	1·3	0.05	10.1	light	manual
ITP system	Detection by a video signal on a CRT		0.2~508	0.2·0.25	0.01	10.01	light	manual
Perkin Elmer system	Detection by a slit-type photo-electric micrometer and a linear encoder		0.5~254	0.5	0.1	10.1	light	auto
Light interfero-meter type II (Nikon)	Detection by a slit-type photo-electric micrometer and a laser interferometer		pitch 10~150	pitch 10	-	-	light	auto
LCM (Olympus)	Detection by a pin-hole type photoelectric micrometer and a laser interferometer		1.5~150	1.5·3	0.1	10.1	light	auto
Telecomparator (Hitachi)	Measurement on a TV monitor		0.5	0.5·0.5	0.01	-	light	manual
NBS (Experiment)	Measurement by SEM/interferometer		-	-	0.01	-	light	manual
Lampus (reflection type) (Nikon)	Automatic measurement by a microcomputer using reflected laser light	size meas.	0.8~100	0.8·2	0.01	0.05	light (laser)	full auto
		position meas.	~150 mm	0.8·2	0.01	0.3		
EB mask in-spection system (VLSI Coop. Lab.)	Automatic measurement by a minicomputer using electron beam	size meas.	0.5~25	0.5·0.5	0.001	0.03	electron beam	full auto
		position meas.	~125 mm	0.5·0.5	0.01	0.15		

Only size inspection system (Screw micrometer through NBS)

Size and position inspection system (Lampus, EB mask inspection system)

1) Detection is done by using the fact that outlines of defect patterns are more complex than those of normal patterns and have many small criss-crossings [4.8,9].

2) Defect extraction is done by using spatial filtering [4.10].

3) Defect detection is done by using laser-light diffraction at the defect [4.11].

In the case of items 2 and 3, a high defect-inspection speed is possible because the amount of transmitted information mentioned in Sect.4.1.1b can be decreased by spatial defect extraction without a computer treatment of the defect information. However, they are not always unique and thus the method is incomplete because not all the defects concerned can be extracted by means of items 2 and 3.

The method of item 1 is effective in speeding up the information treatment, but there are problems with the information transmission (Sect.4.1.1b) from a mask to the computer. Moreover, it is not guaranteed that noticeable defects are fully extracted.

As mentioned above, the method of extracting defects from their features is less reliable in mask-defect inspection when only few experimental data are available. Thus, one must adopt a data-comparison method where defects are decided by comparing experimental data with the design as one of the basic defect-inspection methods. In this method, the measurement accuracy depends upon the alignment of design data and the data of the mask under inspection, because the data at an equivalent point should be compared. The alignment is started at a standard coordinate by standard-mark detection. This method is divided into two aspects by way of making standard data. One is to obtain the data by scanning a neighbouring chip synchronously to a chip under inspection, assuming that all the chips on a mask have the same patterns and have different defects (compare the two-chips method [4.12]). This method is only effective under the above-mentioned assumption.

The other aspect is to use design data as a standard (compare the design-data method [4.13,14]), where the problem is to store and transmit a lot of design data. To solve the problem, design data should be condensed before they are stored and transmitted to an inspection point, where defect inspection is done after converting them into standard data. These techniques still require many problems to be solved.

b) List of Mask Defect Inspection Systems

At present, only one system is commercially available. Other systems have been reported using an electron beam as in the case of a mask dimension-inspection system (Table 4.3). Some examples will be discussed in Sect.4.2.2.

Table 4.3. List of defect inspection systems [4.7] (Part 1) (including experimental systems)

System name	Defect display	Detection method	Detection of common defects	Min. defect size	Inspection time per 5·5 cm	Detection means	Comments
Metallograph microscope	Magnified (~100 - 400×) image	human eyes	in-complete	2 μm	~30 min, down to 2 μm defects	light	much data-scattering by operator
Mask analyzer (Nikon)	150 times defect image displayed by color variation	human eyes	impossible	2~3 μm	~30 min per 2 masks	light	be used as a comparator
Laser applied system MDI (Toshiba)	Red color display of defects extracted by spatial filtering	human eyes	in-complete	~1 μm	~20 min, down to 2 μm defects	light (laser)	-
Fujitsu	Spatial filtering by laser light diffraction	photo-multiplier auto	in-complete	2 μ	-	light (laser)	experimental system
Bell Lab.	Display by computer analysis of laser light diffraction	photo-detector auto	in-complete	-	-	light (laser)	for mask inspection of hybrid IC
Partial feature extraction (Toshiba)	Character display of extracted defects	auto	in-complete	1 μm	-	light	-
SEM (Fujitsu)	Count display of detected defects	auto by scattered E beam	in-complete	0.3 μm	2 h, down to 0.3 μm defects	light	experimental system

Inspection by feature extraction

Table 4.3. (continued)

System name	Defect display	Detection method	Detection of common defects	Min. defect size	Inspection time per 5·5 cm	Detection means	Comments
AMIS (Bell Lab.)	Map display of defects detected by laser signal comparison	automatic detection	im-possible	2 µm	5 min, down to 2 µm defects	light (laser)	practically used in Bell Lab.
5MD1 (Nikon Auto. Control)	Map display of defects detected by photo multi-plier output signal	automatic detection	im-possible	0.75 µm	3 min, down to 2 µm defects	light	–
KLA	Map display of defects detected by memory comparison	automatic detection	im-possible	2 µm	3 min, down to 2 µm defects	light (laser)	–
Mitsubishi electric	Map display of char-acterized defects by colored 3 chips com-parison	automatic detection	im-possible	–	–	light	experimental system
CAD data method (NEC)	CRT display of defect position	automatic detection	possible	10 µm	30 min, down to 10 µm defects	light	experimental system
Perkin Elmer	Coordinate data display of defect position by density signal com-parison	automatic detection	possible	–	–	light	experimental system
EB mask in-spection system (VLS Coop. Lab.)	CRT display of defects detected by design data comparison	automatic detection	possible	0.5 µm	–	electron beam	experimental system

Making data used in inspection

Using stored design data

Inspection by data comparison

4.2 Mask Inspection Systems

4.2.1 Mask Pattern Inspection System

a) Dimension Inspection System Using a Light Beam

The laser interferometric coordinate measuring machine model 21 type II
[4.15,16] is a fully automated system for measuring pattern-position coordinates, pitch dimensions and fine-line width by using a He-Ne laser beam
(Sect.4.1.2a). Its basic principles are: A concentrated light spot scans along
the surface to the object to be measured and the scattered light beam due to
reflection at the edge of the object is sensed by a detector as an edge signal.
Then, the line width of the object is measured using the spot scanning length,
and the coordinates are measured by the amount of travel of the interferometer
stage. So, this system can be used not only for optical masks but also for the
wafer itself. A schematic diagram of the measuring system is shown in Fig.4.3.
As can be inferred from the figure, a He-Ne laser beam is expanded by a col-
limator, reflected by a vibrating mirror, and then introduced into a moving
prism through lens L3. The moving prism is mounted on the travelling stage
with a corner cube and the stage is moved perpendicularly to the optical axis
at the speed of 0.5 mm/s to scan the beam spot on the X-Y stage. The distance

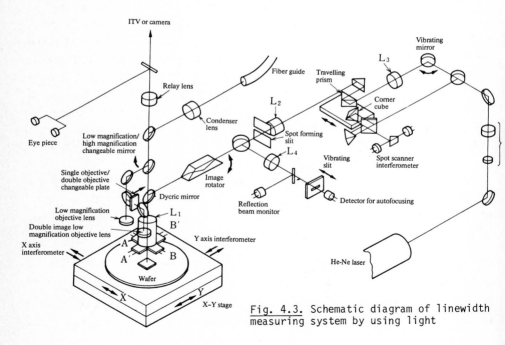

Fig. 4.3. Schematic diagram of linewidth
measuring system by using light

of the prism moved is measured precisely by a He-Ne laser interferometer. The reduction rate of this optical microscopic system is 25, but the beam spot movement in the X-Y stage is doubled due to the two reflections of the travelling prism. Accordingly, the spot movement of the X-Y stage gives 1/12.5 of the value of the travelling prism measured by interferometer. This method is capable of measuring more precisely than the old optical microscopic method which measures the magnified image by slits only.

The slit aperture has the purpose to extend the light spot up to 2 μm in one direction by diffraction for a spot of 0.5 μm in diameter on the sample surface. By this method, the roughness of the patterns can be measured by averaging once. As the scanning direction of the X or the Y axis can be switched by rotating the image by 45 degrees, which results in a rotation of 90 degrees of the beam-scan direction, the long axis of the rectangular beam through the aperture slit is always perpendicular to the scanning direction.

The mirror vibrates in the arrow direction at 3 kHz to bend the beam scanning direction; it can be moved with 0.1 μm steps on the sample surface.

The detection signal is modulated at 3 kHz and the S/N ratio of the signal is recovered by a simultaneous rectification.

As described in Sect.4.1, this instrument involves the automatic focussing mechanism to eliminate the roughness and the inclination of the sample surface.

This sort of instrument using light has a smaller focussing depth. In Fig.4.3, the reflected light spot through the L4 lens makes an image in the opening port of the vibrating slit. As this slit vibrates at 50 Hz with respect to the arrow direction, the output of the detector for autofocussing becomes maximal. It reaches twice the signal when the beam focal point agrees with the vibrating center of the slit. But if the focal point is out of the vibrating slit, the detected frequency becomes equal to the vibrating frequency, but the phase differs by 180 degrees with respect to either position of the vibrating direction. The autofocussing mechanism is accomplished by detecting the focal point of the objective lens.

All steps of this procedure are fully controlled by a small built-in computer (HP-21 series, 64 kB memory). The measured values, such as the average and the dispersion, are displayed on a TV screen and can be printed if necessary. Table 4.4 lists the major specifications of this system.

b) Dimension Inspection System by Using an Electron Beam

Block Diagram, Specification. Figure 4.4 shows the block diagram of the dimension inspection system (VL-MII) [4.17-19] using an electron beam. This system consists of an electron-optical system such as an electron gun, electro-

Table 4.4. Major specifications of the laser interferometric coordinate measuring machine model 2I type II

Pattern size measurement	Measurement sample	Mask and wafer
	Measurement rage [μm]	0.8 - 100
	Minimum measurement pattern [μm]	0.8 · 2
	Reproducibility [μm]	0.05 (3 σ)
	Minimum output unit [μm]	0.01
	Resolution [μm]	0.7 - 0.8
	Measurement time [pts/h]	800
	Computer	HP system 100
Pattern position measurement	Measurement range [μm]	100 - 150
	Accuracy [μm]	±0.13 (3 σ)
	Reproducibility [μm]	0.3 (3 σ)
	Spot size [μm]	0.8 · 2
	Measurement time [pts/h]	400

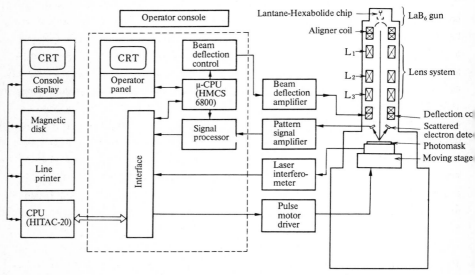

Fig. 4.4. Schematic diagram of linewidth inspection system by using electron beam

magnetic lenses, a backscattered electron detector, an X-Y moving table, a pumping system, a control system such as a small computer, an operator console, and a beam deflection amplifier.

The operator console includes a microcomputer display on the operator panel for processing the input signal from the detector and to send the input signal

184

to the small computer transmitting instructions back to the operatore via an operator panel.

This dimension inspection system has the following features: i) Due to the availability of a small spot size, the system allows a highly accurate size measurement of patterns in the submicron range. ii) The beam positioning is possible over a large area by electrical control so that it is easy to operate with a short response time. iii) An electrical image display permits an easy and quick operational parameter setting.

This system has a three-stage electromagnetic lens permitting a beam spot size of 0.2 μm in diameter or smaller. The objective lens has a long focal length and thus allows a long working distance (distance between the main plane of the objective lens and the specimen surface) of 135 mm. This construction makes possible an electron-beam scanning for a 5 mm square area with high speed. Therefore, it does not require a frequent mechanical stage movement, which helps to reduce the measurement time.

Electron Optics. The electron gun utilizes a single-crystal lanthanum hexaboride (LaB6) emitter with a gun brightness of 10^6 A/cm^2 at an accelerating beam voltage of 20 kV. Electrons emitted from the LaB6 cathode are forming a small focussing point called the cross-over point in the gun. They are aligned to the axis of the electromagnetic lenses by a set of beam alignment coils and finally arrive at the mask. The three-stage electromagnetic lens (L1-L3) demagnifies the cross-over image by a few hundredth, which means a beam spot size of 0.1 - 0.2 μm on the specimen. There are two sets of beam deflector coils for each of the X and Y axes (a total of four coils). These coils, when energized by saw-tooth wave current, scan the beam over the specimen area. A portion of the electron beam is backscattered when it strikes against the specimen with the pattern, and is detected by a backscattered or reflected electron detector. The backscattered detector utilizes a silicon Schottky junction. The detected electrons are amplified in the detector by nearly 10000 times and are used as an input signal for the signal amplifier. The specimen with its cassette is mounted on a specimen stage with an X and Y traverse mechanism. The specimen may be taken out automatically with its cassette through a prepumping chamber. The specimen stage is driven in X and Y by pulse motors mounted outside the vacuum. Its position is monitored by a laser interferometer and controlled by computer to allow precise positioning better than ±3 μm. The He-Ne laser interferometer has a resolution of 0.008 μm which is 1/80 of the laser wavelength. The electron optics is composed of an electron gun, a column and a specimen chamber. The gun is evacuated by an ion pump which attains a vacuum of 10^{-8} Torr, as is the column (10^{-7} Torr), and the specimen chamber is pumped by an oil diffusion pump (10^{-6} Torr).

Control System. The main computer - a HITAC-20 - has a 64 kB word memory. In addition, a disk drive (4.9 MB memory) is used for input (key board and card reader) and for output (console display and line printer). The operator console is composed of a microcomputer (HMCS 6800), an operator panel, a special keyboard, a beam reflector controller, a signal processor, and an interface for HITAC-20. It allows the operator direct access to the system and to preset operating conditions of the system for data acquisitions.

This dimension inspection system has two operating modes, one for presetting the position of inspection and other conditions of inspection, and the other one is for data acquisition under the preset conditions with its data processing. The operator console is used exclusively for the first mode.

Figure 4.5 shows pictures on the operator panel. Figure 4.5a is an entire view of a chip. The white cross mark is a cursor indicating the point of in-

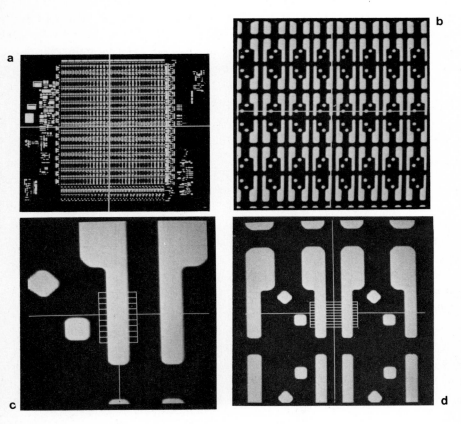

spection. Figure 4.5b is a magnified view of Fig.4.5a. It is available at the touch of a button with the center of the cross mark properly maintained. Figure 4.5c is available by pressing the gate key on the operatore console. It shows a number of white vertical lines around the cross mark. These lines indicate the length of line scans that will be done for the inspection. The horizontal separation of these vertical lines means that the vertical line scans will be repeated at this spatial separation. This is necessary to carry out an average measurement of the pattern width for several points. The number of lines shows the number of measurements that will be done. The length of vertical lines, their separation and number can be preset via the special key board.

Figure 4.5c shows the preset condition for measuring the white specimen pattern for 10 times changing the scanning position each time and averaging the measurements.

Figure 4.5d exhibits another preset condition for measuring the two white specimen patterns in the same way as for Fig.4.5c.

Upon the operator's approval via the key board, these measurement conditions are memorized in the computer. The measurements can be done for all chips specified in the measurement mode and the measured data will be stored in the computer.

Measured Results and Specifications of the System. Table 4.5 shows the general performance of the dimension inspection system using an electron beam. The data were taken with a chromium mask. The pattern-width inspection time is about 10000 points/hour which is outstandingly quick due to the use of an electron beam. The chip-arrangement inspection time is 1000 points/hour, a relatively low value. This is due to the mechanical motion required in measuring a large distance.

4.2.2 The Mask-Defect Inspection System

a) Optical Mask-Defect Inspection System

The optical mask-defect inspection system (5 MD 23 [4.17]) utilizes an illumination spot due to an electron beam on a CRT screen. The spot is deflected by the well-known flying-spot scanner. It is focussed on two masks via two

---

Fig. 4.5. Mask pattern image on operator console CRT. (a) Whole chip image (b) Magnified image (magnified in center of cursor) (c) Magnified image (automatically measured the white pattern linewidth surrounded by gate) (d) Magnified image (automatically measured the three line width of white, black and white surrounded by gate)

Table 4.5. Specifications for a dimension inspection system by using electron beam

Minimum pattern width	0.5 μm
Number of inspection patterns	max. 25000 points
Reproducibility for pattern size measurement	±0.03 μm (3 σ)
Reproducibility for pattern position in chip measurement	±0.2 μm (3 σ)
Measurement time for pattern size	10 cm mask (10000 pts/h)
Chip number of pattern position between-chip measurement	max. 1000 chips
Reproducibility for pattern position between-chip measurement	±0.2 (3 σ)
Measurement time for pattern position between-chip measurement	10 cm mask (1000 marks/h)
Mask size	max. 12.5 cm
Beam scan	Digital scan
Deflection control	16 bit D-A converter
Resolution for deflection step	0.025 - 0.2 μm
Scanning speed for deflection	2 - 32 μs/step
Moving method for sample stage	Step and repeat by pulse motor
Max. moving distance of sample stage	X: 150 mm, Y: 110 mm
Moving speed of sample stage	max. 10 mm/s
Detection accuracy for sample stage position	0.008 μm (by laser interferometer)
Electron gun	LaB_6
Accelerating voltage	20 kV
CPU	HITAC - 20 (64 kB) with disk (4.9 MB)

separate optical microscopes. The optics is designed to scan the same corresponding area on the two masks. The transmitted light beams are received and amplified by two separate photomultipliers. Mask defects may be detected by the difference signal of the two amplifiers. This system corresponds to the "two-chip comparison" mentioned in Sect.4.1.3a.

Figure 4.6 shows a sketch of the system. A 12.5 cm CRT is used for the flying spot scanner. The wavelength of the light emitted from the CRT is 5250 Å. The electron-beam deflections of the CRT are at 31.5 kHz (horizontal) and 75 kHz (vertical). It is the same scanning system used in a commercial TV system.

The field of view can be varied as in an optical microscope or by aperture-stopping the lens to 0.15 mm square and 0.3 mm square. The scanning line is

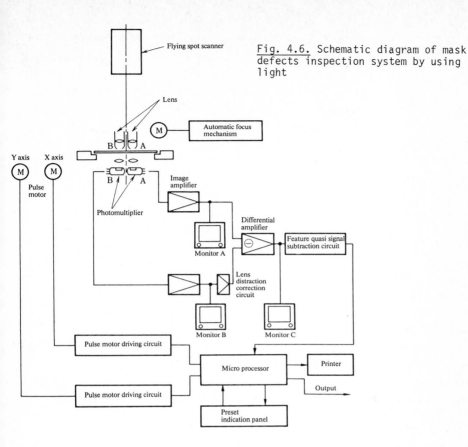

Fig. 4.6. Schematic diagram of mask defects inspection system by using light

always 0.375 μm for the small field (former) and 0.75 μm for the large field (latter). The two optical microscopes (A and B) are independently auto-focussed by air micrometer readings. The distance of the lenses is adjustable in the range of 13-26 mm. Positioning of the lenses may be made manually so that the relative positions of the scanning lines on the two chips may correspond to each other.

The two signals coming from the two chips are detected and amplified by photo-multipliers. They are distortion-corrected and connected to a differential amplifier. Its output signal carries the compared information of the two chips and may be used as the chip defect signal. This signal is displayed on a TV monitor screen.

When there is no defect in the two chips (or the two chips are identical), the two monitors (A and B) show the same image, and a monitor C does not show any. When there are some differences in the monitors (A and B), the monitor C shows an image composed of the difference of the two chips, which represents some defects.

In practice, fine deviations of the patterns on the masks, stage drift, or noise components in the signal (particularly associated with peripheral areas of the pattern, the so-called quasi-defect signal) often appear. This quasi-defect signal needs to be corrected through a correction circuit prior to entering the microprocessor for recording. Positional coordinates for the defects may be entered up to 768 point and can be given to a printer or X-Y recorder, etc. The microprocessor (8080) does some statistics on defects and mask-stage control. This instrument permits detection greater than 0.75 μm at a speed of around 1 cm^2/s.

b) Mask-Defect Inspection by an Electron Beam [4.20,21]

Defect Inspection by Comparison with Standard Data. This system utilizes a comparison with the design data described in Sect.4.1.3a. For this purpose, it is necessary to correct or compensate the alignment error during measurements, the expansion or distortion of the pattern due to temperature rise or lithographic distortion. This experimental system has incorporated a blanking area around the periphery of the pattern by which slight differences may be eliminated and more serious defects may be detected.

Figure 4.7 shows this system. A finely focussed electron beam of 0.2 μm diameter or smaller is scanned over the mask. Its backscattered-electron signal is used for inspection in a binary code after noise subtraction. The noise subtraction is done by positioning the binary-coded picture elements in three lines and three columns. When the central picture element differs from any of the eight adjacent elements, the central picture element is regarded as a noise and therefore it is cancelled.

For the alignment purpose of standard and inspection data, a position mark on each chip is detected prior to the measurement so that both coordinates may be identical. Therefore, both standard and inspection data have correct relative positions and there is no need for reconsidering the positional error for defect judgement.

The defect inspection is done by comparison of the design and inspection data for each picture element. When there are some disagreements, some defects may be suspected. This inspection program has some blanking area around the periphery of the pattern. This area has been incorporated after a series of experiments by changing the picture elements 0 through 4 around the pattern. The experiments performed with some conductive masks having defects of various sizes indicated detection of defects up to 0.2 μm.

190

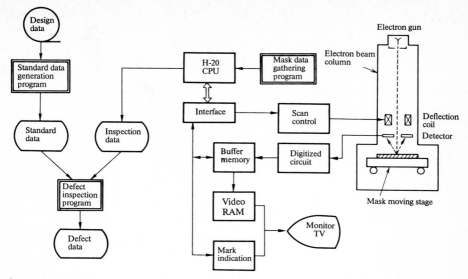

<u>Fig. 4.7.</u> Schematic diagram of mask defect inspection experimental system by comparison with standard data

Defect Inspection by Features. This defect-inspection method belongs to Sect.4.1.3a and is of the type of feature extraction. It is composed of a commercial scanning electron microscope with a minicomputer hooked up. Figure 4.8 illustrates the principle of the system. Electrons emitted from the gun pass through lenses and a deflector system and scan over the mask in form of a finely focussed beam. Secondary electrons emitted from the mask are corrected by the detector. The signal is digitized and fed to a minicomputer for filtering, feature extraction, CDCM (Counter Direction Counting Method) and defect inspection. The primary electron beam has a diameter of 0.5 μm resulting in an image resolution of about 0.2 μm.

The filtering is due to noise subtraction associated with binary coding, and is similar to the method described in Sect.4.2.2b.

In the feature extraction, the system operates to check whether the picture element does not belong to the pattern and eliminates it when any one of the eight adjacent picture elements is unequal to zero. The same picture element is regarded good as a part of the pattern when any one of the eight adjacent elements is zero.

In the CDCM, the system looks at data through a square window larger than the defects. This square window is slightly smaller than the minimum line width so that the normal pattern through the small window has only features in one direction, but defects smaller than the minimum line width have a complicated feature in the small window. A defect in the pattern is judged by this abnormal feature, having structure in more than two identical directions with the figure in the window.

This system permits detection of defects up to 0.3 μm.

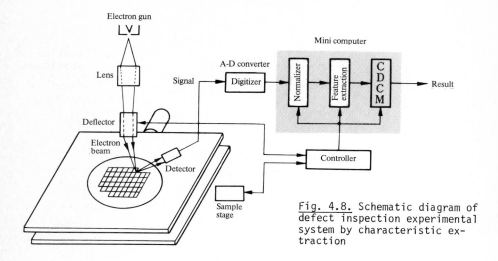

Fig. 4.8. Schematic diagram of defect inspection experimental system by characteristic extraction

5. Crystal Technology

Czochralski grown (CZ) Si wafers presently used for VLSI are of high quality, but still contain oxygen and carbon as residual impurities. These impurities are responsible for the formation of microdefects. Especially, oxygen suppresses the wafer warpage and deformation during various heat treatments in device-fabrication processes. Thus, these impurities have strong influence on the device-production yield. Therefore, the understanding of their behaviours, and both the analysis and the control of the impurities are of importance.

Growth striation in Si crystals, which is a main factor of nonuniformities in the crystal, has recently been improved, but the striation is still difficult to be suppressed, and it is one of the factors to reduce the yield. Therefore, it is desired to establish a method to control the striation.

As for the mechanical processing, bow and warpage of wafers, we re-examined systematically and carried out various experiments from the point of micro-patterning, and obtained interesting and useful results.

On the other hand, in the field of epitaxial crystal growth, the studies of low-pressure epitaxial growth and molecular beam epitaxial growth are gaining significance in the reduction of autodoping effect and the precise control of the epitaxial layer growth from the view of micro-structure device fabrication. In this chapter, we first describe the present status, problems and future trends in impurity research, then the wafer mechanical processing, wafer bow, warpage and deformation, thereafter microdefects and, finally, epitaxial growth related to Si crystals for VLSI technology.

5.1 Overview

In the last decade the minimum line width of VLSI has been decreasing from the 10 μm to the 1 μm order and the number of device elements increased up to $5 \cdot 10^5$ on a chip, as described in Chap.1. The wafer diameter became larger and larger from 50 to >100 mm, and even 125-mm diameter wafers have been introduced in actual processing lines.

For VLSI fabrication, the wafer needs to have two important characteristics. For fine lithography there is the geometrical form requirement and its stability throughout all thermal processes. Controlled physical, metallurgical and crystallographic properties are process requirements that came from advanced devices with fine geometry and very large scale integration and the need of a high production yield.

For some time, it has been misunderstood that the former characteristics depend on the mechanical processing and the latter on the crystal growing process only. But, our research made it clear that both characteristics are interrelated. This means that the Si wafer for VLSI fabrication use must be developed systematically under the consideration of VLSI fabrication processes.

In this section, we have no space to go into details of the crystal growth and fabrication process. Due to the needs of considering all the wafer production processes, they are reviewed briefly [5.1-6].

5.1.1 Crystal Growth and Machining Process

The process steps from raw material, quartzite, to Si single-crystal growth are schematically shown in Fig.5.1 with the impurity level change in each step.

In general, the Si crystal used in LSI or VLSI device is grown by the Czochralski (CZ) method and crystals grown by the float zone (FZ) are not common.

In the case of the CZ method, the fused quartz (SiO_2) crucible and the high-temperature active Si melt are in contact at their interface, and the melt acts as solvent for SiO_2. Oxygen being one of the components of SiO_2 dissolves in the melt as a solute. The oxygen concentration changes by convection and diffusion in the melt, and the evaporation at the open surface of the melt. Oxygen is incorporated into the Si single crystals at the melt-crystal interface and the effective segregation coefficient at the crystallization is estimated to be 1.25. Then, the presence of convection, the strength and direction of the convection and the temperature at the interface influence the concentration and distribution of the impurities.

Carbon in Si comes from the carbon hot-zone material, crucible holder, heater and thermal insulator, as shown in Fig.5.1. Mainly, carbon monoxide originating from the above-mentioned carbon hot-zone material dissolves into the melt and is transported to the melt-crystal interface by diffusion and convection in the melt. The effective segregation coefficient at the interface is 0.07. Sometimes, the carbon in the crystal originates from the raw material.

194

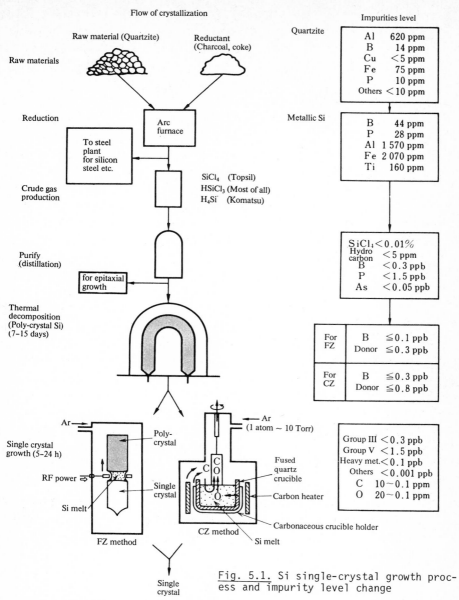

Flow of crystallization

Impurities level

Raw materials
Raw material (Quartzite)
Reductant (Charcoal, coke)

Quartzite

Al	620 ppm
B	14 ppm
Cu	<5 ppm
Fe	75 ppm
P	10 ppm
Others	<10 ppm

Reduction
Arc furnace
To steel plant for silicon steel etc.

Metallic Si

B	44 ppm
P	28 ppm
Al	1 570 ppm
Fe	2 070 ppm
Ti	160 ppm

Crude gas production
$SiCl_4$ (Topsil)
$HSiCl_3$ (Most of all)
H_4Si (Komatsu)

Purify (distillation)
for epitaxial growth

$SiCl_1$ <0.01%
Hydro carbon <5 ppm
B <0.3 ppb
P <1.5 ppb
As <0.05 ppb

Thermal decomposition (Poly-crystal Si) (7–15 days)

For FZ	B	≤0.1 ppb
	Donor	≤0.3 ppb

For CZ	B	≤0.3 ppb
	Donor	≤0.8 ppb

Single crystal growth (5–24 h)
Ar
RF power
Si melt
Poly-crystal
Single crystal
FZ method

Ar (1 atom ~ 10 Torr)
Fused quartz crucible
Carbon heater
Carbonaceous crucible holder
CZ method
Si melt
Single crystal

Group III <0.3 ppb
Group V <1.5 ppb
Heavy met. <0.1 ppb
Others <0.001 ppb
C 10~0.1 ppm
O 20~0.1 ppm

Fig. 5.1. Si single-crystal growth process and impurity level change

Generally speaking, axial carbon concentration does not agree with the calculation based upon the natural freezing condition for either raw material or atmospheric contamination. It is usually fitted to the calculation assuming raw-material contamination, dissolved from the atmosphere and evaporation.

The FZ method does not use a carbon heater, and not even a carbon hot zone, as shown in Fig.5.1. There are big differences in oxygen and carbon impurity concentrations, and this introduced differences in several characteristics and properties to be discussed below. Other added and residual impurities show up in axial distributions and give effective segregation coefficients for natural freezing in the CZ method or the zone melting in the FZ method.

There is a big difference between both methods in the cooling-temperature history of the crystal ingot. It is due to the difference between melt volume, existence of a heater, and the design of the furnace. It also exists between CZ-grown crystal ingots due to differences in crystal diameter, structure of the puller, charge volume, hot-zone structure and materials, recharge, pulling speed and crystal-growth length. This problem will be discussed in Sect.5.4.

There is not such a big difference in the thermal decomposition process except for low-level special impurities that came from the purification system and raw materials from the manufacturer. There is a difference of charged materials in each crystallization process. For the CZ method, polycrystalline lumps are charged in the crucible, and for the FZ method, the initial polycrystalline charge is equal to about one half of the diameter of the growing crystal. There is also a possibility of contamination of heavy metals in the process of dividing into lumps of the thermally decomposed rod. But, this process is not clearly understood.

In the epitaxial growth technique which is important in device production, the incorporation of impurities from the substrate into an epitaxially-grown layer (the so-called autodoping) is a big problem and the suppression of the autodoping is strongly desired. From this viewpoint as well as that of precise control of the epitaxial growth, low-pressure epitaxial growth and molecular-beam epitaxial growth are being investigated with good results [5.8]. The future of these growth technologies looks promising.

In the VLSI fabrication process, the micropattern is drawn or replicated on Si wafers. To do so, wafers with highly flat surfaces are needed to obtain sharp patterns. Wafer-related problems such as slicing, bowing, warpage and deformation have not been studied thoroughly. The mechanism, causes, and behaviour as well as the definition of these technical terms and the measuring methods have not been investigated in detail, but are the subject of current research. Especially, the effect of both dissolved oxygen and precipitated oxygen on the wafer deformation is being studied and interesting results are obtained.

The process of mechanical wafering is illustrated in Fig.5.2. Recently, research reports on this process have been published, but not in large numbers

Wafering process	Operation in process	Work loss
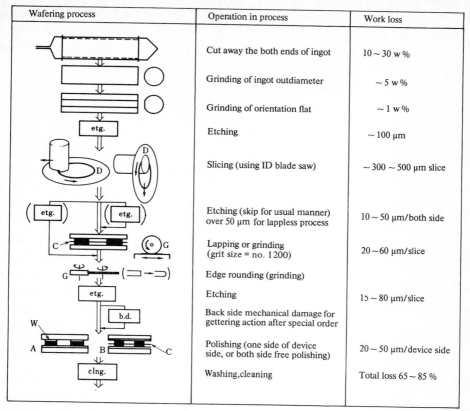	Cut away the both ends of ingot	10 ~ 30 w %
	Grinding of ingot outdiameter	~ 5 w %
	Grinding of orientation flat	~ 1 w %
	Etching	~ 100 μm
	Slicing (using ID blade saw)	~ 300 ~ 500 μm slice
	Etching (skip for usual manner) over 50 μm for lapless process	10 ~ 50 μm/both side
	Lapping or grinding (grit size = no. 1200)	20 ~ 60 μm/slice
	Edge rounding (grinding)	
	Etching	15 ~ 80 μm/slice
	Back side mechanical damage for gettering action after special order	
	Polishing (one side of device side, or both side free polishing)	20 ~ 50 μm/device side
	Washing, cleaning	Total loss 65 ~ 85 %

Fig. 5.2. Mechanical wafering process for Si

The figures in the right column are approximate ones for the removal or loss of Si crystals. These numbers are not the same for each Si wafer supplier, and some users require special specification for certain steps. These differences in removal affect the wafer characteristics at higher temperatures.

There is a big difference of the removal value between the device sides, namely the mirror-polished surfaces and their back sides. The maximum-use ratio of removal values for front to back side reaches 5~6 for commercial wafers. This difference depends on the wafer form, convex or concave, for a certain wafer manufacturer. In Sect.5.3.4, an example is shown that in a convex wafer, on which devices are fabricated, with residual damage on the back side, the damage influences the thermal deformation.

A wafering process without a lapping step is called a lapless process. This lapless process was applied to machining the 125-mm-diameter wafer in this study, but it is not the principal process for commercially available wafers.

The method is not unsuitable for super-flat wafers used in the VLSI fabrication process; it requires careful wafering, an advanced etching method and a free polishing process for both sides. (Refer to Fig.5.23 showing a histogram of flatness for latest wafer data.)

The wafer periphery is grounded by a diamond wheel to the given form. The grit size used in the diamond wheel is in the range of 200-300 lines. This diamond grit size gives deep residual damage that cannot be removed by light chemical etching, 15 - 40 μm in both wafer sides. This thickness is not sufficient to remove the grinding damage perfectly. Then, the residual damage causes frequent slips at the wafer periphery in thermal treatment. The residual damage is removed by etching deeper than 40 μm on one side. This problem will be discussed in Sect.5.3.4 on thermal treatment. The etching depth relates to residual damage on wafer surfaces and large wafer warpage in heat treatment.

A standard 125-mm-diameter wafer with 630 μm thickness has 18 g in weight but the total loss in the wafering process is almost 50 g per wafer. To reduce this weight loss is not a subject of this book, but it is a problem very important for energy and resource saving especially for solar cell production.

5.1.2 Impurities

Recently, semiconductor technology, especially impurity control technology, has advanced greatly, but the impurity control is still an essential problem, and studies on impurities are being actively carried out. With the development of large scale integration of Si devices, the control of doped impurities, residual impurities, and contaminant impurities is required. Especially, the nonuniform impurity distribution leads to scattering on resistivity and defect distributions, therefore it is one of the causes for lower production yield of LSI. Nonuniform distribution such as growth striation has been investigated and improved, but the precise control of the distribution has not been performed and remains to be solved.
The sources of the residual and contaminant impurities are believed to come from dust in the air, Si raw material, and relevant materials such as chemical reagents, water, gas, quartz materials and heater.

Ultra purification of ambient and various relevant materials is needed, and also the development of advanced analytical techniques is needed. If the contamination is not perfectly avoided, the application of gettering technique which removes the contaminants from the active region of the devices during the device process, is effective. Therefore, at present, various gettering techniques are being studied. It is well known that the main residual impurities in CZ Si crystals are oxygen and carbon. The concentration and distribution of these impurities are to some extent controlable.

As for oxygen, the average concentration is around $1 \cdot 10^{18}$ cm^{-3}, and it is possible to control it from $2 \cdot 10^{17}$ to $20 \cdot 10^{17}$ cm^{-3}. As for carbon, it is possible to reduce it to below $4 \cdot 10^{15}$ cm^{-3}. Oxygen in silicon, as will be described later, is useful in some cases and harmful in others; the effect of oxygen is very large. The oxygen behaviour is very complicated, that is, the oxygen effects the growth of microdefects, the gettering action by microdefects, the resistivity variation (by the formation of oxygen donor) and the heat treatment-induced wafer warpage. Oxygen can be correlated to many problems to be solved. It is known that the measured values of oxygen solid solubility and the diffusion coefficient in silicon are different. From the viewpoint of precise control of oxygen impurity in Si, these deviating values should be corrected. Especially, these values at relatively low temperatures (450 - 800°C) have not been measured yet. Reliable measurements are strongly desired.

For the quantitative measurement of oxygen and carbon concentrations in silicon, the infrared absorption method is generally used because of its simplicity. Direct measurement of the concentrations by this method alone, however, is not possible. Generally, the infrared absorption coefficient is corrected for the particular concentration by using a conversion coefficient derived from the concentration value obtained by other methods such as the activation analysis, or the vacuum fusion analysis.

The conversion coefficient, especially for the oxygen concentration, has not been standardized, and sometimes confusion occurs when comparing the oxygen concentrations. Therefore, the determination of a more reliable conversion coefficient is strongly desired and presently being carried out.

It is known that the presence of carbon impurity in silicon more than the solid solubility leads to detrimental effect on the device characteristics. It has recently been demonstrated that the generation of heat-treatment induced microdefects is related to carbon impurity even less than the solid solubility. Differing from oxygen impurity, the carbon impurity concentration should be as low as possible. Furthermore, it has been shown phenomenologically that carbon impurity affects the generation of the oxygen donor. Its detailed mechanism as well as the oxygen-carbon interaction, however, have not been clarified yet. A more advanced technique is required for the analysis of these impurities.

Also, the re-examination and improvement of conventional techniques, and the development of new techniques are being carried out with some good results. For example, Fourier transform spectroscopy in infrared absorption, the photoluminescence (PL) technique to characterize Si crystals and the isotope analysis in ion microanalysis are successfully employed. In addition to the above-mentioned impurities, point defects such as vacancy and interstitial Si atoms, and their complex are believed to play important roles in defect formation, diffusion phenomena, impurity interaction and oxygen donor formation. These problems related to the point defects are being actively studied, but not thoroughly clarified and remain to be solved because of the difficulties of both direct measurement and observation [5.7].

5.2 Impurities in Si Crystals

5.2.1 Outline

CZ Si crystals, which are mainly used for VLSI production, contain some residual impurities with nonuniform distribution. These impurities show complicated behaviour after various heat processes for the device fabrication and, consequently, largely effect the VLSI production yield. In order to control the behaviour of these impurities, both quantitative analysis of the impurity content and analytical characterization of the impurity-related defect are needed. In general, the resistivity measurement and the infrared absorption method are widely used for the analysis. Recently, the PL and ion-microanalysis methods have newly been employed, and good results are being obtained.

As the behaviours of impurities and defects in the crystal are complicated, it is not easy to understand them as a whole. Therefore, characterization with only one method is not enough, but systematic characterization with various methods is needed.

For the quantitative measurements of the residual oxygen and carbon impurities in Si crystals, the dispersive-type infrared spectrometer is widely used. In this case, however, it is difficult to measure precisely the impurity content of thin wafers with conventional 350 - 400 µm thickness because of optical interference. Therefore, wafers with 1 - 2 mm thickness should be prepared specially. Furthermore, it takes a long time for the spectrometer to scan the wavelength range and, therefore, it is not easy to measure a lot of samples. Contrary to this type of spectrometer, the Fourier transform spectrometer, which is being in use along with the development of computers, can eliminate the optical interference component and has the advantage of quick measurement of lots of wafers. At the VLSI Cooperation Laboratory, therefore, the oxygen and carbon concentrations of both as-received and heat-treated wafers were measured by using a Fourier transform spectrometer (JEOL JIR-40). Based on results by this measurement and the others, the behaviours of the oxygen and carbon were analyzed as a whole.

PL measurement has not been actively used for Si crystal characterization because of several reasons such as its weak intensity. After careful consideration, however, the PL method was employed to detect minute impurities and defects in Si crystals. The method was found to be highly sensitive and a very useful tool for the characterization of Si crystals. Although there remain some problems to be solved, the method is expected to be widely used.

By the above-mentioned optical method, it is difficult now to measure the impurity distribution profile in the micro-region. Such a micro-analysis, however, is needed for the VLSI research and development. For such a purpose, an ion-microanalyzer (IMA) is suitable and widely used. As for the oxygen and carbon impurity analysis, however, the background level is so high that meaningful measurement is difficult.

Recently, however, the detection of the order of 10^{15} atoms/cm^3 of these impurities has been possible through some technical improvements. Therefore, one can now obtain the spatial distribution of these impurities which was nearly impossible to know before. Thus, new information on the oxygen and carbon impurities in silicon crystals is being expected. Of the above-mentioned analytical methods, future common problems to be solved are high sensitivity, high precision, high resolution, easy operation and low cost of the instruments as well as the preparation of standard samples for the quantitative analysis.

Growth striation, a main feature of a nonuniform impurity distribution, appears as resistivity variation or distribution variation of microdefects in wafers. Since the nonuniformity leads to the decrease of the VLSI production yield, this is a big problem. Based on the study of the origin of the striation, several methods to suppress the striation are being investigated and carried out. But, this problem has not been fully solved yet.

Oxygen impurity in silicon crystals precipitates by a high temperature heat treatment and causes various microdefects. Also, it is well known that oxygen impurity forms donors (the so-called oxygen donors) by a low-temperature heat treatment. Recently, from the precise resistivity control point of view, and also for the deeper understanding of the oxygen impurity behaviour, this oxygen donor problem was brought up again. Formerly, the characterization of the oxygen donor was mainly carried out by the resistivity measurement. Recently, however, direct characterization of the donors by the PL method became possible, and much information is being obtained.

5.2.2 Fourier Transform Infrared Spectroscopy

Impurities in crystals induce optical absorption in the infrared region, especially if they are in the electrically active state. The absorbed energy is related to the energy level of the impurity. Lines appear around 30 μm in the optical absorption spectrum of Si crystals at liquid helium temperature for crystals containing shallow impurities as B and P. Absorption lines of around 20 μm are observed for relatively deep impurities such as Al and Ga. As for oxygen (O) and carbon (C), although they are electrically inactive, absorption lines appear at 9 μm (1106 cm^{-1}) and 16.5 μm (607 cm^{-1}), respectively, as shown in Fig.5.3. They are due to the local mode vibrational modes associated with interstitial O and substitutional C, respectively [5.9a,b]. In this way, an analysis of impurities in crystals can be made by means of absorption spectroscopy.

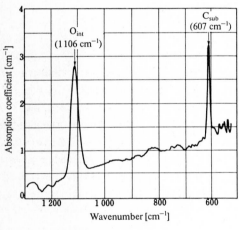

Fig. 5.3. Infrared absorption spectrum of Si crystal at room temperature

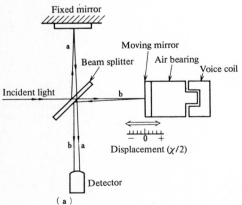

(a)

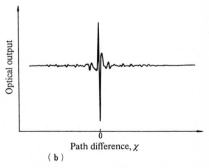

(b)

Fig. 5.4. (a) Schematic illustration of Fourier transform spectroscopy. (b) Interferogram

In general, the absorption spectrum is obtained by using the dispersive spectrometer. The wavelength is scanned and the transmitted light intensity is measured. In Fourier transform spectroscopy, on the other hand, the absorption spectrum is obtained by Fourier transformation of the interference curve measured by a double-beam interferometer (Michelson interferometer). Figure 5.4a shows the schematic illustration of the Fourier transform spectroscopy. The incident light is separated into two beams by the beam-splitter. One beam is reflected by the fixed mirror, and the other one is reflected by the moving mirror. The two beams are again superimposed by the beam-splitter and then collected by the detector. Its signal is an interference curve as a function of the optical path difference (Fig.5.4b). This curve is called an interferogram. The moving mirror is suspended by air bearings and moved rapidly by a voice coil. The interferogram is accumulated in order to reduce noise, and then Fourier transformed. Figure 5.3 is an example of the absorption spectrum by the Fourier transform spectroscopy.

The Fourier transform spectroscopy has two great advantages [5.10]. The first is the through-put advantage which comes from the fact that this method needs neither slits nor dispersive elements such as a prism and a grating. A further advantage results from the fact that the measurement can cover the whole wavelength region at a time. Therefore, the spectral resolution is far better and the time required for the measurement is much shorter, in comparison with those of a conventional dispersive spectrometer. Furthermore, the spectrum can be measured very easily by using a mini-computer to perform interferometer operation, Fourier transformation, and various data processings. The resolution $\Delta\nu$ and the maximum wavelength region ν_{max} were given in wavenumber units

$$\Delta\nu = \frac{1}{\alpha D} \quad , \quad \nu_{max} = \frac{1}{2\delta}$$

where D is the maximum path difference, α is the correcting factor, $1 < \alpha < 2$, depending on the measurement condition, and δ is the sampling interval of the interferogram.

Today infrared spectroscopy is widely used for the quantitative analysis of O and C in Si. The peak heights of the absorption lines due to C and O vary in proportion to their concentrations. Lattice absorption normally overlaps the O and C lines. In order to eliminate this contribution, a difference method is used. Hereby a reference sample being free of O and C and of the same thickness as the unknown sample is placed in the reference beam of the double-beam spectrometer. The conversion coefficient obtaining the O and C concentrations from the absorption coefficients are determined by ASTM [5.11]. In Fourier transform spectroscopy, the difference spectrum can be obtained in the single-beam mode by storing the spectrum of the reference sample beforehand and then subtracting it from the spectrum of the unknown sample. Fourier transform spectroscopy also enables us to correct the difference in the thickness between the reference sample and the test sample.

The detection limit becomes higher for thicker samples. However, too thick samples reduce the transmitted light because of the increase in the lattice absorption. This is a disadvantage for the measurement of C, since the lattice absorption is strong. Therefore, the thickness of the sample is usually around 2 mm. It is very convenient if the measurement can be done on an ordinary wafer (about 400 μm thick). However, the path difference causes interference between the light a, penetrating the wafer, and the light b, reflected at the inter-faces (Fig.5.5a). In this figure, n and d are the refractive index and the wafer thickness, respectively, and multiple reflection is neglected. This ef-fect introduces interference fringes on the spectrum, which makes it difficult to read the absorption peak height correctly. It is impossible to eliminate the interference fringes in a conventional dispersive spectrometer, but it is possi ble in Fourier transform spectroscopy as follows. The interference components appear as spikes at a path difference of $2nd$ in the interferogram, as shown in Fig.5.5b. The interference fringes can be eliminated if the spike is neglected in Fourier transformation. Such software is available in recent Fourier spectro meter systems. If we measure 400 μm thick wafers at a resolution of 4 cm^{-1}, the condition of D < $2nd$ is satisfied; no spikes appear in the interferogram. There fore, interference fringes are not introduced in the spectrum. So, the Fourier transform spectrometer enables the measurement of O and C concentrations in conventional Si wafers nondestructively. This is the main reason for the exten-sive use of the Fourier spectrometer in the silicon industries.

The detection limits for O and C in a 400-μm thick wafer at room tempera-ture are $1 \cdot 10^{16}$ and $2 \cdot 10^{16}$ cm^{-3}, respectively. They can be improved to

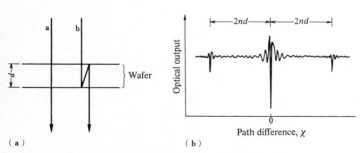

Fig. 5.5. (a) Interference between penetrated beam and reflected beam. (b) Interferogram in the case of (a)

$2 \cdot 10^{15}$ (for O) and $4 \cdot 10^{15}$ cm^{-3} (for C) by cooling the sample down to liquid nitrogen temperature and by putting two wafers together.

Oxygen precipitates produced by heat treatment show the characteristic absorption lines. The absorption due to crystobalite appears at 1125 and 1120 cm^{-1} and that for amorphous SiO$_2$ at 1040 cm^{-1}.

The interference due to the interface reflection is utilized to measure the sample thickness. The interference causes the spikes in the interferogram, as shown in Fig.5.5b. The sample thickness can be accurately determined from the path difference. This is the principle of the conventionally used film-thickness gauge (FTG).

5.2.3 Photoluminescence Spectroscopy

When a crystal is irradiated by light whose energy is higher than the band gap of the crystal, excess electrons and holes are excited. Electrons and holes can come to recombination through various electronic states, partially by the radiative recombination in combination with light emission. This emission is called photoluminescence (PL). Electronic states in crystals can be investigated by spectroscopic analysis of PL.

The photoluminescence from Si was not applied to the characterization until recently. This comes from the fact that the PL efficiency is low because Si is an indirect band-gap semiconductor. In addition, the measurement of PL is difficult since the radiation is in the near infrared region. However, many works have recently been carried out on PL due to shallow impurities, such as B, P, In, Tl, As, Sb, Bi, Li, etc., taking advantage of the recent progress in the measurement techniques [5.12]. Based on these results, PL of Si has been applied to the characterization of impurities in Si [5.13]. Today, this technique is the most sensitive method to the shallow impurities. Furthermore, this technique enables an accurate assessment of trace impurities which have not been analyzed by any other techniques. The advantages of the PL technique are (i) high sensitivity (the detection limit is shown in Table 5.1), (ii) capa-

Table 5.1. Estimated detection limits of the PL method for various impurities

Impurity	Detection limit [cm^{-3}]
B	$5 \cdot 10^{10}$
P	$1 \cdot 10^{10}$
Al	$1 \cdot 10^{11}$
As	$5 \cdot 10^{11}$

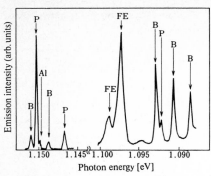

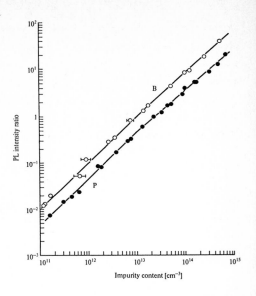

Fig. 5.6. PL spectrum of Si crystal with high resistivity (liquid helium temperature)

Fig. 5.7. Calibration curves to determine B and P concentrations in Si from PL analysis

bility of quantitative analysis for B and P, (iii) capability of the analysis of compensating impurities, (iv) nondestructive and contactless method, (v) high spatial resolution, and (vi) unnecessity of special sample preparation

Figure 5.6 demonstrates the PL spectrum from an FZ crystal with a resistivity of 1.8 ± 0.5 kΩ·cm. It was taken by the irradiation of an Ar ion laser on the sample immersed in liquid helium. The laser beam spot on the sample surface was 2.5 mm at $1/e^2$-intensity-point with an incident power of 300 mW. The observed spectral lines are due to the exciton and the multiple exciton complex bound to neutral impurities. They are labelled with the chemical symbol of the associated impurities, where FE denotes the PL due to the free excitons. This spectrum clearly shows that the crystal contains B, P and Al. The PL intensity ratio between free and bound excitons is related to the impurity concentration (Fig.5.7) [5.13,14]. This relationship was obtained by plotting the intensity ratio as a function of the impurity concentration in a log-log scale.

The problem of this method is that no techniques are available at present to calibrate the PL intensity ratio with an absolute concentration in the concentration range below 10^{15} cm^{-3}. Therefore, the neutron-transmutation-doping (NTD) technique has been used to make calibration-standard samples for P [5.14]. The controlled and accurate doping was performed with using ultra-high-pure crystals as a starting material so as to be free of background impurities. The transmuted P concentration is calculated from the thermal neutron dose. As for B, the samples exclusively doped with B were prepared by the vacuum multipass FZ method. The elimination of the other

impurities can be realized by this method, because the segregation and evaporation effects for B are the smallest of all impurities in Si. The B concentration of these samples was determined by the precise resistivity measurement.

The B and P concentrations in the sample shown in Fig.5.6 can be obtained
to $7.0 \cdot 10^{12}$ and $4.5 \cdot 10^{12}$ cm^{-3}, respectively, using the calibration curves.
This result agrees with the value of $(N_A - N_D) \sim 2.4 \pm 0.7 \cdot 10^{12}$ cm^{-3}, calculated
from the resistivity. The accuracy of this method is estimated within an uncertainty of $\pm 30\%$.

The PL method is now in practical use, for example, in the analysis of
residual impurities in high-resistivity materials for detectors and high-
power devices, in the study into the cause of the abnormal resistivity, and
in the characterization of segregation and evaporation effects in purifying
processes.

5.2.4 Ion Microanalyzer

The ion microanalyzer (IMA) or secondary ion mass spectrometer (SIMS) is an
apparatus which is capable to perform high-sensitive surface analysis by detecting secondary ions emitted from sputtered surface of a specimen in a high
vacuum [5.15-17]. SIMS can provide useful information on hydrogen in solids,
isotopic effects, and three-dimensional distribution of impurities in a submicrometer resolution [5.18].

Improvements in design, manufacturing, and evaluating techniques are required for the successful development of VLSI devices. For instance, perfect
control of impurities in silicon is absolutely necessary to develop actual
VLSI devices and, hence, there will be an increased demand for more detailed
knowledge regarding the distribution of impurities in silicon. For the purpose of analyzing the three-dimensional distribution of impurities located
in a submicrometer region, SIMS is a most powerful apparatus available in
the semiconductor industries.

A basic application of SIMS is to measure compositional profiles of impurities implanted into silicon, and a number of reports has been published
on this subject. The detection limit of phosphorus (^{31}P), which has been
about 10^{19} cm^{-3} owing to the presence of interfering mass as ^{30}Si^{1}H, is now
decreased to a level as low as $1 \cdot 10^{16}$ cm^{-3} by the newly developed high-mass-
resolving technique utilizing a revised ion optics [5.1].

Carbon and oxygen are still the prevailing impurities in CZ silicon single
crystals, which are the starting materials for VLSI devices. The effects of
these impurities on crystal defects and mechanical properties of silicon

have gradually become clear. Conventionally, infrared spectroscopy has been used to detect low concentrations of carbon and oxygen in silicon, but this method cannot give the three-dimensional distribution of carbon and oxygen in silicon. Knowledge regarding the distribution of carbon and oxygen will be necessary for optimizing manufacturing parameters of VLSI devices, and detection of these elements by SIMS is discussed next in more detail.

The secondary ion yields of carbon and oxygen in silicon are high enough and these elements can be detected as low as $\sim 10^{15}$ cm^{-3} [5.15-17] by the SIMS technique. However, a high background signal due to these elements restricts the detection limit to much higher values than the signal limit mentioned above. Actually, the background levels of common oxygen and carbon in conventional SIMS systems are $1 \cdot 10^{20}$ and $3 \cdot 10^{19}$ cm^{-3}, respectively. Carbon and oxygen are present even in high vacuum. Typical residual gases as H_2O, N_2, CO, CO_2, and CH_4 can be detected in a sample chamber of a conventional SIMS system which has been evacuated down to $1.5 \cdot 10^{-8}$ Torr. The number of residual gas molecules impinging upon the specimen in the chamber is estimated to be $10^{12} \sim 10^{13}$ s^{-1} cm^{-2}. These impinging carbon and oxygen atoms can be adsorbed on the surface to produce a high background signal which may hide the intrinsic signals of these elements in silicon in the SIMS analysis.

A simple method to overcome the problem is to attach a shroud cooled by liquid helium with a cryogenic refrigerator surrounding the sample holder, which may reduce the background rate one order of magnitude. To modify the sample chamber to be evacuated independently with an oil-free pumping system such as a cryogenic pump will be much more effective for the analysis of carbon and oxygen. Another method to avoid the high-background problem is to measure stable isotopes of these elements intentionally doped into silicon. Owing to the low natural abundance of the isotopes ($^{13}C \sim 1.108$ %, $^{18}O \sim 0.204$ %), much lower detection limits of carbon and oxygen in silicon can be obtained. It is possible to estimate successfully the behaviour of low-concentrated carbon and oxygen in the order of 10^{15} cm^{-3} in silicon by this method [5.2].

An example for the high sensitive detection of carbon and oxygen is illustrated in Fig.5.8, which shows the redistribution of implanted carbon and oxygen after furnace annealing [5.3]. It is clear in Fig.5.8 that complicated multiple peaks appear in the atomic profiles of all oxygen, whereas only carbon doubly implanted with oxygen in silicon shows the same kind of stratified structure. Differences in the atomic behaviour of carbon and oxygen in silicon can be estimated successfully by the present method.

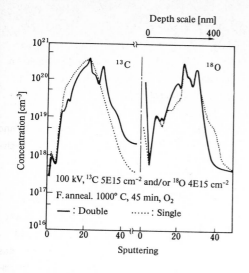

Fig. 5.8. ^{18}O and ^{13}C concentration profiles in ^{18}O and ^{13}C doubly implanted (100) silicon after 45 min furnace annealing at 1000°C in dry O_2 ambient. The dotted lines show the atomic profiles obtained in the singly implanted substrates for the reference

It is thus possible to lower the detection limits for carbon and oxygen in silicon by modifying the experimental procedure of the SIMS analysis, and the availability of SIMS will be much increased in the semiconductor industries.

5.2.5 Striation [5.6,19,20]

A wafer shows some kinds of crystal defect distributions which can affect VLSI fabrication. Figure 5.9 gives an example of spreading-resistance measurement across a CZ-grown (100) 76-mm diameter wafer. As shown in this figure, there are big differences in the resistivity distribution across wafers, depending on the growth conditions. It is known that the wafers with big differences in many peaks and valleys show a low VLSI fabrication yield. The variations influence the oxygen donor generation and disappearance, too, as will be described in Sect.5.2.6. Then a careful heat treatment is required at a low temperature of around 450°C.

Many striae are observed as shown in Fig.5.10. They result from etching CZ and FZ as-grown 76-mm diameter Si crystal wafers sliced off parallel to the growth axis. These striae indicate the difference of etching rate for different positions. They include etch pits row, which cannot be observed by the X-ray Lang topograph technique. Infrared photoelastic observation shows the striae which have the same form as shown in Fig.5.11. The striae contrast is decreased by lowering of the carbon concentration (less than about 10^{16} cm^{-3}) in the CZ-grown Si wafer which has a high oxygen concentration at its solubility limit. For an FZ-grown crystal, with a very low oxygen concentration,

Fig. 5.9

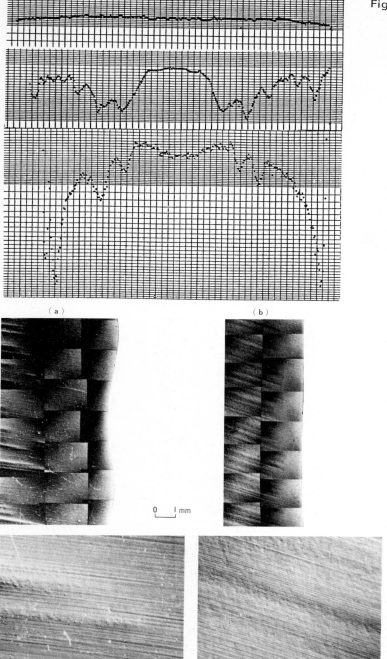

(a)   (b)

0 ⌊ mm

0   0.5   ⌊ mm

Fig. 5.10

Fig. 5.11. IR photoelastic observation for a CZ-grown crystal shoulder, vertical slice. Crossed polarizers. (There are two different pitch striations, 500 ~ 1000 µm and ~ 5mm.)

striae are observed with sufficient contrast at a high carbon concentration (10^{17} cm^{-3}) level.

Figure 5.12 shows the X-ray section topographs for as-grown and heat-treated wafers. Using other methods, it is confirmed that black spots are dislocation loops and precipitates. The striae of spot rows show the same form and different spacing as compared with Figs.5.10 and 11.

There is a strong interrelation between the methods of growth and the striae spacings. For a CZ-grown wafer, there is a difference in the spacing between new and old (before ca. 1970) wafers. The spacings are classified roughly as

CZ-grown crystal   30-60 µm, ca. 500 µm, and mm-order cycles for
                                       30-60 µm striae
FZ-grown crystal   ca. 50 µm and ca. 500 µm.

These are growth striations pointed out by MURGAI et al. [5.21].

It is known that the origin of striation is the impurity distribution change caused by an alteration of the growth conditions around the liquid-solid interface. Then, the striation form depends on the form of the liquid-solid interface and the wafer cutting direction of the ingot. The striation shown in

---

Fig. 5.9. Examples of spreading resistance profile for a CZ-grown (100) 76 mm diameter wafer (after 475°C heat treatment)

Fig. 5.10. a,b. Comparison of etching striation of a vertically sliced wafer for (a) CZ-grown and (b) FZ-grown crystals

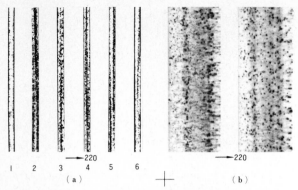

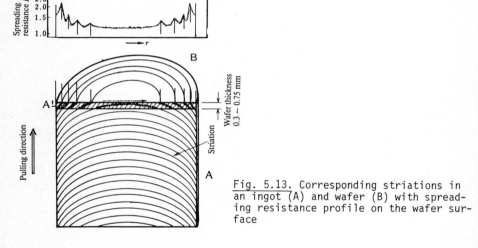

Fig. 5.12. a,b. Section topograph of a heat-treated CZ-grown wafer (thickness : 600 μm). [a] 1: as grown crystal, 2: 820°C, 3: 940°C, 4: 1070°C, 5: 1160°C, 6: 1245°C heat treatment + 1050°C heat treatment in wet O$_2$. b) enlarged photograph of 2 in (a), the striation pitch: 150 ~ 200 μm]

Fig. 5.13. Corresponding striations in an ingot (A) and wafer (B) with spreading resistance profile on the wafer surface

Figs.5.10 and 11 represents the observation on the A-plane in Fig.5.13. The section topograph shown in Fig.5.12 corresponds to the thin section A in Fig.5.13. Figure 5.9 shows the impurity distribution corresponding to the spreading resistance change across the wafer center on surface B of wafer A'.

The conditions of growth that relate to the striation and its width for the two growing methods described above are as follows.

1) Average crystal growing speed v_0 [mm/min]
2) Actual crystal growing speed v [mm/min]
3) Relative rotation speed between crystal and liquid n [rpm]
4) Cycle of temperature change and its amplitude, n_f [Hz] and $\pm\Delta T_f$ [°C]

5) Rotational speed of convection in the melt n_c [rpm]
6) Temperature difference by convection $\pm\Delta T_c$ [°C]
7) Convection mode in melt and its change (center→ periphery or pheriphery→ center) M
8) Vaporization coefficient change $\Delta\varepsilon$ (0: vaporize, <0 and C: contaminate, >0)
9) Oxygen solubility change at the crucible-melt interface ΔC_0
10) Cyclic pulling rate change for the diameter-controlling feed back variable n_p [rpm]
11) Offset value between the pulling rotation center and the furnace temperature center ΔT_0 [°C]

We do not succeed in making clear a total correlation between the factors mentioned above and the striation structures. But those two types of crystal growing methods have shown the following correlations for their striation pitches:

CZ-grown crystals: 30-60 μm pitch (v_0/n),
ca. 500 μm pitch (v_0/n_c),
modulation cycle of 30-50 μm pitch (v_0/n_p)
FZ-grown crystals: ca. 50 μm pitch (v_0/n)
ca. 500 μm pitch (v_0/n_c).

After the computer simulation and model experiment, the changes of the impurity concentration along the direction of growth are caused by the effective distribution coefficient k_{eff} and the interface concentration of impurities (C_{OL} and C_{CL}) both modulated by vaporization coefficient, oxygen solubility change, crucible-melt interface area change and convection mode change depending on the growing volume rate g.

The wafer for VLSI is required to be of a composit method because the uniformity of several properties and the demand for high yield are necessary for processing different device types and process steps. As a consequence, there is no unique solution when to stop the convection in a crucible.

5.2.6 Oxygen Donor

When CZ Si crystals are heat treated at a relatively low temperature, around 450°C, it is well known that donors are newly formed from dissolved oxygen impurities [5.22]. Much work has been carried out to elucidate this problem. Although the true nature of the donor has not been fully understood, several models such as famous KAISER's SiO_4 model [5.23] have been proposed. When CZ

Si crystals pass through a temperature range at around 450°C during cooling af
ter crystal growth, or when Si wafers are subjected to a heat treatment proces
at around 450°C, oxygen donors (10^{16} cm^{-3}) are formed resulting in a resistivity change. It is well known that the oxygen donors disappear by annealing at
temperatures over 500°C, typically around 650°C. Complete controlling of such
a donor killing process, however, is difficult. For example, in the case of
crystal growth with a resistivity of more than 50 Ω-cm, it is not easy to
control precisely the resistivity with good reproducibility. It is understood
that the difficulty is due to the imperfect control of the generation of the
donors during crystal growth and cooling. The temperature difference between
the center and the periphery of the ingot increases with increasing ingot
diameter; as a result, the oxygen donors are nonuniformly generated. Investigation on the oxygen donor in today's dislocation-free CZ Si crystals showed
[5.24] that the reduction rate of the donor by a 550°C annealing is far smalle
than that obtained by KAISER et al. [5.23]. To describe this donor reduction
behaviour, the following reaction is proposed

oxygen donor + X → electrically neutral complex;

here, X is assumed to be contained, to a great extent, in the old crystals, bu
only a few are contained in today's crystals. As described before, it is known
that short-period annealing at around 650°C reduces the oxygen donors, and lon
period annealing produces again oxygen-related donors [5.25]. KANAMORI et al.
[5.26] have investigated further this behaviour, and concluded that the nature
of the donors generated by 650°C annealing differs from that formed by 450°C
annealing. This donor was called as a new donor (ND) [5.26]. Its characteristics are as follows.

1) ND is generated by annealing of CZ Si crystals at temperatures of
   550-800°C.
2) ND is not generated in float-zoned (FZ) crystals, but is generated in
   FZ crystals into which oxygen is diffused at a high temperature.
3) ND is easily generated in carbon-rich crystals, but there is no origin
   in carbon.
4) The maximum concentration of ND is about $1 \cdot 10^{16}$ cm^{-3}.
5) ND generation can be accelerated by pre-annealing at 450-550°C.
6) Concerning the 800°C annealing, it can be stated that

   (i) ND is not generated by an 800°C annealing alone, but is produced by
       800°C annealing with a pre-annealing at 450-550°C;
   (ii) ND generation is suppressed by 800°C pre-heating;
   (iii) an ND, once generated, is stabilized by 800°C annealing.

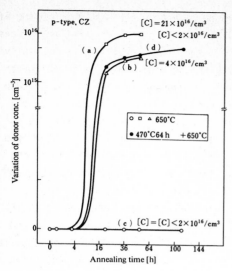

Fig. 5.14. New-donor concentration variation versus annealing time at 650°C for typical p-type CZ wafers with various carbon concentrations. (Curves a, b, and c: without preannealing; curve d: with preannealing at 470°C for 64 h. The same wafer was separated into two pieces for curves c and d)

The effect of the carbon concentration and the pre-heating conditions on the ND-generation behaviour is shown in Fig.5.14. A two-step reaction model has been proposed to interpret the donor formation mechanism [5.26], but the true nature of ND is not fully clarified, and further investigation is needed.

On the other hand, a characteristic PL spectrum relating to the new donor is found, as shown in Fig.5.48c [5.27]. From these spectra, it was also proved that ND is different from the donor generated by 450°C annealing. A measurement of the resistivity cannot distinguish these two kinds of donors, but using the PL method this can be done easily. Therefore, the PL method is a useful tool for investigating the oxygen-related donors.

Recently, the neutron transmutation doping was carried out even on CZ Si crystals. In this case, the ND generation is expected by annealing, which is necessary to recover the radiation damages in the crystals. Thus, the ND problem covers a wide field of problems to be used especially in the direction of controlling the NDs in the near future.

5.3 Wafer Bow and Warpage [5.29,30]

5.3.1 Definition and Measurement Methods

Deviating from the focussing planes in writing or printing instruments leads to the out-of-focus problem. Superfine patterning is achieved in the same manner as taking a fine photograph using a photographic camera. The allowed deviation is of the same order as the minimum line width in the pattern. For example, the deviation tolerance for a 1.5 μm minimum line width is required to be 1.5 - 3 μm. This limit must be kept in the imaging area of any optical patterning instruments. Then, the one shot type printer requires a tolerance for the whole wafer area, and the step and repeater must have the same tolerance for the one step printing area.

Today the X-ray lithography is based on contact, soft contact or proximity exposure techniques except the SOR (Synchrotron Orbital Radiation) exposure system, because of the large divergency and the focus size on the target of the X-ray source. For these exposure techniques, especially the proximity technique, the local wafer unevenness is without influence. The defocussing of the printing line is due to the wafer's unevenness of over 1 μm. These problem will be considered for the optical proximity lithographic system.

Discussing the wafer bow and warpage, the following definitions are necessary (Fig.5.15):

1) Bow: The average surface form at unchucked free-state raw wafer. Bow includes an elastic and/or plastic deformation caused by work damage, etc.
2) Warpage: Elastic and/or plastic deformation induced by processing including heat treatment. In many cases, they are combined.
3) Flatness: Local deviation from an average wafer surface form.
4) Parallelness: The average separation between mirror and back surfaces being constant.
5) NTV (Nonlinear Thickness Variation): Sum of the maximum peak and valley absolute values from the standard plane at the chucking state, or dif-

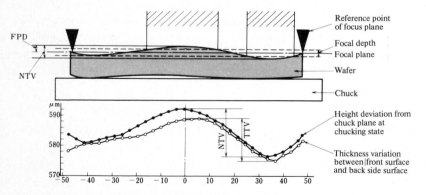

Fig. 5.15. Relation between FPD, NTV, TTV of chucked wafer, focal plane at explosure state and its focal depth. (Hatched region shows out of focal depth)

ference of maximum peak and minimum valley values from the chucked surface.

6) FPD (Focal Plane Deviation): Maximum deviation from the focal plane defined at the corresponding chucking state using the printer or the writer.

7) Standard plane:
   (i) Most probable plane: The plane defined by the least square method for all measured points in an inscribed regular square on a measured wafer. This definition is applied to the free and chucked surface.
   (ii) Standard plane behind lithographic instruments: Plane across more than three points behind the lithographic instrument used on the measured wafer (generally, three points taken on the periphery of the wafer making a regular triangle).

8) Sign of bow and warpage: Depending on the site of center of curvature: + sign is taken for the case that the center takes place on the device (mirror) side. Corresponding to a concave wafer.

9) Bow and warpage form: Typical examples are shown in Fig.5.16 with the examples of the thickness variation.

10) Local thickness variation, LS (local slope): The local slope is the maximum value of the difference of the maximum and minimum values related to all square meshes corresponding to the exposure of area by the step-and-repeat process on the wafer. The standard area is chosen as $10 \cdot 10 \text{ mm}^2$ square but recently a larger square is chosen.

NTV, FPD and LS, as described above, are quantities that describe lithography machines in use. These values change for each standard plane characteristics of such a machine. In this discussion, the standard plane is utilized in the above Definitions 7 (i) and (ii), and the main standard plane in 7 (i). For a small deformation, the differences between the standard planes chosen are negligibly small. At large deformations, the difference between standard plane definitions cannot be neglected.

There are many methods for measuring bow and warpage. These methods are classified and described by their characteristics in Table 5.2.

It was pointed out that the most important aspect in this study of wafer bow and warpage is the separation of the kinds of deformation and their signs with the absolute values, their physical origins for elastic and/or plastic deformation and their in situ observation. These separations are made by a combination of several instruments shown in Fig.5.17. Especially, for the thermal warpage, there are some techniques based on the deformation due to a long working distance, the photoelastic macroscope, and the large-diameter

Table 5.2. Wafer bow and warpage measurement methods

Classification	Method, Name	Rough surface	Mirror surface	Crystal plane	Contact	Distance [mm]	Measurement value / Sensitivity	One point	Multi point	Line	Multi line	2-dimensional	Commercial	Diversion	Automated	CPU control	Special merit and notice
Mech. method	Stylus Technique	o	o		o	contact	z 0.02 µm	o		o			o				Suitable for as-sliced surface
	Spherometer	o	o		o	contact	z ~1 µm	o		o			o				
	Ring spherometer	o	o		o	contact	z ~1 µm	o					o		o		
	3-ball jig	o	o		o	contact	z ~1 µm	o					o		o		ASTM and JEIDA Standard, not adapt to VLSI wafer
Electr. method	Capacitance method	o	o			0.5~2	z 0.1 µm	o	o	o			o				Multi-point instrument reported
	Eddy current method	o	o			0.5~2	z 0.1 µm	o	o	o			o				Average data for sensor area
	Ultrasonic method	o	o			10~100	z ~0.1 µm	o				o	o			o	
Optical method	Focussing technique	o	o			0.3~5	z ~1	o		o			o			o	Fit to periphery form measurement
	Light section method macro/micro	o	o			0.3~3/10 100	z 1/~10 µm		o				o			o	Long working distance
	Autocollimater method		o			100~10000	o 1~30 s	o					o		o		Required sample rotation
	" scanning method		o			100~300	o 1~30 s			o			o		o		
	" multi scanning method		o			100~300	o 1~30 s		o	o			o		o		
	Multi-divided laser beam method		o			300~2000	o 2~30 s		o				o		o		Long working distance
	Oblique incident interferometer		(o)			0.01~0.1	z 1~10 µm					o	o		o		
	Two beam interferometer		o			0.3~500	z 5~20 nm					o	o	o			
	Multi-beam interferometer		o			~0	z 1~5 nm					o	o	o			
	Moiré method	o	o			0.01~1000	z 5~100 µm					o	o		o		Suitable for as-sliced and lapped surface
	Optical micrometer	(o)	o			~1	z ~1 µm	o					o				
	" multi sensor	(o)	o			~1	z ~1 µm		o				o		o		
	" multi sensor scanning type	(o)	o			~1	z ~1 µm			o	o	o	o		o	o	
X-ray method	Lang camera method			o		5 100	o ~5 s			o	o	o	o		o		Required to take photographic technique
	Double crystal method			o		5 100	o ~0.2 s			o	o	o	o		o		Required to take photographic technique
	Laue method			o		20 100	o 1~2 min	o					o				

218

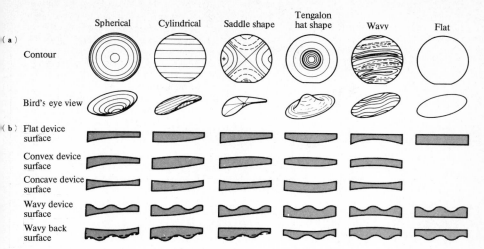

Fig. 5.16. a,b. Wafer bow, warpage and thickness variation. (a) Typical shape of bow and warpage, and (b) thickness variation of front and back surfaces combined

Relation of wafer surface and crystal plane	External form	Crystal plane	Other measurement and observation
	F	F	
	C	F	
	C	C	
	C	C	
	C	C	*Slip observation* X-ray topograph, Etching (with microscope), IR-photoelasticity.
	C	C	*Observation and measurement of both side surface* X-ray topograph, IR-photoelasticity, etching, surface analysis.
	C	C	*Observation and measurement of bi-metal structure* X-ray topograph, IR-photoelasticity, surface analysis.

F: Flat   C: Curved

Fig. 5.17. Combined measurement for search of wafer bow and warpage

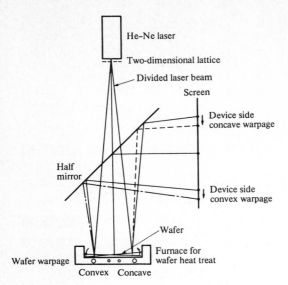

Fig. 5.18. Schematic drawing of wafer warpage in situ measuring instrument (a He-Ne laser beam is devided into several beams and the light beams are guided on a wafer surface put in the furnace. Reflected devided laser beams by a wafer surface move on the screen according the wafer bow sign, convex or concave)

automatic Lang camera method for X-ray topographs. The long working distance deformation measuring instruments are shown schematically in Fig.5.18. Their characteristics are

1) contactless measurement;
2) high measuring and data processing speed for multi-points on a deformed wafer;
3) separation of curvature sign and sorting.

5.3.2 The Slicing Condition

Convex wafers keep a high throughput in the contact aligning process. On measuring the wafer bow with sign for many wafer lots, one finds a distribution of mixed convex and concave forms as can be inferred from Fig.5.19. The throughput of an aligner decreases by these different wafer signs, and the yield is affected by the mixing.

Printing a fine pattern, it is required to keep the wafer flatness below 5 μm or even less than 3 μm desirable for the check-testing of fine-pattern printing instruments. In the past, it was believed that a convex wafer was made flatter than a concave one by chucking. But, experiments showed that concave wafers are corrected by vacuum and electrostatic chucking the bow and warpage similar to the mechanical chucking case. Another of our research works led to the result that high thermal warpage is generally found for convex rather than concave wafers commercially available. Fortunately, superfine

220

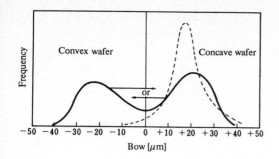

Fig. 5.19. Bow distribution of a raw wafer [(——) commercial wafer, (---) requirement]

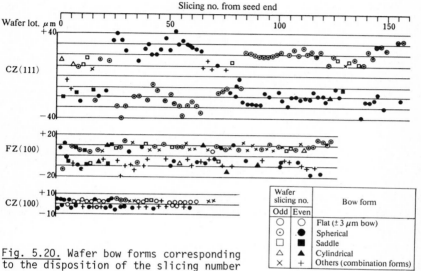

Fig. 5.20. Wafer bow forms corresponding to the disposition of the slicing number from seed end. [(+) concave, (-) convex, wafer lot number represents a combination of growing methods, CZ or FZ, and wafer directions, (111) or (100)]

Wafer slicing no.		Bow form
Odd	Even	
○	○	Flat (± 3 μm bow)
⊙	●	Spherical
□	■	Saddle
△	▲	Cylindrical
×	+	Others (combination forms)

lithographic systems generally require a mechanical, vacuum or electrostatic chucking structure.

From this point of view, it is clear that for wafers made by mechanical processing of a silicon ingot, a bow sign control technique is required. But, this mechanical processing technique is not discussed.

To clarify this problem, we made the following experiment. Nine lots of slices were numbered from seed end to tail end, and the given number was not skipped by any mechanical processing. The nine lots contained crystals CZ and FZ grown in the <100> and <111> directions by three silicon venders. The numbered slices were polished with alternating surfaces, seed side or tail side, after the odd or even integers of the slice numbers. There was a total of near-

Fig. 5.21. Average blade rim deviation during Si slicing. (Si ingot diameter: 76 mm; slicing condition: Blade: 6" ID, 16" 5/8 OD, blade rim speed: 1100m/ min., slicing speed: 25mm/min)

ly 1000 slices. Part of the slices measurements is shown in Fig.5.20. From the results, some conclusions can be drawn. The slice bow sign was determined in the slicing process and the wafer bow sign followed that of the polished side. Then, a mirror-polished wafer sign control should be realized by keeping the polished side of the slices.

The bow slicing originates from (i) the buckling of the slicing blade at a critical slicing pressure, and (ii) the initial deformation of the blade rim. The main origin (ii) is due to a static and dynamic blade rim deviation. This fact was found from the measurement of a static and dynamic blade rim deviation by using an eddy current deviation meter attached to the diamond blade slicer. Figure 5.21 shows some of the dynamical blade deviation measurements. The deviation was kept to some ten slicings of a 76-100 mm diameter silicon ingot. An inappropriate blade-rim lubrication induces a change in the slice bow sign, and the absolute value could be stabilized after few ingot slicings. This result corresponds to the experiment on the alternatively polishing side change.

Some wafers have shown an increasing absolute value of wafer flatness by chucking. This phenomenon was observed for the following cases: (i) distinct saw marks were observed on the back sides, and (ii) convex or saddle-type wafers.

The reasons for Case (i) are shown in Fig.5.22, and those for Case (ii) are explained by leaping of non-chucked parts of the wafer due to buckling. A local deviation from the average shape was noticed in a 125-mm diameter wafer. This deviation was originated by some air bubbles or dust particles in the adhesive layer between slice and base plate for polishing. Such defects appeared in convex wafers frequently.

Measures to eliminate the above-mentioned defects in polishing a wafer are as follow:

1) a careful control of the diamond blade-rim bending;
2) sufficient etching and lapping before polishing;
3) a light lapping pressure
4) a concave form control by mechanical processing, and
5) the application of both side free polishing, and others.

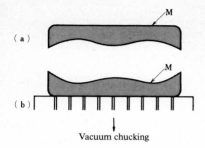

(a)

(b)

Vacuum chucking

Fig. 5.22. a,b. Schematic drawing of the mirror surface deformation caused by chucking of an uneven back side. Vacuum chucking; (a) Free state, and (b) vacuum chucking state

Fig. 5.23. Bow histogram of a 125 mm diameter wafer (about 100 wafers for each lot)

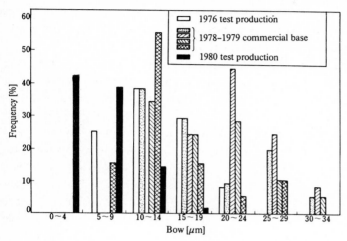

Histograms of bow distribution for several lots of 125-mm diameter wafers are shown in Fig.5.23. The open column corresponds to the first test run of 125-mm diameter wafer lots made in 1976 by the following mechanical processing. After deep etching, slices were polished on the top side surface without lapping. All these wafers were of concave forms, and their average bow value was + 13.2 µm. This is a good value for recent 125-mm diameter wafers commercially available. Hatched and filled columns are wafer lots from 1977-1981.

Flatness and bow values of this test-lot wafers were not sufficient for fine-pattern printing. After the observation mentioned above there was a trial of both side free polishing under a light lapping force. The results are shown in Fig.5.24. Figure 5.24a illustrates the initial bow and flatness of the starting wafer in a bird's eye view and Fig.5.24b is a result of a light lapping pressure both side free polishing technique applied on the wafer mentioned in Fig.5.24a. The NTV and parallelism of the initial wafer were remedied from 36 µm and 6 µm to 10 µm and 2 µm, respectively, after 50-µm-layer removing on both sides by this polishing technique.

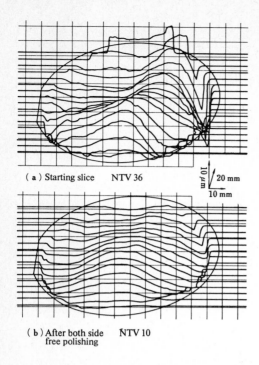

(a) Starting slice   NTV 36   10 μm / 20 mm   10 mm

(b) After both side   NTV 10
free polishing

Fig. 5.24. Bow from change of a starting slice and the consequence of both side free polishing (starting slice bow: NTV 36 μm, wafer diameter : 125 mm)

5.3.3 Warpage of Silicon Wafers in Heat Processing

The deformation of silicon wafers generated during LSI processings can be mostly classified into the following three groups, i.e.,

1) elastic deformation due to the thermal expansion coefficient difference by formation of thin films such as silicon dioxide, silicon nitride, and aluminum,

2) elastic/plastic deformation due to a lattice parameter difference by epitaxial growth, and high-concentration impurity diffusion,

3) elastic/plastic deformation due to thermal stresses generated in high-temperature furnace processings such as oxidation, and diffusion.

In this section, the deformation due to thermal stresses, the so-called warpage, which strongly influences LSI production yield, will be discussed.

High-temperature furnace processing of silicon wafers often produces sufficient thermal stresses generating slip and dislocations. The stresses arise during temperature transients. For an enhanced productivity, the wafers are stacked vertically in diffusion or oxidation boats. When such a boat is inserted into or pulled out of a high-temperature furnace, a temperature gradient [5.30,31] arises between the center and the edges of the wafers. The resulting stresses [5.32] can be sufficient to generate dislocations inducing plastic deformation (warpage). The temperature gradient or its resulting thermal stress depends on a construction of wafer row on the boat (wafer diameter, spacing of wafer to wafer, number of wafers, and heat capacity of the boat), and also on furnace-operation conditions (furnace temperature, temperature distribution around the furnace front portion, and boat speed).

Although in a practical furnace operation, slow insertion and pull-out of the furnace, or the closed boat method [5.33] is applied, warpage is often observed.

When one deals with this deformation problem, it is necessary to get detailed information on the thermal stress. HU [5.32] has simulated by computer the thermal stress for specific conditions, i.e., rapid cooling of wafers from high temperatures. A general theory on thermal stress, which is feasible to be applied to practical furnace operations, is developed here, on the basis of Hu's theory, and compared with experiments [5.34].

If the longitudinal temperature distribution $T_F(x)$ near the furnace front and boat speed v_b are assumed, the effective temperature T_e of a wafer due to external heat radiated from the furnace tube is given by $T_e(t) = T_F(x + v_b t)$, t being the time. In this case, the radial temperature distribution $T(r,t)$ of a wafer is expressed by the following differential-integral equation:

$$c_p w l \frac{dT(r,t)}{dt} = 2\varepsilon\sigma \left[-T^4(r,t) + \int f(\mathbf{r},\mathbf{r}')T^4(\mathbf{r}')d^2\mathbf{r}' + f_0(r)T_e^4(t) \right]$$
$$- Kl\nabla T(r,t) \quad , \tag{5.1}$$

where c_p denotes the specific heat, w the density, l the wafer thickness, σ the Stephan-Boltzmann constant, K the thermal conductivity, and ε the heat emissivity. The distribution function $f(r)$ for heat emission/absorption is given by

$$\left. \begin{aligned} f(\mathbf{r}) &= \frac{1}{\pi H^2} (1 + |\mathbf{r}|^2/H^2)^{-2} \\ f_0(r) &= 1 - \int f(\mathbf{r}')d^2\mathbf{r}' \end{aligned} \right\} , \tag{5.2}$$

where $H = d/R$ (d being the spacing of the wafer to wafer and R the wafer diameter).

Figure 5.25 shows a comparison of theory and experiment for the boat speed of 0.35 cm/s. The deviations for lower temperatures between theory and experiment are presumably due to heat convection effects in the furnace tube.

The stress distribution in a wafer can be determined by using the above-calculated temperature distribution. The wafer radial (σ_r) and angular (σ_t)

225

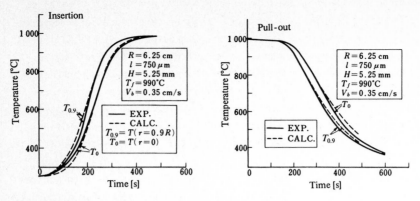

Fig. 5.25. Wafer temperature change with time

stresses are, respectively, given by

$$\sigma_r(r) = \alpha E\left[\int_0^R T(r')r'dr' - \frac{1}{r^2}\int_0^r T(r')r'dr'\right]$$

$$\sigma_t(r) = \alpha E\left[-T(r) + \int_0^R T(r')r'dr' + \frac{1}{r^2}\int_0^r T(r')r'dr'\right]$$
(5.3)

and, particularly, if the radial temperature distribution in a wafer is of parabolic type, i.e., $T(r) = T_0 + \Delta T(r^2/R^2)$,

$$\sigma_r(r) = \frac{1}{4}\ \alpha E \cdot \Delta T\ (1 - r^2/R^2)$$

$$\sigma_t(r) = \frac{1}{4}\ \alpha E \cdot \Delta T\ (1 - 3r^2/R^2)$$
,
(5.4)

where α is the linear thermal expansion coefficient, and E the Young's modulus.

Figure 5.26 shows the loci of both angular stress at the wafer edge and radial stress at the wafer center against temperature under the same conditions as in Fig.5.25.

Next, the experimental results of warpage during high-temperature furnace processings will be described [5.35]. A number of heat cycles is carried out in a practical LSI production. Simulating this situation, the heat treatment was repeated under the same furnace operation conditions. Figure 5.27 shows typical examples of warpage against the number of heat cycles.

From this figure, it can be found that the warpage increases monotonously with the number of heat cycles, and that the warpage is enhanced as the temperature is elevated. The former can be explained via the well-known dislocation multiplication mechanism: the dislocation generation is enhanced due to lowering of the critical yield stress with the number of heat cycles when dis-

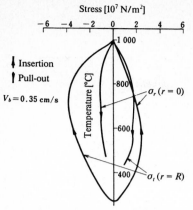

Fig. 5.26. Locus of stress versus temperature

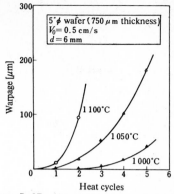

Fig. 5.27. Warpage vs. number of heat cycles

Table 5.3. Summary of warpage shape

Items	Shape	
	Saddle	Bowl (convex, concave)
Warpage generation	Insertion	Pull-out
Dislocation distribution	Periphery	Center
Thickness over diameter ratio	Large	Small
Heat process condition	High temperature and low boat speed	Low temperature and high boat speed

location generation occurs once during the heat cycle [5.36]. The latter is due to an increase of dislocation generation, being enhanced by increasing thermal stress with elevated temperature. The same tendency is observed if the boat speed becomes higher, and the spacing of wafer to wafer becomes narrower.

Through many of these experiments, it is found that the warpage shape can be divided into two types, i.e., saddle and bowl. Table 5.3 shows the summary of warpage shapes corresponding to the generation conditions. From these results, it is inferred that elastic buckling deformation occurs in the shape of a saddle for the wafer insertion period to the furnace, and of a bowl for the pull-out period. In order to verify this prediction, the deformation was directly measured as a function of the wafer temperature difference between the center (T_c) and the periphery (T_p), which is simulating experimentally the insertion/pull-out stage of the furnace processings. The results are

227

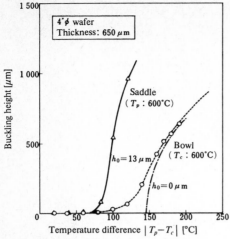

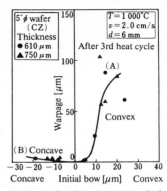

Fig. 5.28. Buckling deformation against temperature difference

Fig. 5.29. Relation between initial bow and warpage

shown in Fig.5.28. The buckling deformation occurred as the temperature difference between wafer center and periphery exceeds a critical value, and its shape is determined by the function of the central temperature versus the peripheral one. There is a saddle when the central temperature is lower than the peripheral one, and a bowl vice versa. This corresponds to the insertion and pull-out stage, respectively. Also, the shape of the buckling deformation coincides with that of the warpage. In the case of a bowl-type buckling deformation, the shape of convex or concave is determined by the bow shape of the initial wafer, and also the warpage shape does agree with the initial bow one, as shown in Fig.5.29.

Next, the theoretical analysis of the buckling deformation will be described 5.37 . If the radial temperature distribution in a wafer is rotationally symmetric, the wafer deformation is supposed to be a rotationally-symmetric bowl within an approximation for the case when the temperature of the wafer periphery is lower than that at the center, the one-dimensional problem. On the other hand, when the temperature of the wafer periphery is higher than that at the center, the analysis yields the two-dimensional problem because the stable deformation becomes of the saddle type. The former bowl-type deformation is analyzed here. If a wafer with internal stresses $\sigma_r(r)$ and $\sigma_t(r)$ becomes deformed as $\zeta(r)$ (ζ being the vertical displacement from a standard plane), the elastic energy U of the system can be expressed by

$$U = \int_0^R \{F_1[\zeta'^2, \sigma_r(r), \sigma_t(r)] + F_2(\zeta'', \zeta')\}r dr \quad , \tag{5.5}$$

228

where F_1 indicates the plane distortion term, which is expressed by a nonlinear second-order algebraic equation with respect to ζ'^2, and F_2 means the bending deformation term which is represented as second-order equation about ζ'' and ζ'. When $\sigma_r(r)$, $\sigma_t(r)$ are given, $\zeta(r)$ minimizing the right-hand side of (5.5) corresponds to the quasi-stable state of a wafer. As the lowest-order approximation, $\zeta(r)$ may be written by

$$\zeta(r) = \frac{1}{2} x \left(\frac{r}{R}\right)^2 \left[1 - \frac{1}{2}\gamma\left(\frac{r}{R}\right)^2\right] \quad . \tag{5.6}$$

By determining γ so that the bending moment at the wafer periphery is zero, and substituting it into (5.5), U can be written by the next fourth-order equation as a function of x

$$U = a(x^2 - x_0^2)^2 + b\xi(x^2 - x_0^2) + c(x - x_0)^2 + d \quad , \tag{5.7}$$

where a, b, ξ, c, and d are independent values of x, and x_0 is an initial bowl value. Therefore, there is an x value to minimize U.

In the special case of an initially flat wafer, i.e., $x_0 = 0$, we have

$$x_{min} = \sqrt{\left\{\max\left(\int_0^R [K_r(r)\sigma_r(r) + K_t(r)\sigma_t(r)]d_r - K_0, 0\right)\right\}} \quad , \tag{5.8}$$

where K_r and K_t are determined by a function of ζ, and K_0 is a constant being proportional to $(1/R)^2$. From the implication of (5.8), the non-deformation state is stable for lower stresses, and if the shear stress exceeds a critical value, (σ_c), the deformation magnitude increases proportionally to $\sqrt{(\sigma - \sigma_c)}$. Further, when σ_r and σ_t satisfy (5.4), its magnitude is proportional to $\sqrt{(\Delta T - \Delta T_c)}$, where $T_c = \Delta T_m(1/R)^2$, (ΔT_m: constant).

The dashed and the broken lines in Fig.5.28 correspond to the case of an initially flat wafer, and the case including an initial bowl value of the wafer used in the experiments, respectively. Good agreement is obtained between theory and experiment.

Although the above simple approximation cannot be applied to the model for the saddle-type deformation, the experimental result suggests that the deformation magnitude is proportional to the same as the bowl type.

To summarize the above results, the wafer warpage during furnace processings is explained by the fact that buckling deformation occurs due to the wafer temperature gradient in the insertion/pull-out period, minimizing the elastic/plastic energy by generation of dislocations and slip, which fix the deformation free of external stress.

5.3.4 Effect of Mechanical Damage on Thermal Warpage

Thermal warpage of a wafer becomes observable when the magnitude of thermal stress in the wafer during its loading/unloading periods for thermal treatments exceeds the level at which dislocation nucleation and multiplication occurs. Thus, the extent of warpage is closely related to the mechanical strength of the wafer.

Factors determining the mechanical strength of a wafer can be classified in two categories, extrinsic and intrinsic. The intrinsic factors include, for example, the initial dislocation density, and the concentration and states of impurity atoms. The initial dislocation density is not an important factor in the present-day commercial wafers, except for the problem of the so-called microdefects such as swirl defects in floating zone (FZ) wafers. Impurity effects are discussed in Sect.5.4.

One of the major extrinsic causes is mechanical damage on the wafer's surfaces. Indeed, dislocations are transmitted from surface flaws such as indentations by a hardness tester, scratch traces by a diamond stylus, etc., when a wafer is heated to several hundred degrees Centigrade. Transmission electron microscopy observations have revealed that these damaged regions accompany a small region of high dislocation density. Thus, these damaged regions act as dislocation sources to initiate plastic deformation.

The front surface on which electronic devices are localized later on is usually very carefully prepared to eliminate such damage. On the other hand, the backside is sometimes intentionally damaged, for example, to utilize gettering effects.

An effect of the backside condition on the wafer's thermal warpage is shown in Fig.5.30. The backside was first lapped with #1200 abrasive powder and then etched chemically to different depths. Thus, wafers with different degrees of backside damage were obtained. The wafers were processed by a series of heat-

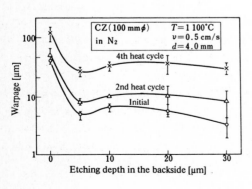

Fig. 5.30. Etched amount vs thermal warpage of wafers with one lapped surface. (v: loading/unloading velocity, and d: wafer separation in the wafer row)

treatment cycles to study the effect of the damage on thermal warpage. As shown in the figure, the effect is negligible in wafers being etched to more than 5 μm. Similar results are obtained using FZ crystals, where the amount of warpage gradually decreases after the initial rapid decrease up to 5 μm of etching depth. To remove the effect of slicing damage, a chemical etching of more than 30-40 μm is required.

The distribution of dislocations in a wafer would follow that of the magnitude of the resolved shear stress for each slip system, if the wafer is homogeneous [5.37]. It can be calculated for any combination of {111} planes and <110> directions using the thermal stress components described in Sect.5.3.3. The actual distribution, however, is strongly influenced by many factors such as the positions of surface flaws.

In the case of wafers without surface damage, during the wafer loading period dislocations mainly generate near the wafer periphery and propagate along {111} planes inclining toward the surface. Furthermore, they form rows of dislocations with the <110> Burger vectors inclining toward the surface, too.

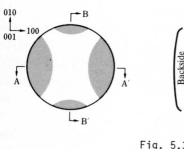

Fig. 5.31. Typical dislocation distribution and sectional view of a thermally warped wafer with mechanical damage on the backside

In wafers with backside damage, many half-loop shaped dislocations nucleate at damage centers. Dislocation generation accelerates in the peripheral regions where the surface curvature is concave to the damaged side. This is because, if elastic buckling precedes the dislocation generation, the concave side exercises the larger stress (Sect.5.3.3). Thus, in <100> wafers with a damaged surface, dislocations can distribute in a two-fold symmetry, as shown in Fig.5.31 (right). The wafers warp in a "trough" shape in contrast to those which warp in a "saddle" shape. The local curvature is large where more dislocations are produced and small otherwise.

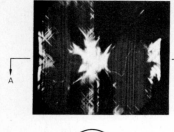

Fig. 5.32. X-ray topograph *above* and sectional view (A-A') of a thermally warped CZ wafer in which a plastic deformation mainly occurred in the central part. (100) wafer. (High dislocation density regions are shown in bright contrast. In this case Bragg condition over the whole wafer is not satisfied because of the wafer deformation)

Dislocation generation near the center of a wafer is liable to occur during the unloading period. This is because in this period the temperature is higher (i.e., the critical stress for dislocation generation is lower) in the central part than in the peripheral part of the wafer. The generated dislocations are normally responsible for vital changes in device characteristics, if they end in the front surface and cross especially the p-n junctions of the devices. For this reason, a good control of the cooling period during the thermal treatment is important. The wafer warps in a shallow cone shape with the maximum curvature in the dislocated central region. This is illustrated in Fig.5.32. Elastic buckling precedes the dislocation generation also in this case, so dislocations are mainly created on the concave side of the central region. High-density dislocation generation and large warpage increase, if the concave side is mechanically damaged [5.38].

Lastly, effects of repeated heat cycles are briefly discussed. In a conventional silicon process a wafer undergoes a series of heat cycles. If dislocations are produced in an early stage, plastic deformation easily proceeds in the following heat cycles. Figure 5.33 shows the change in warpage during a repeated heat-cycle experiment. The amount of warpage increases rapidly after several cycles. This is because dislocation multiplication and propagation may occur at a lower stress than its generation. Thus, wafers should be handled with care so that mechanical damage and dislocations may not be created in an early stage.

5.3.5 Oxygen Effects on Wafer Thermal Warpage

The strength of a crystal is sensitive to the impurity concentration and its character in the crystal, i.e., whether it is located substitutionally or interstitially, or in precipitates. Here the effects of groups III and V im-

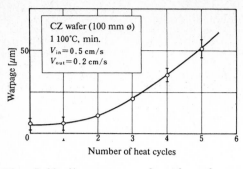

Fig. 5.33. Warpage as a function of the number of heat cycles

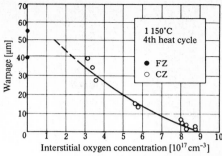

Fig. 5.34. Warpage after repeated heat cycles as a function of initial interstitial oxygen concentration. 75 mm φ, (100) wafers

purities are briefly mentioned and then those of oxygen which is principally a neutral impurity in CZ-grown silicon.

The velocity of dislocation motion in a covalent crystal is influenced by solid solution of electrically active impurities. In group III or V impurity-doped silicon crystals, the dislocation velocity is known to differ from that in non-doped ones, in a temperature range where the doped impurity concentration exceeds the intrinsic carrier concentration. In the case of n-type impurities, the dislocation velocity is increased, whereas in that of p-type impurities the results vary among different investigators. Effects on yield stress of the group III and V impurities are not clearly understood, either. One of the major reasons for this lack of coincidence, especially in CZ crystals, is that the effects of neutral impurities have not yet been fully clarified.

The major one among such neutral impurities is oxygen in CZ crystals. Effects of dissolved oxygen on mechanical properties of silicon have been studied extensively. For example, a comparison of CZ crystals with FZ ones [5.39,40] and oxygen denuded layers on the surface region of CZ crystals [5.41] have been reported in connection with dislocation behaviour.

A result of warpage measurements on repeated heat treatments of various CZ and FZ crystals is shown in Fig.5.34. In all the tested CZ crystals oxygen atoms were interstitially dissolved and no essential precipitation was detected both before and after the heat treatments [5.42].

Possible mechanisms for the interstitial oxygen effects would be: (i) interaction of individual oxygen atoms with dislocations, (ii) interaction of small clusters of oxygen atoms with dislocations, and (iii) oxygen precipitation at dislocation sites as well as dislocation nuclei. The Mechanism (i) seems unlikely to work because of the high Peierls potential in silicon. HU tried to

233

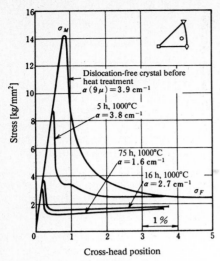

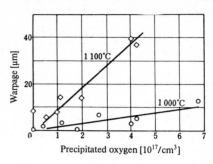

Fig. 5.35. Change in the stress-
strain curve of a 1000ºC heat-
treated CZ-grown silicon [5.00]

Fig. 5.36. Effect of oxygen precipitation on the wafer thermal warpage.
100 mmφ, (100), CZ-grown wafers.
(Loading/unloading velocity: 0.5 cm/s.
wafer separation: 4 mm)

explain the results of microindentation experiments mainly by the Mechanism
(ii) [5.41]. The results shown in Fig.5.34 were obtained under conditions
similar to those used in ordinary processes, where the fairly high temperature
allows oxygen atoms to diffuse a long distance and their concentration is above
the solubility limit, at the same time. Under such conditions the third mechanism will also work.

The oxygen impurity effects on the yield strength of silicon have been extensively studied, too [5.43-45]. PATEL showed (Fig.5.35) that long-term heat
treatment of CZ crystals at about 1000°C results in the reduction of infrared
absorption due to interstitial oxygen, which coincides with the reduction of
the yield strength [5.44]. Oxygen atoms agglomerate to form large ($\stackrel{\sim}{\gtrless} 1$ μm) precipitates which often accompany dislocation loops generated by the prismatic
punching mechanism. The yield-strength reduction easily occurs by the multiplication of these dislocations.

Such crystals are deformed easily by thermal stress. Figure 5.36 shows an
example of the oxygen precipitation effect on thermal warpage. The wafers had
the initial interstitial oxygen concentration of about $9.5 \cdot 10^{17}$ atoms/cm^3 with
a negligible amount of precipitated oxygen. The wafers were heat treated at
1050°C for various times to obtain wafers with various amounts of precipitated
oxygen. Then the wafers were processed by a series of repeated heat cycles in

234

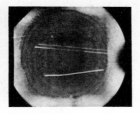

Fig. 5.37. Typical X-ray topographs of wafers without *left* and with *right* oxygen precipitation. (100 mmφ, (100), CZ-grown wafers)

a wafer row. The warpage was actually increased approximately in proportion to the increase in oxygen precipitation.

Different states of oxygen impurity also cause different dislocation distributions. This can be clearly seen in Fig.5.37. The right and left photographs show typical X-ray topographs of dislocated wafers with and without oxygen precipitation, respectively.

The topographs illustrate the case where the effect during the heating period is dominant. Dislocations in the crystal without precipitation are generated mainly at the wafer periphery and form distinct slip lines. In contrast, in the wafer with precipitation small dislocation loops are formed at precipitates growing by thermal stress.

Besides the precipitated oxygen amount, size, density and distribution of precipitates are also major factors to determine the extent of warpage. Wafer which have been processed by a relatively low-temperature ($\simeq 800°C$) precipitation heat treatment are sometimes more warpage resistant than those which have received such a treatment at a higher temperature, or even than as-grown crystals.

5.4 Thermally Induced Microdefects

5.4.1 The Definition of Thermally Induced Microdefects and Their Detection Technique

a) Characteristics of Microdefects

As a result of improved crystal-growth techniques, silicon crystals are nearly perfect dislocation-free crystals in these days. However, this fact does not mean that those crystals are defect free. Microdefects have first been observed in an FZ-grown crystal, being of the as-grown state [5.46]. They did appear as shallow pits especially in dislocation-free FZ crystals and did

Fig. 5.38. Thermally induced microdefects in CZ-grown wafer revealed by the etching technique, where microdefects appear as a swirl pattern. Upper and lower halves show front and back surfaces of the wafer, respectively

not in crystals containing dislocations. These phenomena are called "swirl defects" because they appear on the wafer surface as swirl patterns (Fig.5.38). On the other hand, the swirl defects had been observed several years ago even in CZ-grown wafers which were of the as-grown state, but they have disappeared recently due to the improvement of the crystal growth technique. However, they can be induced during thermal treatments at temperatures around 1000°C, to be described later. It is well known that CZ wafers are mainly used in LSI processes involving a rich variety of heat treatments at temperatures of around 1000°C. These thermally induced microdefects do not appear only on wafer surfaces but also in the bulk after thermal treatments [5.47-49]. They are of strong influence on device properties and performances [5.50-52]. This is a very serious problem for the whole LSI technology because nearly all the processes involve repeated thermal treatments at temperatures of around 1000°C.

The thermally induced microdefects have intensively been studied for several years. As a result, it is ascertained that the microdefects are originated in the precipitation of oxygen that is included in CZ crystals and comes from the quartz crucible during crystal growth. Therefore, the oxygen concentration of the CZ crystal is comparable to the solubility of the oxygen in the silicon crystal at the growth temperature. Accordingly, the concentration is supersaturated in the crystal at those temperatures of heat treatments used in LSI processes. These heat treatments result in precipitation of the oxygen inducing a rich variety of microdefects to be described later.

The characteristics of the microdefects vary with the temperature of the relevant heat treatment as follows [5.53]. Very small-sized oxygen precipitates and dislocation dipoles are induced at temperatures between 650° and 800°C, as shown in Fig.5.39a. At temperatures of about 1000°C, oxygen precipitates and punch out dislocations are formed besides stacking faults, as demonstrated in Fig.5.39b. When the heat-treatment temperatures are above 1150°C, large-sized stacking faults and precipitates of polyhedron are introduced (Fig.5.39c).

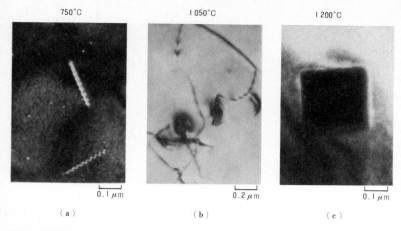

750°C   1050°C   1200°C

0.1 μm   0.2 μm   0.1 μm

(a)   (b)   (c)

Fig. 5.39. a-c. Thermally induced microdefects observed by transmission electron microscopy (TEM). Figures (a-c) show microdefects induced by heat treatments at 650-850°C, around 1000°C, and above 1150°C, respectively

Microdefect densities increase with a decreasing of the heat-treatment temperatures, as can be inferred from Fig.5.40. The absolute value of the density
varies with wafer to wafer owing to the crystal quality of the relevant wafer;
this tendency can always be noticed in all the wafers. The formation of the
microdefects depends on not only the oxygen concentration of wafers, but also
on various factors such as the relevant heat-treatment conditions, carbon concentrations in the wafers, and the thermal history of the relevant crystal from
which wafers are made. The complexity of the thermally induced microdefects requires wide-spread investigations by various techniques in the near future.

b) *Infrared Absorption Spectroscopy*

Microdefects produced in CZ crystals during heat treatment are due to the precipitation of residual oxygen. Possibly carbon is related to the oxygen precipitation (Sect.5.4.2a). Therefore, the infrared absorption technique is indispensable for the characterization of microdefects, since infrared spectroscopy
can analyze oxygen, carbon and oxygen precipitates (Sect.5.2.2). It is convenient to use the Fourier transform spectroscopy, because an accurate, rapid
and effective analysis can be made on a wafer by this technique.

Figure 5.41a shows the variation of interstitial oxygen, when a commercial
CZ-grown crystal for process use is subjected to an isochronal anneal for 64 h
at a temperature ranging from 450° to 1250°C [5.54]. The crystals are classified into two groups depending on the variation of the oxygen. In group I
crystals, the interstitial oxygen is reduced greatly, and correspondingly pre-

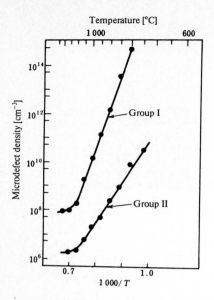

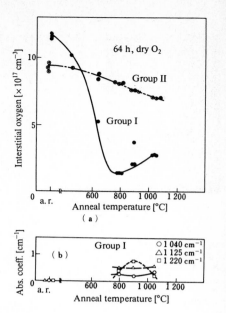

Fig. 5.40. Dependence of thermally induced microdefect density on the heat-treatment temperature. Group I and II show wafers in which a high density of microdefects are and are not easily induced by a heat treatment at a temperature around 1000°C

Fig. 5.41. (a) Interstitial O concentration as a function of annealing temperature for two groups of Si crystals. (b) Absorption coefficient of Si-O complexes as a function of annealing temperature

cipitates, such as cristobalite and amorphous SiO_2, can be observed (Fig.5.41b). In group II crystals, on the other hand, the reduction of interstitial oxygen is small and the precipitate-content is below the detection limit. This shows that the generation of microdefects is more significant in the first group than in the second one. Similar results are obtained by other characterization techniques, which will be described in the following sections.

In order to inquire about the origin of the difference in the generation of the microdefects between both groups, the influence of carbon has been investigated, as shown in Fig.5.42 [5.55]. This figure shows the oxygen reduction rate (the ratio of reduced interstitial oxygen to the initial interstitial oxygen) after an annealing at 650°C for 64 h as a function of carbon in CZ-grown crystals before annealing. The oxygen precipitation becomes remarkable for higher carbon concentrations. Effects of oxygen and carbon on the generation of microdefects have also been examined by annealing experiments on oxygen-diffused FZ-grown crystals with various carbon concentrations [5.56]. Effects of the thermal history have been analyzed by two-step annealing experiments [5.54]. It has been clarified that the generation of microdefects due to

238

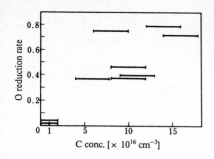

Fig. 5.42. Dependence of initial C concentration on interstitial O reduction
rate caused by an anneal at 650°C

oxygen precipitates depends on the residual carbon in the initial crystals as
well as the thermal history during crystal growth.

c) Measurement Techniques Utilizing X-Ray Diffraction

X-ray diffraction techniques are widely used to investigate crystal defects
because this is a non-destructively operating method. Of X-ray diffraction
techniques, X-ray topography (section and traverse topography) [5.57-59],
anomalous transmission measurement [5.60,61], intensity measurement of transmitted-diffracted X-ray beam, and diffuse scattering [5.62] have been applied
to the analysis of microdefects. For large-size microdefects, X-ray topography
is useful but this technique cannot reveal small-size defects less than one
micrometer because of its resolution of several micrometers. For small-size
microdefects, intensity measurements are powerful. However, this technique
does not reveal characteristics of the defects without a detailed analysis
of the result. This method has an advantage for a rapid survey of defects or
for the detection of the quality variation of crystals among many wafers. The
measurement becomes more powerful if the density and the characteristics of
defects are investigated by another technique such as an etching technique or
transmission electron microscopy (TEM) in advance.

It should be noted that the condition of $\mu_0 t \geq 10$ is preferable for the anomalous transmission measurement, μ_0 being the absorption coefficient of X-ray
and t the thickness of the sample wafer, whereas $\mu_0 t \leq 1$ is desirable for the
measurement of the normal transmitted-diffracted intensity. The measured X-ray
intensity is strong when the defect density is low for the former case and
vice versa for the latter case.

A representative result is shown in Fig.5.43 [5.54,63] concerning five different wafers supplied from five different vendors being used as starting samples. The sample wafers were annealed at 1050°C for 64 h in a dry O_2 atmosphere.
Subsequently, transmitted diffracted X-ray intensities were measured, as shown
in the figure with solid circles. For reference, unannealed wafers were also

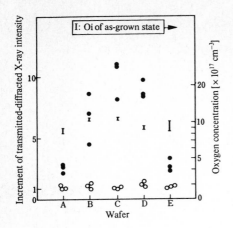

Fig. 5.43. Transmitted-diffracted X-ray intensity as a function of crystal quality of a starting wafer where open and solid circles show unannealed and annealed wafers, respectively. It is to be noted that high X-ray intensity shows a high density of microdefects, and vice versa

investigated and indicated in the same figure with open circles. As a result, the X-ray intensities were constant among wafers when they were unannealed. However, the intensities increased due to the annealing under conditions described above. That means that microdefects were induced by annealing. Moreover, the intensities correlating to the annealed wafers were different among one another. The difference depended on the vendor from which sampled wafers were supplied. In conclusion, it is ascertained that the degree of formation of thermally induced microdefects varies from one wafer to another, even though starting wafers are similarly dislocation-free crystals.

d) Transmission Electron Microscopy

Since microdefects in silicon are generally very small (less than a few micrometers), transmission electron microscopy (TEM) is the most powerful method to observe them [5.64]. TEM observation of a crystalline defect is based on the contrast analysis of diffraction phenomenon electrons transmitted through a crystal. When the defect is relatively large ($\gtrsim 500$ Å), it can be observed under the condition that only one diffracted wave excites (two-beam condition) [5.65]. In order to identify the individual defect correctly, it is necessary to observe the defect under not only one diffraction but also various diffraction conditions, and the diffraction image must be analyzed by using the dynamical diffraction theory [5.65-68]. Therefore, it is indispensable to equip a goniometer for sample tilting in TEM. When the defect (such as a dislocation etc.) density is measured under the two-beam condition, the diffracted beam must be selected to be able to show all the defects, because of the disappearance of the defect contrast under $g \cdot b = 0$ (g: diffraction vector, b: Burger's vector of the dislocation). The sample thickness is calculated from the number

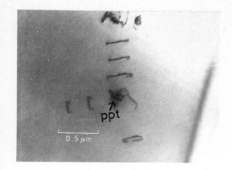

Fig. 5.44. Microdefects in CZ-grown silicon annealed at 1050°C for 64 h. (Punch-out dislocations are generated from a precipitate)

of equal thickness fringes and the defect volume density is deduced on the assumption that all the defects recognized uniformly exist in the sample film. Then, one must attend to the deduction of sample thickness at the film edge like a wedge.

Figure 5.44 shows a TEM micrograph of defects induced by thermal annealing at 1050°C for 64 h in CZ-grown silicon (400 excited bright field image). An oxide precipitate is indicated by ppt, and punch-out dislocation loops into the <110> direction from the precipitate can be observed. When the defect is very small (smaller than 500 Å) and its strain field is faint, one had better use the weak-beam method [5.69], that is the method observing the dark field image under the off-Bragg condition. Figure 5.45 shows a weak-beam image of ultramicro-defects generated in CZ-grown silicon annealed at 800°C for 64 h. Bright dots and lines in Fig.5.45 show oxide precipitates (about 80 Å ⌀) and dislocation dipoles, respectively. We can directly observe the ultramicro-defects as small as 40 Å by using the weak-beam method.

Crystalline defects are often generated in a limited area in a wafer during the device manufacturing process. Then it is very difficult to make a thin sample for TEM observation of the limited area. In this case, it is possible to make a uniform thin film of wide sample area using a special apparatus [5.70] and to cut off an interesting place from the film for TEM observation.

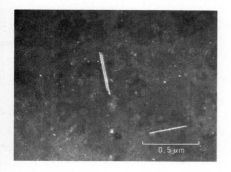

Fig. 5.45. TEM micrograph of microdefects in CZ-grown silicon annealed at 800°C for 64 h. (Weak beam image)

When one wants to observe an individual microdefect, it is difficult to make a thin sample for TEM and to get good resolution in X-ray topography. On the other hand, the preferential etching generally used is a very easy technique with relatively high resolution. Table 5.4 shows the representative etch solution and its characteristics to study crystalline defects. In order to detect microdefects by the etching method, the etched surface must be cleaned enough, since a contamination or an oxide layer left on the etched surface could become an etch hillock or an etch pit after the etching. The defect character cannot be correctly identified by the etching method only, so it is necessary to make a correspondence between the defect character and the etch pit or hillock. The defect density is calculated on the assumption that there is the same number of defects with the etch pits in the etched region. However, as the etch rate depends upon surface features, resistivity, surface damage of the sample and so on, the etch rate must be adjusted to the sample.

In the case of observation of thermally induced microdefects in CZ-grown silicon, the defect density is different between surface and bulk of the wafer. Near the surface region, a defect-free layer called denuded zone is formed. Therefore, one must attend to the depth of the etched layer, and it is useful to observe the cross section of the sample. For a silicon wafer, a (110) cleavage plane is flat and smooth, regardless of the strongest (111) cleavage plane. Figure 5.46 shows an interference micrograph of a (110) cleaved surface etched by a Dash solution for 50 min from CZ-grown silicon wafers annealed at 1050°C for 64 h in nitrogen ambient. Line and dot-like de-

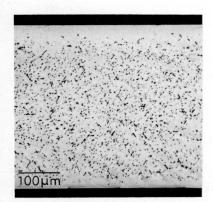

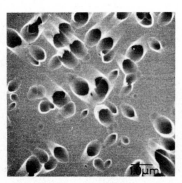

Fig. 5.46. Dash etching/optical micrograph of (110) cleaved surface in CZ-grown silicon anneal at 1050°C for 64 h in nitrogen ambient

Fig. 5.47. SEM micrograph of Dash-etched surface in CZ-grown silicon annealed at 700°C for 64 h

Table 5.4. Etch solution and its feature

Name	Composition	Etch rate [µm/min]	Comments
Dash etch	HF 10 cc HNO_3 30 cc CH_3COOH 120 cc	$\simeq 0.1$	Applied to all surfaces. Etch rate is slow. Fit for observation of microdefects.
Sirtl etch	HF 100 cc CrO_3 50 g H_2O 100 cc	$\simeq 1$ ($<111>$)	Fits for (111) and (110) plane, but not for (100) plane.
Secco etch	HF 100 cc $K_2Cr_2O_7$ solution (0.15 mol %) 50 cc	$\simeq 1.5$	Etch rate is fast. Etched uniformly with ultrasonic agitation. Gives elliptical etch pits.
Wright etch	HF 60 cc HNO_3 30 cc CrO_3 solution (5 mol %) 30 cc $Cu(NO_3)_2$ 2 g H_2O 60 cc CH_3COOH 60 cc	$\simeq 1$	Applied to all surfaces. Etched uniformly. Gives sharp etch pits.
Schimmel etch	HF 100 cc CrO_3 solution (1 mol %) 50 cc	$\simeq 2$	Etch rate is fast. Etched uniformly without ultrasonic agitation. Gives elliptical etch pits.
Sailor's etch	HF 60 cc HNO_3 30 cc Br_2 0.2 cc $Cu(NO_3)_2$ 2.3 g Dilute 10:1 with water before using	$\simeq 0.4$	Show a/6 $<112>$ partial dislocations but not stair-rod dislocations.

fects correspond to stacking faults and precipitate-dislocation complexes, respectively. The defect-free region is obviously formed near the surface.

Figure 5.47 shows a scanning electron micrograph of a Dash-etched surface from CZ-grown silicon annealed at 700°C for 64 h. Each etch pit, tailing to the $<110>$ direction, corresponds to a dislocation dipole shown in Fig.5.45.

However, small hillocks, indicating microprecipitates shown in Fig.5.45, cannot be observed. Then, it is found that the microdefects, which are very small, cannot be detected by the etching method. Therefore, in order to study microdefects, it is necessary to use not only etching but also other technique

f) Photoluminescence Spectroscopy

The photoluminescence (PL) technique has recently been developed but not fully been established as a standard technique. However, this technique has been reported to be quite sensitive to microdefects in comparison with other techniques [5.71-73]. The measurement methods were described in Sect.5.2.3. No special sample preparation is necessary for the PL measurement. Samples are immersed in liquid helium or liquid nitrogen and are excited by an Ar ion laser. PL from the sample is spectroscopically analyzed. The effects of defects in crystals upon PL are as follows:

(i) Defects act as radiative recombination centers introducing new emission

(ii) Defects act as nonradiative recombination centers reducing the other emission.

(iii) Defects introduce strain in the crystal causing the spectral shift and/or deformation.

Therefore, information on defects can be obtained by investigating how the PL spectral pattern and intensity are changed in comparison with the PL from defect-free crystals.

Figure 5.48 shows the typical PL spectral pattern at liquid-helium temperature observed in heat-treated CZ crystals [5.72]. Figure 5.48a is the PL spectrum from the crystal before heat treatment. Here, the PL lines associated with dopant impurity B are dominant (A type). Figure 5.48b is the B-type pattern observed in the crystal after an annealing at 450°C for 64 h. The TD components in the spectrum have been identified to be due to thermal donors through the experiment using oxygen-diffused FZ crystals [5.71]. It can be determined definitely whether thermal donors are produced or not by the appearance of the TD components. These A- and B-type patterns are commonly observed in commercial CZ-grown crystals, regardless of their vendors and crystal growth conditions. However, the CZ crystals are classified into two groups, depending on the PL spectral change caused by an annealing at temperatures higher than 600°C. In one group (group I), the PL pattern is drastically changed, as shown in Figs.5.48c and d, where the PL intensity is greatly reduced. On the other hand, the PL pattern is not changed from the A type by an annealing at $\gtrsim 600°C$ in another group (group II). This corresponds with the behaviour of oxygen analyzed by infrared spectroscopy

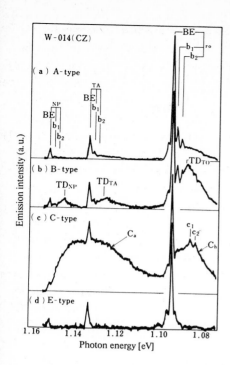

Fig. 5.48. PL spectral change at liquid-helium temperature caused by heat treatment. (a) as-received, (b) 450°C-, (c) 600°C-, (d) 900°C- annealed in dry O_2 for 64 h

(Sect.5.4.1b [5.73]). It should be pointed out that the PL signal is quite sensitive to the thermally induced defects. For instance, the magnitude of the maximum PL intensity change due to the heat treatment is more than a factor of 1000 at liquid helium temperature (more than 5000 at liquid nitrogen temperature), which is far greater than that of the infrared absorption change (a factor of 50).

The carbon concentration is higher in group I crystals than in the other one (Sect.5.4.1b). The effect of carbon on the defect formation will be described in detail in Sect.5.4.2. C-type patterns appear in group I crystals after an annealing at around 650°C (Fig.5.48c). The C-type pattern also appears in group II crystals, if the crystals are pre-annealed at 450°C followed by an annealing at around 650°C. In correspondence with the appearance of the C-type pattern, an increase in the donor concentration is always observed. These results are a strong evidence for the idea that the C-type spectrum is due to new donors as described in Sect.5.2.6.

5.4.2 The Role of Oxygen and Carbon for the Formation of Thermally Induced Microdefects

a) Thermally Induced Microdefects in Wafers Where the Carbon and Oxygen Concentrations Are Controlled

It is ascertained that thermally induced microdefects are originated in the oxygen precipitation. The relevant oxygen is incorporated into the crystal with a high density from the quartz crucible during crystal growth. However, a range of the oxygen concentration, by which thermally induced microdefects are introduced by heat treatments, has not been established. Besides the oxygen concentration, the effect of other impurities such as carbon atoms which are contained with a rather high density ($10^{16} \sim 10^{17}$ cm^{-3}) in the silicon crystal is not yet understood. These items must be investigated without fail in order to know the mechanism by which thermally induced microdefects are formed. That is, it must be clarified whether the oxygen in the crystal can precipitate without fail or cannot under the only condition of the supersaturation. The role of other impurities, such as carbon, which is believed to be the most likely nucleation center of swirl defects in an FZ-grown crystal [5.74,75], must be clarified, too.

Based on this background, a basic experiment using FZ wafers was done, where the oxygen and carbon concentrations were artificially controlled [5.58, 76]. For starting, three kinds of wafers shown in Table 5.5 were chosen. They were of different carbon concentrations in the as-grown state. These wafers were subjected to an oxygen-diffusion process at the temperature of 1300°C. The resultant wafers had the oxygen profiles shown in Fig.5.49 as a function of the wafer depth. These wafers were put into a two-step annealing process where a 800°C annealing for $1 \sim 64$ h was followed by the 1050°C annealing for 64 h. After these procedures, thermally induced microdefects in the bulk were investigated using both the X-ray diffraction technique and an etching technique. In the X-ray diffraction technique, transmitted-diffracted intensities

Table 5.5. The oxygen and carbon concentrations in the starting FZ wafers

Samples	Interstitial oxygen concentration [$\cdot 10^{16}$ cm^{-3}]	Substitutional carbon concentration [$\cdot 10^{16}$ cm^{-3}]
A - 84	<1	<2
B - 124	<1	7
C - 42	<1	12

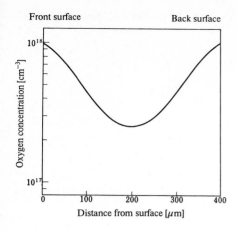

Front surface   Back surface

<u>Fig. 5.49.</u> Oxygen concentration profile in the bulk of the wafer after an oxygen diffusion process at 1300°C in a dry O_2 atmosphere for 18.5 h

were measured and the increments of X-ray intensity by the heat treatment were shown as functions of both carbon concentrations and annealing conditions (Fig.5.50). As the density of microdefects in the bulk increases, the increment of X-ray intensity increases. This is ascertained and thus the defect density induced by the two-step annealing increases dependent on the carbon concentration in the wafer (Fig.5.50). Figure 5.50b and c show results with no 800°C pre-annealing, and without both the 800°C pre-annealing and the oxygen diffusion process, respectively. These results illustrate that no low-temperature pre-annealing produces a low density of microdefects regardless of carbon concentrations in the wafers.

Without oxygen diffusion, no microdefects are, of course, induced. It should be noted that the microdefects are not easily induced in the low-carbon-contained wafer even if the low-temperature pre-annealing period is long [5.58].

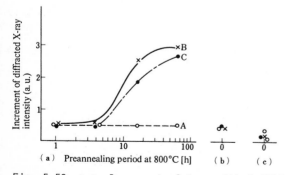

<u>Fig. 5.50. a-c.</u> Increment of transmitted-diffracted X-ray intensity as functions of both carbon concentrations in the wafer and the preannealing period at 800°C. Figures (a) and (b) show the case where the oxygen diffusion is carried out, while (c) the case of no oxygen diffusion

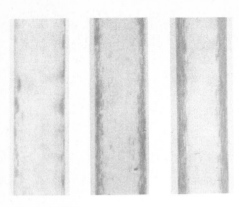

Fig. 5.51. a-c. X-ray section topographs after a two-step annealing
using oxygen-diffused FZ-grown wafers.
Carbon concentrations in (a-c) are
below 2×10^{16}, 7×10^{16}, and 12×10^{16} cm^{-3}, respectively

Fig. 5.52. Thermally induced microdefects revealed by Dash etching
on (110) cleavage plane normal to
(100) surface where an oxygen-diffused wafer is sampled. Sample is
B, shown in Fig. 5.50, where 800°C
annealing is followed by 1050°C
annealing for 64 h after the oxygen
diffusion process

Therefore, it is clear that the thermally induced microdefects are not necessarily induced by heat treatments under the only condition of supersaturation
of the oxygen in the crystal.

Interiors of annealed wafers were investigated by X-ray section topography,
as indicated by Fig.5.51. These topographs show that induced microdefects are
located near the surface where oxygen concentrations are high. The microdefect density depends on the carbon concentration in the wafer as can be seen
in Figs.5.51a-c. Subsequently, the depth distribution of microdefects in the
bulk was investigated in detail by an optical microscope after Dash etching,
in which (110) cleavage planes were etched (Fig.5.52). This figure makes clear
that the defected area is restricted to a surface region of $70 \sim 90$ μm in
depth. This area is correlated with the region where the oxygen concentration
is over $5 \sim 6 \cdot 10^{17}$ cm^{-3}. From this result, the lower limit of the oxygen concentration, where microdefects are induced by thermal treatments at a temperature of 1050°C, can be taken to $5 \sim 6 \cdot 10^{17}$ cm^{-3}. Resultantly, it is clear
that nucleation centers such as carbon atoms are necessary for the formation
of microdefects besides the supersaturated oxygen. This experiment also indicates that the thermal history plays an important role on the formation of
microdefects. In this experiment the thermal history of the crystal was artificially introduced into wafers by adding the 800°C pre-annealing before the
final thermal treatment at 1050°C for 64 h.

b) Formation Mechanism of Thermally Induced Microdefects

In the basic experiment using FZ-grown crystals, both the carbon concentration in the crystal and the thermal history of the crystal are described to be important for the formation of microdefects. Are these items also important in practical CZ-grown wafers used in LSI processes? It has been reported [5.77] that thermally induced microdefects do not depend on the carbon concentration in the wafer when they are induced by a heat treatment at a temperature around 1000°C, frequently used in LSI processes. According to the mentioned experiments, there is not necessarily a correlation between the carbon concentration and the microdefect density when the relevant thermal treatment is carried out at a temperature above and around 1000°C. However, the correlation becomes clearer when the relevant thermal treatment is carried out at a temperature around 650°C inducing a high density of microdefects. That means the oxygen reduction ratio (ORR) by heat treatments at 650°C depends on the carbon concentration in the wafer (Fig.5.53) [5.53,78]. When the ORR is high, a high density of microdefects is being induced, and vice versa. This is because interstitial oxygen in the wafer is precipitated by the relevant heat treatment when microdefects are formed during the heat treatment. These data shown in Fig.5.53 were obtained by a more detailed examination than those of Fig.5.42. Resultantly, it can be said that the formation of microdefects depends on the carbon concentration in the wafer when the heat-treatment temperature is around 650°C.

The formation of microdefects is due to the oxygen precipitation regardless of heat-treatment temperatures and thus the formation mechanism is not affected by the heat-treatment temperature. Therefore, it can be concluded that the formation of microdefects is correlated with the presence of carbon atoms in the crystal as a whole. The question arises why there is no clear correlation between the carbon concentration in the wafer and the microdefect density for a heat-treatment temperature higher than 1000°C. In order to clarify this

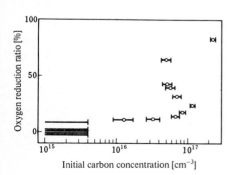

Fig. 5.53. Oxygen reduction ratio by thermal treatment as a function of carbon concentration in the starting wafer where a low-temperature annealing at 650°C for 64 h is carried out

Table 5.6. The oxygen and carbon concentrations in wafers sampled

Samples	Interstitial oxygen concentrations $[\cdot 10^{17}\ cm^{-3}]$	Substitutional carbon concentrations $[\cdot 10^{16}\ cm^{-3}]$
A	11.0	5.1
B	10.2	<0.4
C	7.6	11.4
D	8.8	3.4
E	9.3	<0.4

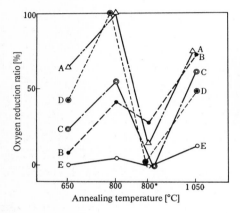

Fig. 5.54. Oxygen reduction ratio by thermal treatment as functions of both the heat-treatment temperature and the wafer quality where 800* shows the result of the two-step annealing (1050°C for 1 h + 800°C for 64 h)

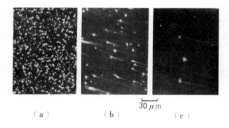

Fig. 5.55. a-c. Dependence of thermally induced microdefects on crystal quality of the starting wafer where the same 1050°C annealing for 64 h is carried out in (a-c). Figures (a-c) are optical photographs of microdefects revealed by Dash etching, where B, C, and E shown in Fig. 5.54 have been sampled, respectively

doubt the following experiments have been carried out. Several wafers with different carbon concentrations given in Table 5.6 were prepared as starting materials. All these samples were quartered and always three pieces of them were separately subjected to isochronal annealings for 64 h in a dry O_2 atmosphere at 650°, 800°, and 1000°C, respectively. The remainder of each wafer was put into the two-step heat treatment, in which 1050°C annealing for 1 h was followed by the 800°C heat treatment for 64 h in a dry O_2 atmosphere. After these procedures, the ORR's were measured, as shown in Fig. 5.54. The relevant microdefects in the bulk of the wafer were demonstrated in Fig. 5.55 sampled for the wafers B, C, and E. A comparison of both figures explains that a high ORR corresponds with a high microdefect density, and vice versa.

The resultant good correlation can be recognized between the ORR and the microdefect density. Furthermore, ORR by the two-step annealing is very low in comparison with ORR obtained by the single-step 800°C annealing. This is a surprising phenomenon because only the 1050°C annealing for a very short period is added to the 800°C annealing in the two-step annealing. This shows that few nucleations of oxygen precipitates occur during the 800°C annealing. Why is a high density of microdefects induced by the 800°C single-step annealing in the case of no high-temperature pre-annealing? It is not described here, but it was observed that thermally induced microdefects are enhanced by adding the low-temperature pre-annealing [5.63,70,78]. These facts show that the formation of microdefects is a result of growth of latent microdefects [5.76,78] being present in the wafer of the as-grown state, when the relevant heat treatment is carried out at a temperature above and around 800°C.

The growth phenomenon of microdefects by the heat treatment can be explained based on the model shown in Fig.5.56 as follows. Before the explanation, it should be pointed out that the nucleus of oxygen precipitates shrinks during the thermal treatment if the size is smaller than the critical size necessary for growth [5.79]. The critical size increases as the heat-treatment temperature increases. Now, it is postulated that a size-density distribution of the latent microdefects can be schematically drawn (Fig.5.56a). When high-temperature annealing is introduced where the critical size is $C-S_1$, a small part of latent microdefects can grow during the relevant annealing. As a result, the induced microdefects have a size-density distribution indicated with

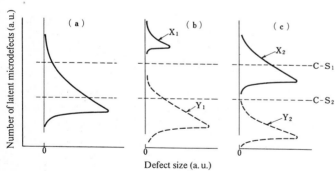

Fig. 5.56. a-c. Size-number distribution of latent microdefects (model). a) The as-grown state, b) The distribution after a high-temperature annealing where X_1 and Y_1 show distributions of defects grown and shrunk, respectively. c) The distribution after a low-temperature annealing where X_2 and Y_2 show distributions of defects grown and shrunk, respectively. It should be noted that $C-S_1$ and $C-S_2$ are critical sizes at temperatures T_1 and T_2, respectively, where T_1 is higher than T_2

X_1 in Fig.5.56b. On the other hand, latent microdefects smaller than the critical size C-S$_1$ shrink during the relevant annealing. They are redistributed as indicated with Y_1 in Fig.5.56b. If the resultant wafer is annealed at a lower temperature than that in the case of Fig.5.56b, a low density of microdefects is induced [5.76,78] because the resultant wafer contains a low density of defects including latent microdefects whose sizes are larger than the critical size C-S$_2$. If a low-temperature annealing is carried out, a high density of microdefects is induced where latent microdefects larger than the critical size C-S$_2$ are grown during the relevant annealing (Fig.5.56c) [5.76, 78]. Therefore, a low-temperature annealing is preceded by the annealing, and an opposite phenomenon occurs (Figs.5.56b and c). That is, a high density of microdefects is consequently induced.

Moreover, the nucleation of latent microdefects depends on the carbon concentration in the crystal. Consequently, the total number of latent microdefects in the crystal is influenced by the carbon concentration in the crystal. The size distribution of the latent microdefects in the crystal is related to the thermal history of the crystal received during crystal growth [5.76,78]. For example, the wafer B shown in Fig.5.54 is of low carbon concentration, but a high density of microdefects is induced by a 1050°C single-step annealing. In this wafer a rather high density of microdefects was also induced by a 800°C annealing and the density was scarcely induced by the high-temperature pre-annealing procedure demonstrated in Fig.5.54. These results make clear that sizes of latent microdefects in the as-grown state are relatively large in this wafer. It can be deduced from these data that the total number of latent microdefects contained in this wafer (B) is small, but that large-size latent microdefects occupy a large part of the total latent microdefects. As a result, a rather high density of microdefects was induced by the 1050°C annealing for 64 h in spite of the low carbon concentration in the wafer. However, it should be noted that the absolute density of microdefects is relatively low, as indicated in Fig.5.40, when the annealing temperature is around 1000°C even if the maximum density is induced. So, it can be concluded that thermally induced microdefects largely depend on the thermal history of the crystal when the heat-treatment temperature is around 1000°C.

c) Suppression of Microdefects During LSI Processes

As mentioned earlier, heat treatments are carried out at temperatures around and above 1000°C in LSI processes. Accordingly, it is necessary that microdefects induced by heat treatments at these temperatures must be diminished. This is possible if the nucleation of the oxygen precipitates scarcely occurs during the relevant heat treatments, as deduced from the model shown in Fig.5.56. One way to do so is as follows. Firstly, the pre-annealing temperature is increased to 1270°C which is higher than that of the previous case. At this temperature, the critical size of nucleus for precipitate growth is very large [5.79] and, consequently, this procedure of pre-annealing can shrink large-size latent microdefects. If all the latent microdefects are smaller in size than the critical size at 1270°C, they shrink into very small sizes by this pre-annealing procedure. Of course, the pre-annealing period must be long enough for the latent microdefects to shrink. Under such a condition a few microdefects are induced by the heat treatment at temperatures around 1000°C, provided the model shown in Fig.5.56 is correct. At the beginning, the starting wafer is halved. One is subjected to the 1000°C single-step annealing for 64 h and the other one is used for the same annealing after the pre-annealing at 1270°C for 2 h. Subsequently, bulk microdefects on the (110) cleavage plane are examined using the Dash etching as shown in Fig.5.57. Figures 5.57a and b indicate that the microdefect density induced by the 1000°C annealing has been decreased substantially by the addition of the 1270°C pre-annealing procedure. Relatively few microdefects are seen in Fig.5.57b even by adding the 1270°C pre-annealing. This is due to the fact that a few latent microdefects are believed to be larger than the critical size at 1270°C. The procedure of high-temperature pre-annealing is not always

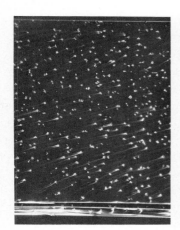

Fig. 5.57. a,b. Effect of high-temperature preannealing on thermally induced microdefects where microdefects are revealed by Dash etching on (110) cleavage plane normal to (100) wafer surface. (a) After 1000°C annealing for 64 h, and b) after 1270°C annealing for 2 h followed by the same 1000°C annealing for 64 h. It is to be noted that the oxygen concentration in the wafer is about 1.2×10^{18} cm^{-3}

useful. If many latent microdefects are larger in size than the critical size at 1270°C, this procedure does not work for the suppression of microdefects. This phenomenon occurs if large temperature fluctuations are induced during crystal growth, as SHIRAI [5.80] pointed out. The size of the latent micro-defects is also enhanced during cooling after crystal growth, if the relevant crystal is kept for a long period at temperatures below 800°C. Therefore, latent microdefects in the as-grown state must be small in size. This can be achieved by controlling the crystal-growth conditions.

In conclusion, the following items must be considered if the crystals are to be grown in order to suppress microdefects induced during the LSI process:

(i) The carbon concentration in the crystal must be below $4 \cdot 10^{15}$ cm^{-3}. This allows a low density of latent microdefects in the as-grown state crystal.

(ii) The temperature fluctuation during crystal growth must be small. This guarantees small-size latent microdefects in the same crystal.

(iii) The cooling period after crystal growth must be short especially at temperatures below 1000°C. This suppresses the growth of latent microdefects during the cooling after crystal growth.

If the crystal is grown under conditions described above, only few micro-defects are induced by the heat treatment at temperatures above 800°C, even if the oxygen concentration in the crystal is a high as $9 \cdot 10^{17}$ cm^{-3}. Even if the above conditions are satisfied, microdefects are induced during the heat treatment when the temperature is around 650°C. This is because nucleation of oxygen precipitates can occur at temperatures around 650°C [5.78]. In this case, carbon atoms act as nucleation centers for oxygen precipitates. Therefore, this nucleation is heterogeneous. It is only emphasized here that the holding time after crystal growth must be short especially at temperatures around 650°C in order to avoid the occurrence of this nucleation phenomenon.

5.4.3 Ion Implantation Induced Defects[1]

Ion implantation techniques utilizing accelerated ions (typically several keV to several hundred keV) has found the most successful application field in the impurity doping of semiconductors. Ion implantation makes possible a highly controllable and homogeneous impurity doping, and is considered to be a very essential process in VLSI fabrication.

---

[1] The space for this section does not allow us to fully explain the ion implantation induced defects. Only a very brief review is given. For further study see the related books [5.81-89] and conference proceedings [5.85,86, 90].

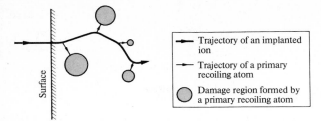

Fig. 5.58. Schematic view of ion-implantation damage formation

Implanted layers, however, need suitable thermal treatments before they can be used as electrically active regions for semiconductor devices. This is because the damage produced during the bombardment of heavy ions causes undesirable effects on electrical properties. The effects include reducing the number of electrically active implanted ions, introducing deep levels, etc. Thus, the generation mechanism and nature of the defects must be understood.

An accelerated ion, passing through a solid, loses energy by successive collisions with atoms forming the solid, and finally stops. This is shown schematically in Fig.5.58.

The energy loss process is customary separated in two major components: nuclear and electronic. The rate of energy loss is denoted by a differential $(-dE/dx)$, known as the stopping power or the specific energy loss. Here E is the ion energy and x is the distance, usually measured along the direction of incidence of the ions. The two components are shown as a function of energy in Fig.5.59.

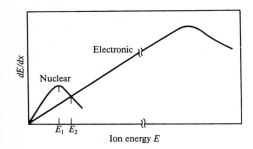

Fig. 5.59. Nuclear and electronic contributions to the energy loss

The energies E_1 and E_2 for various ion species in Si are listed in Table 5.7. It is shown that, in the conventional energy range, popular doping impurities lose energy mainly by the nuclear stopping process, with the exception of the boron.

In the nuclear stopping process the specific energy loss is due to collisions of the ion with an atomic nucleus of the solid (primary collisions).

Table 5.7. E_1 and E_2 (Fig.5.59) in Si

	E_1	E_2
B	3	17
P	17	140
As	73	800
Sb	180	2000
Bi	530	6000

At ion energies below several hundred keV, usually applied in ion implantation, elastic scattering is the dominant process, as compared with inelas-
tic nuclear excitation. The elastic collisions result in an energy transfer
ranging from fractions of a keV to many tens of keV. Provided that a re-
coiling atom receives an energy in excess of a minimum value, about 25 eV,
called the displacement energy, it can leave its lattice site to become per-
manently displaced within the solid. Such a pair of displaced atoms and
vacant sites is called a Frenkel pair and is the simplest form of radiation
damage. In ion implantation, however, lattice atoms recoil from primary collisions usually with kinetic energies far in excess of the displacement energy and penetrate many atomic distances into the surrounding lattice, repeating
a succession of uncorrelated collisions with the lattice atoms. Thus, in heavy-
ion cases, not the simple defects, but damaged regions containing many displaced atoms remain around the pass of the incident ions.

The specific nature of the damage depends on the chemical composition and
the structure of the irradiated material, on the ion species, energy, and dose,
as well as on the irradiation temperature. Figure 5.60 shows the critical dose
at which continuous amorphous layers form in the cases of boron and phosphorus

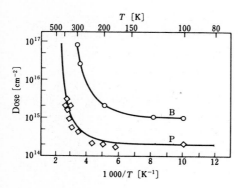

Fig. 5.60. Implant temperature dependence of critical dose for continuous amorphous-layer formation
in phosphorus and boron implanted
silicon

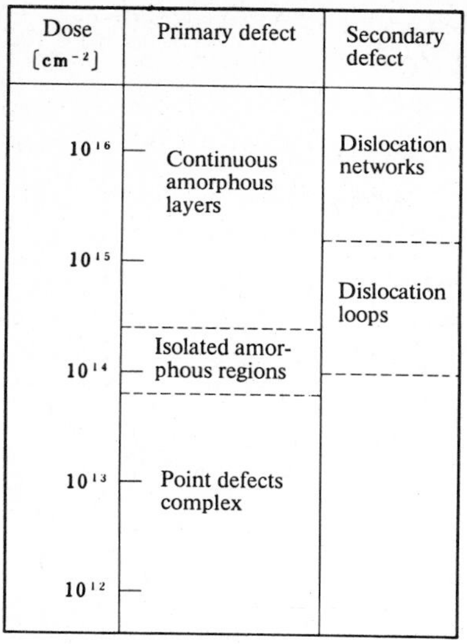

Dose [cm^{-2}]	Primary defect	Secondary defect
10^{16}	Continuous amorphous layers	Dislocation networks
10^{15}		
		Dislocation loops
10^{14}	Isolated amorphous regions	
10^{13}	Point defects complex	
10^{12}		

Fig. 5.61. Crystalline defects in medium-mass ion-implanted silicon (secondary defects are for 50 to 100 keV implants annealed at about 1000°C)

ions in the silicon. Only a lower implant dose is required to form amorphous layers at a lower implant temperature.

The structure of the implanted damage depends on the implanted dose. The relation for the room-temperature implant of the phosphorus and other medium-mass ions in silicon is shown in Fig.5.61. In the dose range below about 10^{14} ions/cm^2 smaller point defect complexes are dominant. They can mostly be annealed out at temperatures lower than 300 - 400°C. In the little higher dose range some of the regions of those defect complexes overlap to form small mutually independent amorphous regions. In the still higher dose range these amorphous regions in turn overlap to form a continuous amorphous layer, which recrystallizes epitaxially at the crystalline-amorphous interface at temperatures above about 600°C. The rate at which the interface proceeds during the solid phase epitaxial recrystallization depends on factors such as the crystal orientation of the interface, species and concentration of the dopant, as well as the annealing temperature.

If the implantation is done at an elevated temperature and the damage zones formed by individual ions are annealed during implantation, continuous amorphous layers are hard to form. In the case that the temperature rise is caused by a high dose rate, the competition between the formation and the annealing of implant damage sometimes results in stable complex defects. Thus, in high dose rate implants a higher density of crystalline defects remains after implantation than in the case of low dose rate implants at the same elevated temperature.

Annealing conditions to obtain specidal device characteristics are determined as a function of many factors such as temperature, ambient, crystal orientation, etc. The annealing temperature is a major factor. A higher temperature is required than to merely recover the crystallinity of the implanted layers, and the nature and the structure of the crystalline defects after the high temperature annealing depend strongly on the implanted ion species.

In the case of phosphorus, dislocation networks are frequently observed. These networks are similar to those observed in high-concentration phosphorus diffused layers, and are of known form due to the elastic strain introduced in the lattice because of the solid solution of the phosphorus atoms (atomic radii difference). In ion-implanted layers, these networks are detected at a lower phosphorus concentration than in conventional diffused layers. A similar tendency is observed for the so-called emitter edge dislocations.

The dislocation-network formation is not so conspicuous in the cases of arsenic and boron implantations.

In the case of the annealed boron implants, stacking fault formation sometimes causes troubles in the medium dose ($10^{14} - 10^{15}$ ions/cm^2) implants being mainly used for the base region of npn bipolar transistors.

These larger-size crystalline defects observed after elevated-temperature thermal treatments are called the secondary defects (the term tertially has also been used for the still larger defects formed after higher-temperature annealing) [5.88].

Defect free layers can be obtained by selecting the annealing conditions for each implant condition carefully.

The gettering techniques have also been applied to obtain implanted layers with still better electrical characteristics. The mechanical backside damage gettering, the intrinsic gettering and other techniques have been utilized for arsenic, boron and other ion-implanted layers and found feasible. Also, the damage formed by ion implantation has been studied as a means of gettering It has been found effective, although it is less favourable from the economica viewpoint.

5.4.4 Gettering

In today's semiconductor technology, it is well known that harmful impurity contamination and crystal defects are introduced into wafers during device fabrication processes. In turn, these contaminations and defects give detrimental influence on device properties and yield. Therefore, it is desirable that the entire wafer or the active region of devices be intentionally made clean and defect free by the removal or the reduction of the impurities and/or point defects responsible for the defect generation. Generally, this process is called "gettering". The gettering in wafer process is, for convenience, divided into two categories, a gettering on the wafer surface (back, or front surface) and a gettering in the bulk. As for the former, the wafer back-side damage process, hydrochloric (or trichloroethylene) oxidation, and the phosphorus gettering process are known as extrinsic gettering. As for the latter, high-density microdefects intentionally distributed inside the wafer are used as capturing centers of contaminant impurities, or point defects. This gettering is called intrinsic, or internal gettering (IG). In the following, examples of the intrinsic gettering are discussed.

As described in Sect.5.4.1, when CZ Si crystals are subjected to annealing at around 1000°C, residual oxygen impurity precipitates and microdefects are generated from the precipitates.

It is well known that when the microdefects are located in the active region of devices, these defects give harmful effect on the device properties. The defects, however, tend to attract and capture disadvantageous impurities (mainly heavy metal impurities). Therefore, if microdefects are intentionally distributed in a wafer, the defects are used for increasing the device production yield. For example, as reported by TAN et al. [5.91], when CZ Si wafers are annealed at around 1050°C for 60h in an oxygen-free environment, a defect-free region, or a denuded zone is formed near the surface because of the out-diffusion of oxygen impurity from the surface region. On the other hand, oxygen in the interior of the wafer produces microdefects and these defects capture the harmful impurities during the process. Thus, the wafer surface region, which plays an important role for a device, becomes microdefect and contaminant impurity free. As a result, device yield has been improved.

It is supposed that this type of intrinsic gettering effectively functioned in the old day's wafer, in which a high density of microdefects can be easily induced by a single annealing at a temperature around 1000°C, without being noticed. This gettering method is applicable to only the old day's wafer. Today's wafer quality has been improved with the development of crystal growth technology, and microdefects have not been easily generated by a single annealing even for the oxygen concentration at the same level as before.

Recently, a new technique has been developed combining a low and a high temperature annealing to induce a high density of microdefects even in a high-quality today's wafer. That is, a low-temperature annealing at 800°C for 16 - 64 h in oxygen atmosphere is added before a high-temperature annealing at 1050°C for 64 h in nitrogen ambient. Using this technique, a significant lifetime improvement was obtained, as shown in Fig.5.62, even if high-quality wafers are used [5.92]. Experimental procedures are indicated in Fig.5.63.

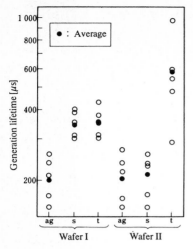

Fig. 5.62. Generation lifetime of MOS capacitors as functions of wafer category and annealing conditions. Wafer I and II indicate
wafer of easy microdefect generation and
high quality wafer, respectively, and ag, s,
and t indicate wafer without annealing (as
grown), with single annealing, and a two-
step annealing, respectively

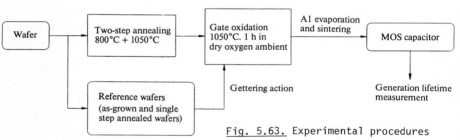

Fig. 5.63. Experimental procedures

The role of the first low-temperature annealing at 800°C is to make the size
of small latent defects[2] large enough for further growth at the following
high-temperature annealing.

By the following high-temperature annealing at 1050°C, the oxygen concentration is reduced near the surface because of the out-diffusion. As a result,
only a few latent defects grow to microdefects. Thus, a microdefect-free region
or a denuded zone can be formed near the surface. On the other hand, in the
bulk region, a high density of microdefects is generated from latent defects
which already grew to some extent by the previous 800°C annealing.

During the subsequent gate-oxidation, these microdefects getter contaminant
impurities (probably heavy metals), and consequently the MOS capacitor life
time increases.

---

[2] Latent defects are believed to be existing in the crystal from the time of
crystal growth and be shrunk or dissolved if a single high-temperature annealing at around 1050°C is applied to the wafer.

In order to make sure that this lifetime improvement is due to both the microdefects in the bulk and the denuded-zone near the surface, the following experiment was carried out. A silicon chunk, in which a high density of micro-defects can be easily induced by a single-step annealing, was heat-treated at 1050°C. Then, the wafers were sliced from the chunk. MOS capacitors were fabri-cated on the wafer and the carrier lifetime was measured. No lifetime improve-ment was obtained in spite of the formation of large-size microdefects, which are believed to have enough ability for gettering. This is because the denuded zone is not formed near the wafer surface.

Later, it was recognized that this gettering technique has a weak point, as described below. Relatively large defects among the tiny latent defects near the surface, which grew larger by the first low-temperature annealing, remain without dissolution by the next high-temperature annealing. In other words, in this case, the denuded-zone near the surface is not perfect. This technique is good enough to getter heavy metals, but leads to the formation of oxidation-induced defects originated from the remaining defects near the surface.

In order to avoid the formation of such an imperfect denuded-zone, an im-proved double pre-annealing technique, with which a nearly perfect denuded-zone and a bulk microdefect region are formed, was proposed and demonstrated successfully, as will be described later [5.93]. In this technique, dissolved oxygen near the surface is diffused out from the wafer by the first high-tem-perature (1050°C) annealing in non-oxygen ambient, and a resultant nearly defect-free region is formed near the surface. And then, subsequent 650°C annealing is carried out. By this annealing, latent defects in the bulk are transformed into nuclei for the formation of the microdefects. By a subsequent high-temperature process (for example, a heat treatment after the ion-implan-tation as described below), a high density of microdefects with gettering abil-ity is formed from these nuclei. By this technique, better results are obtained on recent high-quality wafers in which a high density of microdefects is not easily induced by a single heat treatment at temperatures around 1000°C.

This technique was applied to the ion-implantation oxidation process which is the easiest way to induce surface defects among the various wafer processes.

The experimental procedure and the result of this application are shown in Figs.5.64 and 65, respectively. Representative optical micrographs are shown in Figs.5.65a-d. The front surface and the cleavage plane of the gettered wafer are shown in Figs.5.65a and c, respectively. Those of the control sample with-out the gettering are shown in Figs.5.65b and d, respectively. A high density of oxidation-induced stacking faults ($3 \cdot 10^4$ cm^{-2}) is seen on the control sample surface b, but not on the gettered wafer surface a. However, the bulk micro-

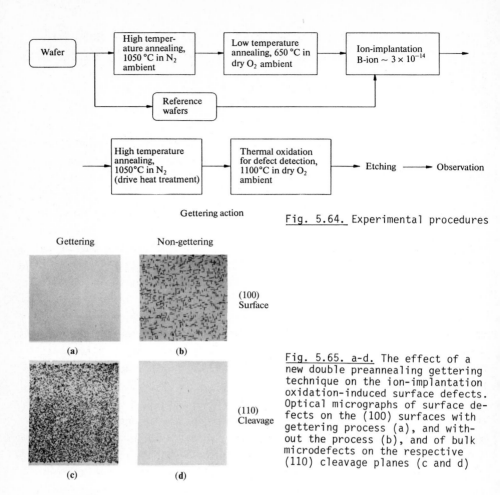

Fig. 5.64. Experimental procedures

Gettering action

Gettering   Non-gettering

(100)
Surface

(a)   (b)

(110)
Cleavage

(c)   (d)

Fig. 5.65. a-d. The effect of a
new double preannealing gettering
technique on the ion-implantation
oxidation-induced surface defects.
Optical micrographs of surface defects on the (100) surfaces with
gettering process (a), and without the process (b), and of bulk
microdefects on the respective
(110) cleavage planes (c and d)

defect density is low in the control sample d, but very high in the gettered
sample c. It should also be noted that a clear denuded-zone occurred, as shown
in the figure. This indicates that a high density of surface stacking faults
corresponds to a low density of bulk microdefects, and vice versa. It is clear
that highly dense bulk microdefects act effectively as gettering sites during
the heat treatment after the ion-implantation.

Furthermore, the improved double pre-annealing technique is tested to as-
certain if it can capture heavy metals such as copper [5.94]. Some wafers
are put into the improved double pre-annealing procedure. After that, a control
sample and a double pre-annealed wafer are simultaneously subjected to a copper
diffusion process at 850°C followed by thermal annealing (for gettering) at
1150°C in an N_2 atmosphere.

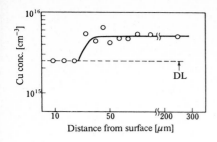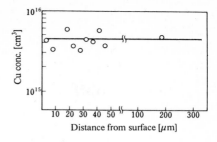

Fig. 5.66. Depth profile of copper concentration in a wafer after gettering with the improved double preannealing technique. DL shows the detection limit of copper atoms in Si crystal by the present NAA technique. Copper concentrations in the denuded-zone are below the detection limit where meaningless low values are counted

Fig. 5.67. Depth profile of copper concentration in the test wafer; not subject to the improved double preannealing before copper diffusion

Following these procedures, a depth profile of the copper concentration in the wafer is obtained with the help of a neutron activation analysis (NAA). As a result, it was found that copper was trapped and accumulated in the wafer bulk and not detected in the denuded-zone near the surface, as shown in Fig.5.66.

The copper concentration in the control sample without the double pre-annealing was found to be nearly flat in the wafer (Fig.5.67). Thus, it has been proved that the technique is effective for the gettering both of surface-induced defects and copper impurity.

In the above-mentioned intrinsic gettering, the bulk microdefects are utilized to capture harmful impurities or point defects. Recently, however, the bulk microdefects begin to be used to capture carriers to prevent light leakage in CCD device, or to suppress soft errors by α-particles in a MOS RAM or CCD memory.

Therefore, in the near future, the application of "controlled microdefects" is expected to be more important.

However, the gettering mechanism is not fully understood yet. It is not clear what kind of defects such as oxygen precipitate itself, the surrounding strain field, and/or punched-out dislocation loops around the precipitate play a role in the gettering phenomena. Also, it is not clear what kind of impurities, point defects and/or carriers are selectively captured by the microdefects. As for the point defects, it is not understood whether the microdefects trap, or emit the defects.

Furthermore, is a continuous formation of the microdefects possible? The impurity capturing ability and capturing duration of the microdefect are important problems especially when the gettering process is carried out in the device fabrication processes. But these problems are not solved yet. With a clarification of these questions, the effectiveness, the controllability, the uniformity, the handling and the cost of the gettering are expected to be greatly improved.

5.5 Epitaxial Growth

5.5.1 Low-Pressure Epitaxial Growth

Much attention has been focussed on the low-pressure silicon epitaxial growth as thin epitaxial method since 1974, because it has a merit of reducing auto-doped impurities from a buried layer (BL) on a substrate. SiH_4 or SiH_2Cl_2 were used as reaction gases [5.95]. Vertical type [5.96], barrel type [5.97,98] and horizontal type [5.99,100] reactors were tried, and the radiation heating method by infrared lamps was mainly used rather than radio-frequency heating.

Here, expitaxial growth at reduced pressure by pyrolysis of SiH_4 using an infrared-lamp-heated horizontal reactor with a double-wall quartz tube [5.101] is described. In the case of lamp-heating from both upper and lower sides of the reactor tube, there is a small difference in temperature (less than 10°C) between a wafer surface and a susceptor in a low-pressure hydrogen atmosphere.

Figure 5.68 compares the distribution of growth rates between atmospheric and reduced pressures. In the case of reduced pressure, the growth rate distribution is better for the lower pressure, although the growth rate decreases Thus, the improvement of thickness uniformity is one of the merits for the low-pressure growth. The ranges of growth conditions are 20 - 100 Torr $(2.7 - 13.3 \cdot 10^3$ Pa) growth pressure, 920° - 1050°C growth temperature, and $1.2 - 2.0 \cdot 10^{-3}$ in the SiH_4/H_2 mole ratio.

The autodoping is classified into vertical autodoping above a BL and lateral autodoping around the BL. For the purpose of studying the aspect of the auto-doping, undoped epitaxial layers were grown on n^- substrates with two types of as-doped BL. One was a single circular BL and the other was a multi-BL of the same circles arranged in two dimensions. And then, the impurity concentration profile was obtained by C-V method after a diode array was formed on the epitaxial layer.

In the case of a single circular BL, the concentration of lateral auto-doping has its maximum value near the interface between a substrate and an

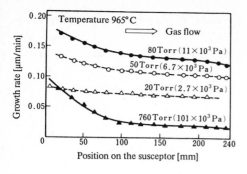

Fig. 5.68. Growth-rate distributions at atmospheric and reduced pressures in the direction of the reaction gas flow

expitaxial layer, and decreases exponentially with vertical distance towards the surface from the interface. The maximum concentrations, i.e., surface concentrations at measured points, also decrease exponentially with the lateral distance from the BL. They are distributed as round contour lines around the BL, because the current direction of reaction gas has less influence upon the impurity distribution in the low-pressure epitaxial layer. The amount of auto-doping can be evaluated quantitatively based on these facts.

On the other hand, in the case of an epitaxial layer grown at atmospheric pressure, the current direction has influence upon an impurity distribution. Therefore, lateral autodoping on the lower stream is more than that on the upper stream, and the impurity concentration is larger by about one order than that in a low-pressure epitaxial layer [5.99].

Figure 5.69 shows the pressure dependence on the maximum concentration of lateral autodoping, at constant growth rate, 25 - 110 Torr (3.3 - 14.6 · 10³ Pa), and 1000°C. It becomes almost an exponential relationship at each pressure

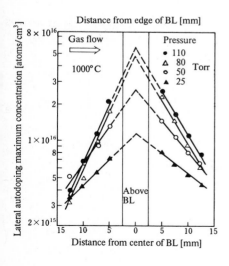

Fig. 5.69. Pressure dependence on lateral autodoping from a single circular BL (growth rate: 0.12 μm/min)

265

between the maximum concentration (surface concentration) and the lateral distance. The peak concentration above BL decreases more for the lower pressure. However, the maximum concentration becomes easy to spread out towards the lateral distance.

When the growth temperature is varied, an exponential relationship is obtained. The peak concentration decreases when the temperature increases. Then, autodoping is also decreased when the growth rate is smaller, or the prebaking temperature is higher and the time is longer at reduced pressure just before growth.

In the case of multi-BL, even if the growth pressure is lower, the maximum concentrations (surface concentrations) are almost constant (Fig.5.70). However, if the extinction coefficient B in the direction normal to the surface is larger, the amount of autodoping is smaller by lowering of the pressure. When the growth temperature is varied, the higher the temperature the larger is the extinction coefficient.

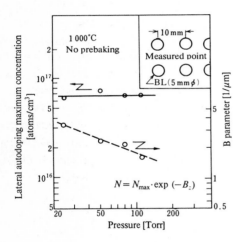

Fig. 5.70. Pressure dependence on lateral autodoping from multi-BL

From the facts mentioned above, it may be inferred that autodoping is reduced because vaporized impurity atoms from BL are rapidly purged in low-pressure growth.

As regards crystal defects in a low-pressure epitaxial layer, the oxidation induced stacking fault (OSF) density increases when the substrate is heated up in vacuum, using both SiC-cotaed graphite and poly-crystalline silicon susceptors. Therefore, better results for crystal defects are obtained by heating up the substrate in reduced pressure. Especially, the substrate surface is badly influenced by release of adsorbed gases from an SiC-coated graphite susceptor [5.97]. It is found that the growth conditions (pressure, temperature, etc.)

are less influenced by crystal defects, if prebaking is carried out at temperatures over 1000°C, 10 min. In the case of favourable pretreatment and handling of the substrate surface, the low-pressure epitaxial layers with very few OSF densities (zero or under 10 cm^{-2}) are obtained.

It is found that the etch pit density in the growth layer decreases for gettered substrates. On the other hand, for reducing OSF, it is important to eliminate the contamination on the substrate surface after a gettering treatment.

Generally speaking, the density of crystal defects in a low-pressure epitaxial layer is as high as that in an epitaxial layer grown at atmospheric pressure. It is not seen that crystal defects increase particularly due to growth at reduced pressure.

5.5.2 Molecular-Beam Epitaxial Growth

Abrupt profiles of the impurities across the epitaxy-substrate interface and arbitrary control of doping levels in the grown layer are required for the fabrication of devices with a finer structure.

The conventional CVD method using SiH_4 gas or $SiCl_4$ vapour, is considered not sufficient to achieve these high qualities of the epitaxial layer, due to autodoping and out diffusion of impurities. On the other hand, Si molecular beam epitaxy (Si MBE) [5.102-105] attracts more and more attention as a novel low-temperature process and has recently produced high-quality-grown layers, because arbitrary control of the impurity distribution and the low-temperature growth (as low as 600°C) are made possible by this process. Figure 5.71 shows the typical growth system of Si MBE. In this system, an electron-gun system is installed to create an Si molecular beam, and effusion furnaces by resistance heating are set to evaporate Sb and Ga beams for impurities doping.

All heated sections are surrounded by liquid-nitrogen-cooled shrouds, because MBE systems are to be maintained with high vacuum $\sim 10^{-9}$ Torr even in a growth period.

One of the major difficulties in Si MBE is the heating method of the substrate. Indirect heating of the substrate by a resistance heater makes a source of contamination gas near the growth surface of samples in an ultra-high vacuum.

The direct heating method of the resistive substrate is often used in an Si MBE system to avoid the above-mentioned contamination. The rectangular silicon substrate is clamped with tantalum clamps at both ends of the samples. The substrate is heated by providing the heating current with the Ta clamps.

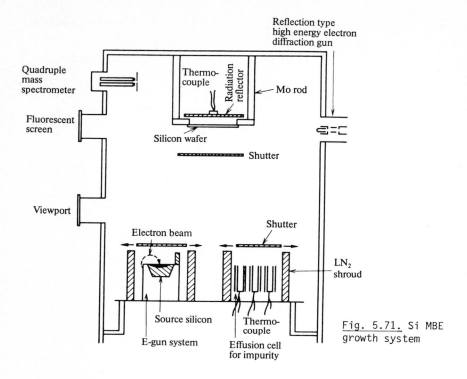

Fig. 5.71. Si MBE growth system

This direct heating method is, however, a barrier for realizing a practical Si MBE system, because it sets serious limits to shapes and dimensions of the silicon samples.

The Si source parts consist of several sections of high-purity polysilicon of resistivity more than 1 kΩ-cm, and are designed carefully to avoid copper contamination which is produced by bombardment of the water-cooled copper crucible surface with stray electrons. The silicon sections surrounding the source silicon section protect the Cu crucible from electron bombardment [5.102].

Hall mobilities of the grown layer on the pn junction between the epitaxial layer and substrate are measured by the Van der Pauw method for evaluation of the grown layer by MBE. Figure 5.72 shows examples of hole mobilities measured by this method in p-type grown layers, where the mobilities in epitaxial layers are found to be almost the same as those of bulk silicon [5.104].

Figures 5.73 and 74 show the abrupt change in impurity profiles near the substrate, which are measured by SIMS (secondary ion spectroscopy) [5.15, 5.106-108]. Interface impurity profiles of grown layers with high Sb and B concentration substrates are shown for a growth temperature of 750°C in Figs.5.73 and 74, respectively.

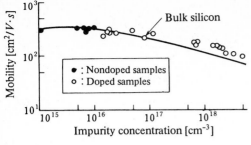

Fig. 5.72. Hole mobilities of MBE grown layer

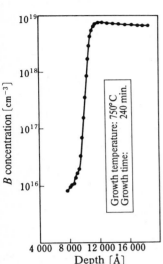

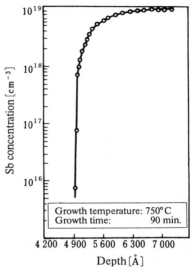

Fig. 5.73. Impurity distribution near the interface of a highly Sb doped substrate

Fig. 5.74. Impurity distribution near the interface of a highly B doped substrate

A problem on impurity doping of Si MBE is the difficulty in introducing high-concentration impurities into the grown layer. Sb doping by the effusion-cell method cannot be controlled enough to a precisely determined level, because the residual Sb vapour inside the growth system prevents doping levels from being controlled finely.

Recently, considerable progress has been made in doping control by a novel system developed by OTA [5.109], which is composed by combining the ordinary MBE chamber with a low-energy ion-implantation section.

Figure 5.75 shows a schematic diagram [5.110] of a growth system with p-type and n-type implantation sections similar to the low-energy implant system proposed by OTA. In this case, each channel of the sections is de-

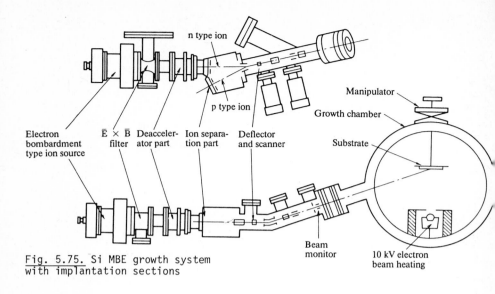

Fig. 5.75. Si MBE growth system
with implantation sections

signed to be exchangeable by another one. Experimental results on arsenic
ion implant MBE were reported by OTA. As^+ ions are mass analyzed by $\mathbf{E} \times \mathbf{B}$
filters and the beams are then deaccelerated to a low energy of 400 - 800 eV
for implantation in the MBE. The ion beams are scanned by a pair of horizontal
and vertical scanning plates as well and mixed with an Si molecular beam
near the substrate.

The system makes it possible to grow silicon epitaxial films with simultaneously introduced implanted impurities. Doping of As impurity in the range
$10^{14} - 10^{18}$ cm^3 was proved [5.109].

At present, Si molecular beam epitaxy is effectively applied to SOS (Silicon On Saphire) systems producing a thin Si film. Growth of more than 1 μm
thick films for an SOS structure in UHV has hitherto been reported, however,
no good experimental result for Si films as thin as 5000 Å has been obtained
[5.110]. Average carrier mobilities in such thin films made by MBE have recently been reported to exceed those by CVD by 20 - 25%, where the back side of a
substrate coated with sputtered Ta film is directly heated by electrical power
supplied to the Ta film [5.111].

6. Process Technology

Although various structures of elemental devices have been proposed for materializing VLSI, they all are based on silicon MOS or bipolar technology. The fabrication process for most of these devices, therefore, consists of lithography using electron beam or photo exposure, etching, diffusion, ion implantation, chemical vapour deposition, thermal oxidation, etc., the same as one used for the conventional LSI. On the other hand, the decrease in the dimension of elemental devices and in the alignment margin requires the high accuracy of processing. Table 6.1 shows typical dimension parameters of MOS device with a gate length of 1 μm. Brief discussions will be given first about the accuracy of processing for fabricating fine structures like those demonstrated in Table 6.1

The accuracy of the structure in the lateral direction of VLSI is determined by the accuracy of resist patterns delineated on each exposure process, by the alignment accuracy between two layers of resist pattern, and by the accuracy of processing, namely, etching, diffusion, etc. The delineation of resist patterns and the alignment accuracy have already been treated in the previous chapters. The self-alignment technique as a technique for avoiding the alignment error is important in the fabrication of LSI and will be still more important in fabricating VLSI. In the self-alignment technique, a structural pattern which was formed by one lithographic and etching process acts

Table 6.1. Dimension parameters of a MOS device having 1 μm gate length

Gate width	1.3 μm
Contact hole diameter	1.3 μm
Junction depth	0.3 μm
Gate oxide thickness	0.025 μm
Field oxide thickness	0.3 μm
Poly-silicon thickness	0.3 μm
Metal thickness	0.3 μm

as a mask for the subsequent pattern definition. Between the former and the latter patterns, there exist some dimension shifts due to etching, diffusion, etc., and, therefore, the shift is required to be controlled precisely. The accuracy of the structure in the vertical direction is determined primarily by the accuracy of layer thickness of thermal oxide, CVD, etc., and by the accuracy of the depth of etching, diffusion, ion implantation, etc. In the following part, a brief survey will be given on the wafer processing technology

Selective etching with an etching mask made of resist or other materials is the most important technique for defining the device structure. For the highly accurate etching, it is required that the material selectivity is high and that the etching bias (difference between edge of the resist pattern and that of the etched pattern, caused by the undercut, etc.) can be controlled. Wet etching techniques using liquid solutions have been widely used in the past, but dry etching techniques without using liquid solutions have been developed recently to improve the accuracy of etching. The barrel-type plasma etching has already been used in the fabrication of LSI, but the planar-type plasma etching which gives the higher accuracy will become primarily important for fabricating VLSI. In Sect.6.1, the planar-type plasma etching technique will mainly be described. The ion beam etching is suitable for fabricating precise patterns and is highly anisotropic in its etching characteristics. This anisotropy offers some new possibilities in the process, but is not so generally used as the plasma etching.

The wet etching has such advantages that the material selectivity is high and the anisotropic etching due to the crystallographic axes is possible for silicon single crystals. Up to the present, it has been used in various proces- ses. It is still difficult to judge today whether the wet etching could be com- pletely replaced by the dry etching or not, and at least in the near future, both of them will be used together in the production line.

In analogy to the dry etching, it might be possible to consider substitu- tions of dry process techniques for wet processes such as the resist coating, developing, removal, the cleaning of wafer surfaces, etc. For the resist re- moval, the plasma etching technique has been materialized by using the oxygen plasma. For other processes, studies are being made also on the plasma depo- sition of the resist material, the plasma development of the resist, the plasma oxidation of silicon, and the plasma cleaning of the silicon surface. However, large improvements will be necessary for these techniques to be introduced into the VLSI processing.

Since the old days, p-n junctions were widely formed by impurity diffusion. In the case of VLSI, it is especially important to suppress the change in the impurity profile due to the internal diffusion and to avoid the shift of the p-n junction during the following heat treatment process, because of the shallow junction and the high density of impurities. To fulfill these requirements, it is usual to lower the temperature, to shorten the time of heat treatment and to use an impurity having a smaller diffusion coefficient. Arsenic recently has been applied because of its smaller diffusion coefficient, instead of phosphorous which had been generally used as an n-type impurity. Solid state diffusion or ion implantation are common for doping with arsenic impurity. Until recently, ion implantation had been applied only for the light doping, for example, controlling the threshold voltage of MOS LSI. As a technique for the formation of p-n junctions, the ion implantation has several disadvantages, especially the long time being necessary for the high concentration doping with an impurity. Furthermore the junction depth is small, and the wafer must always be annealed after the implantation. In the case of VLSI, however, the ion implantation has become important as a technique of forming p-n junctions, because the precise control of the impurity profile is required and the junction depth is small. Both demands can be fulfilled by implantation.

In the case of ion implantation with high dose, the rate of inactivated impurities which do not act as donor or accepter levels even after the annealing becomes higher. If there is an increase of the annealing temperature, the rate of activated impurities extends, too, but the impurity profile becomes wider due to the internal diffusion. The precise control of temperature and time of the annealing, therefore, is a necessity. It is also to be said that the annealing efficiency depends strongly on the atmosphere at the heat treatment. Because the implantation with high dose has some problems, namely the temperature rise of wafers and the low throughput, machines specialized for high dose implantation are being developed.

The heat treatment process is used for sintering the CVD or evaporated films as well as for annealing the ion implanted wafers. Beam annealing techniques including the laser annealing and the electron beam annealing are being studied as methods of annealing wafer surfaces within a short period and without enlarged substrate temperature. These techniques offer a low energy consumption and a short annealing time compared with the conventional annealing in a furnace, because the light or electron beam is focussed on a minute volume of the wafer surface and heats it very locally and transiently. The technique also makes possible processes such as the local melting and recrystallization which are very difficult to do by using only the conventional annealing technique in a furnace. Section 6.2 is going to describe the beam annealing technique, especially done by the laser.

The thermal oxidation is used for growing the gate or the field oxide. For present VLSI devices, the gate oxide will become as thin as 20 to 50 nm. The breakdown of the oxide might not be a large problem as far as the applied voltage follows the scaling rule. A problem appearing in VLSI is a shift in the electrical characteristics of the devices caused by the injection of hot carriers into the oxide. A variation and scattering of the threshold voltage due to other origins must also be avoided. In Sect.6.5 the evaluation method of the gate oxide will be described.

In such a process, the ion implantation onto the field region is followed by the field oxidation, the oxidation is desirable to be performed at the temperature as low as possible for repressing the diffusion of the implanted impurities. The oxidation at the lower temperature is preferable also for suppressing wafers to warp. High pressure oxidation is one of the ways of lowering the temperature of oxidation. The rate of oxidation of silicon is approximately proportional to the oxygen pressure, and this fact is the base for a decrease in the oxidation time or the oxidation temperature by application of the high pressure oxidation. It can be performed experimentally at 150 to 500 atm. as reported, but for practical purposes, high pressure oxidation furnaces are being available for the dry and wet oxidation in the pressure range of 5 to 25 atm. In the wet oxidation using the high pressure furnace, the size of the oxidation-induced stacking faults is smaller than the ones created by the conventional oxidation in the atmospheric pressure provided the same oxide thickness for both cases [6.1]. MOS and bipolar devices fabricated by high pressure oxidation technique have been reported [6.2,3].

The film deposition techniques for poly-silicon, insulator and metal with CVD, evaporation, etc. are widely used in the processes of fabricating gate electrodes, interconnections and the intermediate insulating layers. With decreasing the film thickness and the pattern size, further requirements are the homogeneity of the film thickness and a decrease in the defect density of the film. In addition to improvements in the CVD and evaporation techniques, the film deposition technique utilizing the plasma-assisted chemical reactions in the low pressure and lower temperature atmosphere are being developed. These techniques will be presented in Sect.6.3.

The actual process for fabricating VLSI is a combination of the individual process steps like those described above. The individual techniques are not independent of but influence each other in the fabrication process. In the case of fabricating a partial device structure, therefore, a fabrication process consisting of a combination of various individual techiques must be developed. If we take the interconnection, for instance, various kinds of tech-

274

niques are possible for the film deposition, etching, etc. depending on the material of the interconnection and so on. These techniques are described in Sect.6.4.

Techniques for producing a clean environment and clean materials have been needed since the old days of transistors. Protection techniques against the contamination of soluble materials such as alkali metals or some organic materials were almost established in LSI technolgy today. On the other hand, the contamination due to dust particles in the air, gas, water, or reagent has become required to be controlled over a wider range of the particle size with decreasing the pattern size. In Sect.6.6, the clean environment will be discussed when the number of dust particles having the size of more than 0.1 μm is controlled. The number of dust particles in gas, water and reagent is also required to be decreased corresponding to the cleanness of the environment. In addition to these, techniques are important which prevent the generation of dust particles and remove them in the environment, process equipments and their loading-unloading mechanism. Loading and unloading of the process equipments with wafers and its transfer from equipment to equipment are especially desirable to be automated as well as possible.

6.1 Dry Etching

6.1.1 Dry Etching and Fine Pattern Definition

Fine patterns of VLSI are formed by etching the layer to be etched and covered with the resist patterns which are delineated by using electron beam or other radiations. For fabricating VLSIs, therefore, a highly exact etching technique is an absolute requirement as well as a technique which delineates very precisely the resist pattern itself. For etching a layer, the wet-etching technique had been used ever before. As the density of integration was tending to the LSI-level, however, the accuracy of the wet-etching has become insufficient, and the dry-etching technique making an etching of the layer more precisely has begun to focus the attention [6.4,5].

Dry-etching techniques are divided into two categories. One is the plasma etching in which chemical reactions between the material to be etched and the chemically active species in the gas plasma play a main role. The other one is the ion etching in which the physical sputtering due to the energetic ions accelerated by the electric field plays a main role.

The plasma etching can further be divided into the barrel type and the planar type, and likewise the ion etching into the ion beam etching and the sputter etching.

The barrel-type plasma etching has already been introduced into the production line for LSI processing [6.5]. The planar-type plasma etching making possible more precise etching has started to be introduced into the VLSI processing. Although the ion-beam etching has some problems being not yet solved, the possibility of the application to VLSI processing is under investigation because of its simple etching mechanism and its good controllability. The sputter etching, however, is now almost disregarded because it has no special advantages except for the relatively simple structure of the equipment.

In the following parts, the dry etching technique will be reviewed briefly and discussed mainly on the planar-type plasma etching which focusses the attention most today.

6.1.2 Plasma Etching Equipment

a) Plasma Etching Equipments and Their Characteristics

Plasma etching is an etching method based on the phenomenon that materials with high vapour pressure are formed by chemical reactions between the material to be etched and chemically active species (atoms, molecules, ions, etc.) existing in the low temperature plasma. Such a plasma can be generated by applying the rf power to the low pressure gas volume.

As described before, there are two types of the plasma etching equipments, namely, the barrel type and the planar type. In the barrel-type plasma etching wafers are placed in a low-pressure gas within a barrel-shaped chamber made of quartz or glass. The rf power which is applied to the electrodes attached outside the chamber generates a gas plasma within the chamber and etches the material on the surface of the wafer. In this case, the etching occurs by pure chemical reactions, and so the process passes homogeneously. Therefore, the pattern shift which is caused by the undercut happens in the order of the thickness of the layer to be etched. This limits the accuracy of the pattern definition.

The planar-type plasma etching equipment consists, as shown in Fig.6.1, of a vacuum chamber having two planar electrodes inside, an exhaust system, gas controllers, and an rf power supply. The wafers are placed at one of the electrodes and etched as soon as the plasma is generated. There are two coupling modes between the wafers and the rf power, namely, depending on whether the wafers are placed at the cathode or the anode. In the cathode coupling mode, for example (Fig.6.2a), the wafers are placed at the cathode electrode (rf electrode, usually negatively charged) supplied with the rf power. In the anode coupling mode (Fig.6.2b), on the other hand, wafers are placed at the grounded

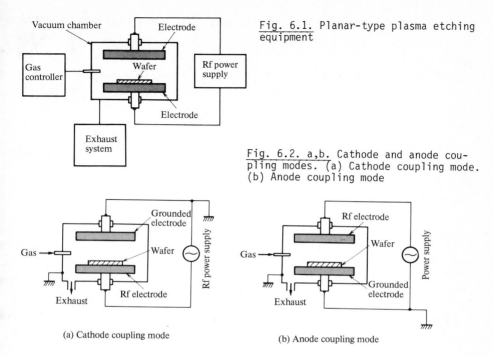

Vacuum chamber
Electrode
Gas controller
Wafer
Rf power supply
Electrode
Exhaust system

<u>Fig. 6.1.</u> Planar-type plasma etching equipment

<u>Fig. 6.2. a,b.</u> Cathode and anode coupling modes. (a) Cathode coupling mode. (b) Anode coupling mode

Grounded electrode
Gas
Wafer
Rf power supply
Exhaust
Rf electrode

(a) Cathode coupling mode

Rf electrode
Wafer
Gas
Power supply
Exhaust
Grounded electrode

(b) Anode coupling mode

electrode (anode electrode). In the case of the cathode coupling mode, the anode electrode is often eliminated.

The planar-type plasma etching has the following advantages in comparison with the barrel-type one.

(i) The anisotropic etching results in a high precision etching because of the etch-rate being smaller in the lateral direction than in the vertical one. For the anisotropic etching, two mechanisms can be considered:
- Due to the electric field in the vertical direction to the wafer surface, active species having electrical charges are supplied much more onto the lateral surface than onto the vertical one.
- The physical sputtering by ions enhances the chemical reactions on the lateral surface.

(ii) The planar-type plasma etching can etch materials etching of which is difficult by using the barrel-type plasma method. This is based on two reasons:
- Active species with a short lifetime can also contribute to the etching process, provided that the region generating the active species is close enough to the wafers to be etched.
- Physical sputtering due to energetic ions enhances the chemical reactions between the active species and the solid material.

277

The application of either the cathode or the anode coupling mode depends on the material to be etched and the gas used for etching. In general, however, the following tendencies are observed: In the case of materials for which the physical sputtering due to energetic ions makes a large contribution to the enhancement of the chemical reaction, as SiO_2 for example, the cathode coupling mode tends to give better etching characteristics (etch rate, etch-rate ratio, etc.). The optimum gas pressure is located in the lower range (1 Pa) for the cathode coupling and in the higher range (10 Pa) for the anode coupling mode. The first one gives the less deposition of materials of the opposite electrode due to the sputtering and the less polymer deposition on the wafer. On the other hand, this coupling mode has the ability for larger crystallographic and/or electrical damages to the surface of the wafer. However, there are no reports dealing with the electrical characteristics of devices fabricated with materials produced with the cathode coupling mode. Even if the physical sputtering generates the crystallographic and/or electrical damages, they could be annealed out by heat treatments in the subsequent processes or could be eliminated by etching off a thin layer of the wafer surface with the wet-chemical etching. Summarizing the discussion, the cathode coupling mode is con sidered to be more advantageous than the anode coupling one.

b) Example of the Planar-Type Plasma-Etching Equipment

The barrel-type plasma etching technique, as described before, has already been introduced into the mass production lines. Equipments being satisfactory for the practical use are commercially available. Contrary to this situation, planar-type plasma-etching equipments which can be used in the mass production line for VLSIs are not easy to find. A plasma etching equipment for practical conditions has to satisfy the following demands besides suitable etching characteristics:

(i) High throughput for production.

(ii) Good homogeneity and good reproducibility in the etching character-istics. The accuracy of the etching is determined finally by these two param-eters.

(iii) Comfortable operation, less trouble and easy maintenance.

It is not easy to find a planar-type plasma-etching equipment which satis-fies all of the conditions mentioned above, especially a good homogeneity and reproducibility, and an easy maintenance as well.

The planar-type plasma etching equipment described in the following part was developed to satisfy those conditions as well as possible. Generally, how-ever, the throughput and the accuracy of the etching are contrary to each othe

Fig. 6.3. Front view of the planar-type plasma etching equipment

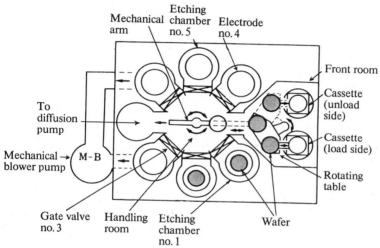

Etching
Mechanical chamber Electrode
arm   no. 5   no. 4

Front room

Cassette
(unload
side)

To
diffusion
pump

Cassette
(load side)

Mechanical → M - B
blower pump

Rotating
table

Gate valve   Handling   Etching
no. 3   room   chamber
                        no. 1

Wafer

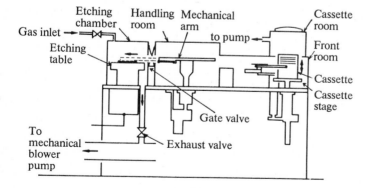

Etching   Handling Mechanical
chamber   room   arm

Cassette
room

Gas inlet →

to pump →

Etching
table

Front
room

M

Cassette

Cassette
stage

Gate valve

To
mechanical
blower
pump

Exhaust valve

Fig. 6.4. Schematic structure of the planar-type plasma etching equipment

Table 6.2. Specifications of the equipment

Materials		SiO_2, Si_3N_4, Si, Al, Mo
Throughput		40 wafers/h
Wafer size		76, 100, 125 mm
Homogeneity and Reproducibility		Within ±5%
Number in batch		1 wafer/batch
Number of chambers		6 chambers
Isolation from the atmospheric air		By high vacuum handling rooms
Exhaust	Main	Mechanical blower pump
	Sub	Oil diffusion pump
Rf source		13.56 MHz, 720 W
Coupling mode		Cathode coupling
Etching monitor		Dual wavelength emission
Control		Microprocessor control

and the optimum condition has to be determined depending on the purpose. In the case of the equipment to be described, the precision is of somewhat greater importance than the throughput because of the equipment being used for the fabrication of VLSIs.

Figures 6.3 and 4 show a front view and schematic structure, respectively, of the equipment. Table 6.2 summarizes the specifications of the equipment briefly. It has the following features:

(i) To improve the accuracy of the etching, one wafer is etched simultaneously in the etching chamber where the etching condition is controlled independently from those in the other ones.

(ii) To increase the throughput under keeping the high accuracy of the etching, the equipment is performed with six etching chambers. Each operating chamber can be shifted to the next chamber station. Thus, the waiting time becomes a minimum for each chamber.

(iii) To improve the reproducibility and make the maintenance easy, the etching chambers are isolated from the atmospheric air by the high vacuum region.

(iv) A cassette to cassette fully automated system.

(v) 5" wafers can be treated.

In the following part, brief explanations will be given about the structure and operation, referring to Fig.6.4. A cassette filled with wafers to be etched and an empty one are set into the cassette rooms of loading and unloading, re-

spectively, which are then evacuated. After evacuation, the cassettes go down
into the front room which is always kept at high vacuum. The wafer handling
mechanism consisting of a mechanical arm and a rotating table, etc. transfers
the first wafer from the loading cassette to the etching chamber no.1. Then,
the gate valve no.1 is closed and the gas inlet valve no.1 and the exhaust
valve no.1 are opened. After a stabilization of the gas pressure, the rf power
is applied to the electrode no.1 and the etching process starts for the first
wafer. The wafer handling mechanism, on the other hand, begins to transfer the
second wafer from the loading cassette to the etching chamber no.2 immediately
after finishing the transfer of the first wafer. Then, after the similar valve
operations as those for the etching chamber no.1, the etching chamber no.2 goes
into the etching operation. The same procedure is followed in turn by the etching chambers from no.3 to no.6. The end signal from the etching monitor of the
chamber no.1 turns off the rf power of the electrode no.1 and closes the gas
inlet valve no.1. After a short period, the exhaust valve no.1 is closed, then
the gate valve no.1 is opened. The wafer handling mechanism transfers the first
wafer from the chamber no.1 to the unloading cassette. The chamber no.1 is then
loaded with the seventh wafer and goes again into the etching operation. Because six etching chambers are operating in different phases of operation as
described above, six wafers are etched effectively within the total period of
the etching time for one wafer and the loading and unloading time for one wafer.
This mode of operation increases the throughput.

The cassette exchange can be done by breaking the vacuum only of two cassette rooms, and the front and handling rooms are always kept at high vacuum
of the order of 10^{-4} Pa. The atmospheric air, therefore, is suppressed almost
completely to go into the etching chambers, resulting in a large improvement
in the reproducibility. This improvement can be explained if we consider that a
part of the poor reproducibility results from the following phenomenon. The
residue which is produced by chemical reactions between active species in the
residual gas and components in the atmospheric air coming into the etching
chamber during the cassette exchange remains on the inside wall of the etching
chamber and affects the etching characteristics at the next etching cycle.

The etching monitor used in this equipment adopts a dual-wavelength, plasma
emission sensing method. Hereby, the intensities of two emission lines with different characteristics from each other are measured, and the ratio or the difference between the two intensities is given as the output signal. Thus, the
output signal level after the end point little changes, even if the etching
condition is somewhat changed. The reliability of the end point detection,
therefore, is strongly improved.

Table 6.3. Examples of performance of the planar-type plasma etching equipment

		Present equipment	Previous equipment
Homogeneity of etch rate 4" wafer SiO_2	Within wafer	±1.8%	5 - 10%
	Wafer to wafer	±2.3%	10 - 20%
	Reproducibility	±2.5%	10 - 20%
	Total	±2.6%	
Homogeneity of pattern width 4" wafer, 7500 Å SiO_2		±0.11 µm[1]	0.2 - 0.4 µm

[1] This value includes the deviation in the width of the resist pattern of 0.08 µm.

Table 6.3 gives examples of the homogeneity-data and the reproducibility obtained from wafers etched by using the equipment described here. These data were obtained by etching wafers in a constant time without using the etching monitor. As can be seen on Table 6.3, the equipment gives the expected good homogeneity and reproducibility. On this table, the deviation in the pattern size of SiO_2 is considerably larger than the expected value. However, it includes the deviation in the pattern size of resist patterns themselves. The deviation of the etching only is presumed to be around 0.05 µm.

6.1.3 Plasma Etching Technique

a) Etching Characteristics and Its Accuracy

By application of the plasma-etching technique to the fabrication of VLSIs, the most important problems are the characteristics of the etching and its accuracy.

The etching characteristics are, to designate them concretely, the etch-rate of the material to be etched and the etch-rate ratio between the etched material and another material. What is required first of all, however, is a suitable etch-rate. If this rate is too small, the etching time becomes too long, and such a process cannot be used in the production.

Two kinds of the etch-rate ratio come into consideration for a material to be etched. One is the etch-rate ratio of this material to the underlying material, and the other one is that to the resist. The etch-rate ratio has, in general, some spreading within a wafer, and the thickness of the layer to be etched is not thoroughly constant. Therefore, if it is wanted to etch completely off the layer to be etched in a part where the etch-rate is a little smaller, the layer is overetched in another part with a larger etch-rate. So, the underlying layer is deeply etched in the latter part. If the etch-rate ratio of this layer to the resist is small, on the other hand, the thickness of the layer to be etched is limited by the thickness of the resist, and the accuracy of etching becomes worse as described later. As a conclusion, there is a second requirement that both the two etch-rate ratios of the layer to be etched to the underlying layer and to the resist are high.

There are two kinds of etching accuracy. One is a shift of the pattern size during the etching, namely the difference between a size of the resist pattern before etching and the pattern size of the etched layer after etching. This shift results from two sources, namely, the undercut and the etching of the resist pattern in the lateral direction. Both phenomena depend on the material to be etched, the resist material, the etchant gas, and the etching conditions. Figure 6.5 shows the phenomena schematically, a) the undercut and b) the etching of the resist pattern in the lateral direction. In the actual case, it is general for both phenomena to appear concurrently with each other.

The shift of the discussed pattern size can be taken into account by the design, when it does not vary. In practice, therefore, the deviation in the pattern size of the layer to be etched is a more serious problem, and this

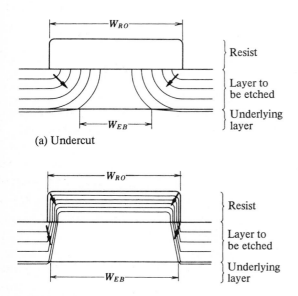

(a) Undercut

(b) Etching of resist pattern in the lateral direction

Fig. 6.5. Schematic illustrations of the shift of the pattern size. (a) Undercut; (b) Etching of resist pattern in the lateral direction. [(W_{RO}) Resist pattern; (W_{EB}) Pattern size of the etched layer after etching]

gives the second kind of the accuracy of etching. The factor which determines this accuracy of etching is not the shift of the pattern size itself, but its deviation depending on the variation of the etching characteristics. In the case of the undercut, as can be understood in Fig.6.5, the small over-etching causes a large variation of W_{EB}, and so a large deviation in the shift of the pattern size. In the case of lateral etching of the resist pattern, as the shift of the pattern size is proportional to the etch rate in the lateral direction and to the over-etching time, the deviation in the shift of the pattern size is limited to a relatively small value.

b) Etching of SiO₂

Etching of SiO_2 is done, as shown in Table 6.4, with using the gas of C_mF_n or C_mF_n added with H as a component. In most cases of pure SiO_2, undercut does not occur. The origin of this fact is considered that the etch-rate of SiO_2 due to the pure chemical reaction is very small, but the chemical reaction proceeds under the assistance of the physical sputtering due to energetic ions and the etch-rate becomes large. The etching technique of SiO_2, therefore, has been investigated mainly to increase, at the same time, both the etch-rate of SiO_2 and the etch-rate ratios of SiO_2 to Si and the resist. The etch-rate ratio of SiO_2 to Si, in general, tends to become larger when the m/n ratio of C_mF_n increases [6.6] or H exists in the plasma [6.7]. The former phenomenon is interpreted by lowering the etch-rate of Si because of C to be remained on Si [6.8] and the latter from the etch-rate of Si becoming smaller because of F-radicals (F^*) which are the main etching component for Si being present in the form of HF.

Figure 6.6 shows, as examples of the etching characteristics, the dependences of the etch-rates and the etch-rate ratios of SiO_2 on the H_2 content in the mixed gas of CF_4 and H_2 [6.9]. With increasing the H_2 content, the etch rate of SiO_2 decreases, but the etch rates of Si and resist decrease more rapidly, so that the etch-rate ratios of SiO_2 both to Si and resist increase. When the H_2 content reaches around 20%, some polymer starts to deposit on the wafer, and the etching becomes unstable above this H_2 content.

Also by covering the electrode with Si, teflon or graphite-consuming F radicals [6.10] or reforming CF_4 with F radicals [6.11,12], it is possible to decrease the number of F radicals and, therefore, to increase the etch-rate ratio of SiO_2 to Si. The similar effect is found with the polymer being previously deposited on the electrode by the C_4F_8 gas plasma, and an etch-rate ratio of about 5 is given with the CF_4 gas plasma [6.13]. Contrary to these phenomena, mixing with O_2 tends to increase the number of F radicals and, therefore, to decrease the etch-rate ratios of SiO_2 to Si and resist.

Table 6.4. Etching characteristics given for the planar-type plasma etching

Material	Gas	Etch rate [Å/min]	Etch rate ratio to other materials	to resist	Reference
SiO_2	CF_4	1000	5 (Si)		[6.11]
	$CF_4 + H_2$	450	35 (Si)	10	[6.74]
		410	12 (Si)	4	[6.9]
	$CF_4 + C_2H_4$	630	4.8 (Si)	4.2	[6.75]
	C_2F_6	900	6 (Si)		[6.12]
	$C_2F_6 + C_2H_4$	800	20 (Si)	8	[6.12]
	C_3F_6	2000	5 (Si)		[6.6]
		750	6.7 (Si)		[6.76]
		750	5.8 (Si)	2.5	[6.75]
	$C_3F_8 + C_2H_4$	700	5.8 (Si)	3.5	[6.75]
	CHF_3	220	11 (Si)	5.5	[6.77]
		450	15 (Poly-Si)	20	[6.78]
		300	20 (Si)	13	[6.14]
	C_4F_8	690	8.6 (Si)	3.2	[6.75]
Si (single)	CCl_4	1000	8 (SiO_2)	1.2	[6.79]
	$CF_4 + O_2$	900	5 (SiO_2)	1.7	[6.79]
	$CF_4 + Cl_2$	430	7 (SiO_2)		[6.80]
Si (poly)	SF_6	4500	>100 (SiO_2)	40	[6.16]
	$SF_6 + He$	1000	20 (SiO_2)		[6.81]
	$CBrF_3$	1200	$6 \sim 8$ (SiO_2)	4	[6.20]
	CCl_2F_2	350	2 (SiO_2)	1	[6.16]
	$CClF_3$	320	2.5 (SiO_2)	1	[6.16]
Al	CCl_4	5000			[6.21]
	$CCl_4 + He$	2000	2 (Si)		[6.23]
	$CCl_4 + O_2$	1500	1.5 (Si)	3	[6.24]
	$CCl_4 + Cl_2$	5000	10 (Si)		[6.26]
	BCl_3	500			[6.82]
	$BCl_3 + O_2$	3000	2 (Si)	2.5	[6.24]
Si_3N_4	C_2F_6	1250	1.4 (SiO_2) 8.3 (Si)		[6.12]
	CF_4	1200	1.7 (SiO_2) 9.2 (Si)		[6.78]
Mo	$CBrF_3$	1000	1 (Si)		[6.20]

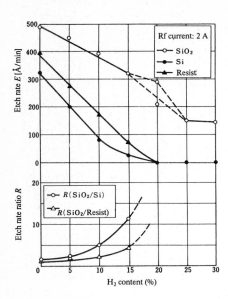

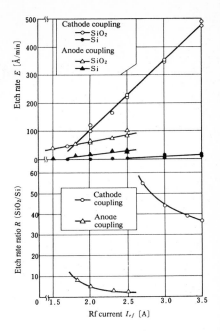

Fig. 6.6. H$_2$ content dependence of etch-rates and etch-rate ratios

Fig. 6.7. Rf power dependence of etching characteristics [6.14]

The etching characteristics of SiO$_2$ depend strongly on the coupling mode, too. This can be interpreted as a result from the physical sputtering due to energetic ions being larger in the cathode coupling mode than in the anode coupling one, because the potential difference between the plasma and the cathode is larger than that between the plasma and the anode. Figure 6.7 shows, as an example, the rf power dependences of the etching characteristics of SiO$_2$ and Si by using the CHF$_3$ gas plasma with two coupling modes [6.14]. As can be seen in this figure, the etch-rate of SiO$_2$ is larger, and the etch-rate ratio of SiO$_2$ to Si is much larger for the cathode coupling mode than for the anode coupling one. That is, for the etching of SiO$_2$, the cathode coupling mode is clearly found to be more superior than the anode coupling one.

The side-wall of the etched pattern of SiO$_2$ is nearly vertical to the surface of the wafer, which is not preferable in many cases of practical device fabrications. The angle between the side-wall and the wafer surface, θ, as is presumed from Fig.6.5b, can be somewhat changed by varying the etch-rate ratio of SiO$_2$ to resist. Figure 6.8 is an example in which the angle θ is changed by varying the gas pressure and thus the etch-rate ratio [6.15]. As

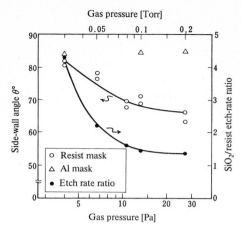

Fig. 6.8. Gas pressure vs. side-wall angle of SiO₂ pattern [6.15]

Fig. 6.9. Cross-sectional view of etched pattern of SiO₂ [6.15] (gas pressure : 0.2 Torr)

the gas pressure of CF_4 mixed with 20% H_2 is increased, the etch-rate ratio of SiO_2 to resist decreases as shown by the symbol ●, which brings about the decrease in the angle θ shown by the symbol o. In this experiment, if Al which is not etched by the above gas is used as the etching mask, the angle θ does not decrease but is kept nearly vertical even with the gas pressure increased, as shown by the symbol Δ. A cross-section of the etched pattern at the gas pressure of 27 Pa is shown in Fig.6.9.

c) Etching of Si and Poly-Si

As shown in Table 6.4, Si and poly-Si are etched by using the gas of CF_4, SF_6 or CF_4 in which some of F are replaced with Cl or Br. Among these gases, CF_4 and SF_6 give a relatively large etch-rate of Si and etch-rate ratios of Si both to SiO_2 and resist. Especially, SF_6 increases the etch-rate ratio. Figure 6.10 is an example of the etching of Si with SF_6 gas and shows the dependence on the gas pressure of the etch-rates of Si, SiO_2 and resist (AZ-1350 J) and the etch-rate ratios of Si to SiO_2 and resist [6.16]. The etch rate of Si, as can be seen in this figure, reaches several thousands of Å/min, and the etch-rate ratios of Si to SiO_2 and resist are in the range of 100 and 40, respectively. CF_4 gives the similar etching characteristics as SF_6 but somewhat smaller values both for the etch-rate and the etch-rate ratios. The addition of O_2 to these gases causes the increase in the etch-rate of Si and in the etch-rate ratio of Si to SiO_2 but the decrease in the etch-rate ratio in Si to resist. In the dependence on the gas pressure, as can be seen in Fig.6.10, a maximum point of the etch-rate of Si exists around 0.1 Torr,

287

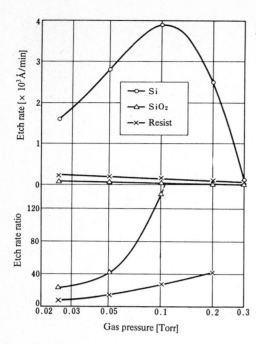

Fig. 6.10. Etching characteristics of Si with SF$_6$ gas [6.16]

and the etch-rate ratios of Si to SiO$_2$ and resist decrease with decreasing the gas pressure. The coupling-mode dependence is small for the etching of Si with SF$_6$ and CF$_4$, namely, the etching characteristics of Si are not so much different from each other between the cathode and the anode coupling mode, except for the etch-rate of Si having the tendency to be larger with the anode coupling mode than the cathode coupling mode. The etching characteristics depend considerably on the frequency of the rf power. Both the etch-rate of Si and the etch-rate ratios are larger with 13.65 MHz than with 400 kHz. The etch-rate of Si given by using these gases is decided under the condition of the supply limited in most cases and, therefore, the loading effect is usually observed [6.17].

In the case of the etching with SF$_6$ or CF$_4$ under the etching condition generally used, the chemical reactions of Si with F radicals play a main role, and the undercut appears. However, as the etching of Si can occur also with ions like CF$_3^+$, an etching having a small undercut can be expected if the condition is found under which the reactions with ions are dominant. One method to realize an etching with a small undercut is to decrease the gas pressure so that the chemical reactions with ions become dominant. Another method is to decrease the concentration of F radicals by using the loading effect, etc. These phenomena have been observed experimentally as follows. In the case of the etching with CF_4, the undercut vanishes at the gas pressure below 40 mTorr. It also vanishes at 50 mTorr where the undercut is normally observed if the surface of the electrode is covered with Si, and appears again by adding O_2 [6.18]. In addition to this, another paper reported that the undercut became smaller if very low frequency of 8 kHz was used for the rf power [6.19].

On the other hand, the undercut becomes generally harder to appear when a gas is used containing Cl or Br. It has been reported that the etching of Si with a pattern width of 1 μm and an etched depth of 3 μm was possible by using $CBrF_3$ gas [6.20]. With these gases, however, there are some problems in that the etch-rate of Si is not so large. Therefore, the etch-rate ratios of Si to SiO_2 and resist are relatively small, namely around 5, and the contamination on the inner surface of the equipment is much more intense compared to SF_6 or CF_4 gases.

In the case of the etching of Si and poly-Si, as mentioned above, the undercut tends to appear if the gas and the etching condition giving good etching characteristics are used. Contrary to this, gas and etching conditions resulting in no undercut tend not to show good etching characteristics. For the time being, therefore, the only way is to determine the optimum condition depending on the purpose after gathering many factors and characteristics related to the etching.

d) Etching of Al

CCl_4 or BCl_3 is generally used for etching Al as can be seen in Table 6.4. In addition to these two, the etching can be done with Cl_2, Br_2, HCl or HBr, and etch rates from 1000 to 1700 Å/min have been reported [6.21]. These gases, however, have some difficult problems and have been little investigated about their etching characteristics.

In the plasma etching of Al, the slow-start phenomenon is usually found. That means that the etching does not start at the moment of applying the rf power but some periods later. Figure 6.11 shows one example where the origin of the abscissa is taken at the moment when the rf power was turned on. It can easily be understood from this figure that the etching process starts after some periods from the origin [6.22]. This delay is considered to be a period during which the natural oxide, Al_2O_3 layer, formed on the surface of Al, is sputtered off by ions like CCl_3^+. This delay tends to become shorter with the rf power increased, but it differs from sample to sample and depends on the equipment. This phenomenon is one of the serious problems in the etching of Al. To overcome it, the etching monitor, which, for example, detects the intensity change of a plasma emission at a certain wavelength, as shown

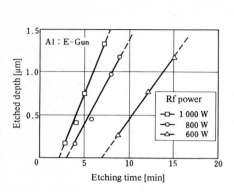

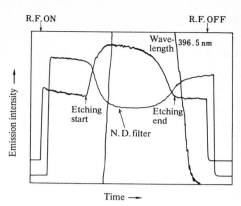

Fig. 6.11. Etched depth vs. etching time for Al [6.22] (CCl_4 + He)

Fig. 6.12. Variation of emission intensity during etching of Al (CCl_4)

in Fig.6.12, is needed for the etching of Al. The time delay between the rf power turn-on and the start of the etching can also clearly be seen in this figure.

In the case of the etching of Al with a CCl_4 or BCl_3 gas plasma, the etch-rate of Al and the etch-rate ratio of Al to SiO_2 is large enough, but the etch-rate ratios to Si and resist are usually insufficient. There are also some problems because of the reproducibility of the etching, which is not so good. Sometimes, residues are found on the wafers, and very often the inside of the equipment is heavily contaminated. As one of the ways of solving these problems, it has been tried to mix various gases.

The mixing of CCl_4 with He or Ar slightly reduced the etch-rate and the degradation of resist and the residue but did not show any remarkable improvements [6.21-23]. The mixing of BCl_3 with O_2 increased the etch-rate of Al by five to ten times, and the mixing with Cl_2 increased it by about five times [6.24,25].

It was also found that the etch-rate of Al increased by about six times, and the etch-rate ratios of Al to SiO_2 and Si achieved almost one hundred and ten, respectively, when Cl_2 was added to CCl_4 [6.26]. In addition to this, it was reported that the mixing of CCl_4 with Cl_2 made the period of delay shorter in the slow-start phenomenon and zero under a certain condition as well as suppressed both the residue and the contamination on the inside of the equipment. Al has been considered to be etched by being scavenged in a compound of $AlCl_3$. In this case, it is presumed that an unsaturated compound like C_2Cl_2 is formed due to the residual C in the plasma and causes the residue and contamination.

The addition of Cl_2 was explained to return the unsaturated compound which will change into complex compounds to CCl_4, which suppresses the residue and the contamination [6.26].

A shift of the pattern size in the etching of Al does not occur in such a simple manner as that shown in Fig.6.5. An undercut is not usually found, but a deposit is found on the side-wall of the pattern being etched, like the re-deposition of the material etched by the physical sputtering in the ion etching [6.22]. In the etching of Al, however, the chemical reactions with the neutral radicals (Cl^*, for instance) are considered to play a main role, from the facts that the undercut comes out in some cases [6.24] and that the etch-rate is as large as several to ten thousands $\overset{\circ}{A}$/min. Based on these facts and considerations, the etching of Al having no undercut can be explained as fol-lows: The redeposition of the material being etched occurs on the side-wall of the etched pattern due to the physical sputtering simultaneously with the etching proceedings, and the redeposited material on the side-wall prevents the undercut.

6.1.4 Future Prospects

The planar-type plasma etching technology has been investigated rigorously, and its characteristics and mechanisms have beomce considerably clear. The results of these investigations have also improved the equipments, and the equipments now being fit for the practical use have appeared on the market. However, this technology seems not to have developed enough that it can be used without problems in mass production lines. It would determine the future of this technology whether there is an ability to materialize the equipment having both high reliability and high throughput or not.

6.2 Beam Annealing

Production of VLSI circuits needs the installment of impurities in a high concentration into thin layers on the silicon surface according to fine and dense device patterns. The ion implantation is being used for such high-grade impurity dopings. In as-implanted crystals, however, the atomic arrangement is distorted in a high degree and impurity atoms are not necessarily located in the substitutional sites in the Si crystal. A furnace annealing has been utilized to repair these defects and set impurities in the proper lattice sites to act as donors or acceptors. When the annealing time is exceedingly long, however, the impurities diffuse in lateral and longitudinal directions and the precision of device patterns is deteriorated. The annealing of a silicon surface by transient heating with lasers, flash lamps or electron beams is therefore being vigorously studied.

6.2.1 Laser Annealing in VLSI Technology

Laser annealing has exhibited an explosive increase of research activities since 1978, triggered by a re-evaluation by Italian scientists of pioneer works initiated in USSR in 1974 [6.27]. The technological background of such a fever is that the laser annealing is characterized not only by the repair of implantation damages but also by the rapid crystal growth, versatile silicide formation and various non-equilibrium effects.

This section chiefly deals with the laser annealing and introduces its basic features [6.28] and effects on the VLSI process. Table 6.5 shows some typical examples of the laser annealing equipments though they are still in a developing stage.

a) Laser Annealing of Ion-Implanted Crystals

Figure 6.13a shows the distribution of implanted As in Si after the annealing by a CW laser [6.29]. In this case, the sample temperature does not exceed the melting point. It was checked that the crystalline perfection has been re-

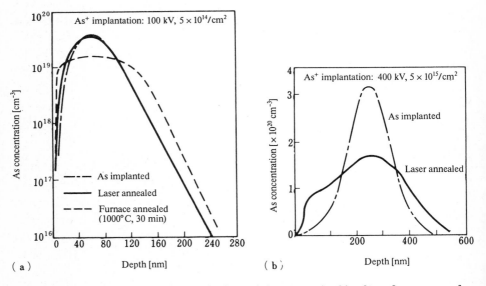

(a) Depth [nm] (b) Depth [nm]

Fig. 6.13. a,b. Distribution of implanted As atoms in Si after laser annealing. (a) CW Ar laser annealing (3.8×10^5 W/cm², 1.4 ms), (b) Q-switched ruby laser annealing (6×10^7 W/cm², 50 ns)

Table 6.5. Specifications of typical laser annealing equipment

Type of lasers	Nd : YAG (Q-switched)	Ar (CW)
Wavelength [µm]	1.06 0.53	0.488
Output	0.15 mJ/pulse	16 W
Mode	TEM_{00}	TEM_{00}
Beam diameter [µm]	100 - 400	25-500
Repetition	2 - 3 kHz	(CW)
Pulse width	100 ns	Beam diameter/scanning speed
Scanning method	Beam scanning (X and Y direction)	Beam scanning or stage movement
Scanning speed [cm/s]	0 - 10	0 - 500 (beam scanning) 0 - 10 (stage movement)

stored and the As atoms have all been activated. Nevertheless, the distribution of As after the laser annealing (solid line) did not change from as-implanted distribution (chain-dot line) which is quite contrary to the distribution after the furnace annealing (broken line). Below the melting point, the diffusion coefficient D of As is at the largest 10^{-14} cm^2/s which results in a diffusion length $\sqrt{Dt}$ of only 0.5 Å during the annealing time (t = 3 ms in this CW case). On the contrary, as shown in Fig.6.13b [6.30] when the Si surface was melted by a Q-switched ruby laser, the As distribution is leveled because D increases to 10^{-4} cm^2/s in molten Si which results in $\sqrt{Dt} = 840$ Å for even shorter t in the pulsed laser annealing (t = 700 ns).

Another feature of the laser annealing is that impurities with the concentration exceeding the solubility limit can be incorporated in the substitutional sites in Si crystal lattices without segregation. This is because the non-equilibrium atomic configuration is frozen in the laser annealing due to its extraordinary rapid cooling rate. This meets the requirement for the VLSI process that the impurities must be doped with higher concentration in a thinner region than the furnace annealing can realize.

b) Laser Annealing for the Self-Alignment Technology

The introduction of the ion implantation in MOS transistor fabrication realized the self-alignment of source and drain regions against the gate. The laser annealing has brought here the realization of much finer patterning without lateral diffusion of implanted impurities. Experimental results have been reported for both poly-Si and Al gates [6.31]. In both cases the gate electrode

shields the channel region from the laser beam and only source and drain regions are irradiated.

c) Crystal Growth by Laser Irradiation

Epitaxial Growth on the Crystalline Substrate. When the ion dose is excessive
in the ion implantation, the Si surface is heavily damaged to almost amorphous
state. The recovery of crystalline perfection by the laser annealing can therefore be regarded as a kind of an epitaxial crystal growth. When the laser annealing involves the surface melting of Si, a rapid crystal growth from the
melt occurs with a growth rate of a few meters/s which is extraordinary high
among the present crystal growing methods (at the highest, a few cm/s in the
ribbon crystal growth). When the laser irradiation does not involve the surface melting, the solid phase epitaxial growth occurs with a growth rate in
the order of 0.1 mm/s.

Following the studies in the crystal growth on single crystal Si substrates,
the laser annealing has been applied to recover the perfection of Si thin films
on sapphire substrates (SOS). It has been reported in this case that the recovery is most perfect when the interfacial layer between Si and sapphire,
which contains many crystalline defects due to lattice mismatch, has been once
made amorphous by the ion implantation and then regrown by the laser annealing.
Figure 6.14 shows the experimental results made by the collaboration of VLSI
Cooperative Laboratories and Electrotechnical Laboratory. By the proper choice
of the energy of implanted ions and the laser power, a higher mobility than
that obtained in the furnace annealing can be achieved [6.32].

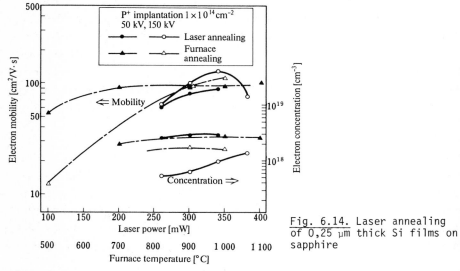

Fig. 6.14. Laser annealing of 0.25 μm thick Si films on sapphire

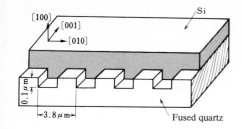

Fig. 6.15. "Graphoepitaxy", the growth of thin, crystalline Si film on the grooved amorphous substrate

Crystal Growth on the Amorphous Substrate. One of the notable features of the laser annealing is that it enables the growth of Si single crystals on an amorphous substrate such as fused quartz, thermal oxide and silicon nitride. H.I. Smith and his collaborators in MIT used fused quartz substrate with grooves on its surface as shown in Fig.6.15. When an amorphous Si layer was deposited with the thickness of about 0.5 μm on it and annealed by the CW Ar laser, a 1.5 mm square single crystal region with the surface orientation (100) has been obtained [6.33]. They call this technology "graphoepitaxy". J.F. Gibbons and his collaborators at Stanford University deposited 0.5 μm thick poly-Si films on a silicon nitride (Si_3N_4) film without any artificial structure. Annealing it with the CW Ar laser, they obtained a single crystal region with the length 30 μm along the laser scanning direction and the width 2 μm. They fabricated an MOS transistor within this region and evaluated its characteristics [6.34]. Though the detailed mechanism of the crystal growth is not known for either case, the extension of these technologies suggests a possibility of stacking single crystalline semiconducting layers and amorphous insulating layers alternately, thus leading to the three-dimensional LSI.

Single crystalline Si layers can also be obtained by laser annealing of poly-Si deposited on the thermal oxide which has through-holes opening in some places allowing direct contact of poly-Si to single Si substrate. Crystal growth started at these through-holes continues, following the laser beam scanning, to the poly-Si layer on the oxide and the entire region becomes single crystalline [6.35]. This technique has been named "bridging epitaxy".

d) Laser Annealing and Interfacial Reactions

Metal–Si Interface. Laser annealing can also be applied to form silicides being used as contact materials in LSIs. Silicides are created when the Si surface is irradiated by the laser beam after thin films of Pt, Pd, Ni, Nb, etc. have been vapour-deposited upon it. In the case of Q-switched Nd : YAG laser irradiation, a layered structure with different phases is formed [6.36]. When a CW Ar laser is used, on the other hand, a single phase silicide layer is formed and its phase type can be selected through the choice of the laser power [6.37]. For example, Pd_2Si is originated for the low power level and PdSi is originated for the high power level. In Pt-Si systems, Pt_2Si_3 is developed in a certain condition which is unstable in thermal equilibrium and is a superconductor with the critical temperature 4 K [6.37].

Si-SiO$_2$ Interface. The Si-SiO$_2$ interface, carefully fabricated by thermal oxidation, is the cleanest and most perfect interface. By a powerful laser irradiation it tends to degrade. It has been found, however, that the radiation damage caused by the electron beam exposure can be annealed by laser irradiation [6.38]. As shown in Fig.6.16, interface charges and states created by the electron beam irradiation can be completely annihilated by laser annealing with the proper choice of the laser power. Electronic traps within SiO_2, however, could not be completely diminished by laser annealing. Taking into consideration the temperature estimation in Sect.6.2.2 that the temperature inside the SiO_2-layer is sufficiently raised to 96% of the Si surface temperature, this suggests that the rearrangement in the SiO_2 network occurs slower than in Si crystal lattices.

6.2.2 Fundamentals of Laser Annealing

a) Temperature Distribution and Variation

The temperature in an Si sample being irradiated by a laser beam can be calculated as a one-dimensional problem if the depth in the sample is sufficiently smaller than the beam diameter. Usually, this assumption is valid because the former is less than 5 μm while the latter is about 50 μm or more. The

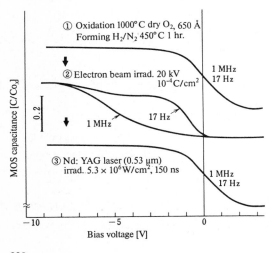

Fig. 6.16. Annihilation of electron-beam induced charges and states at Si/SiO$_2$ interface by laser annealing

sample can then be regarded as a semi-infinite medium and its temperature is obtained by solving the heat diffusion equation [6.39]

$$K \frac{\partial^2 T(z,t)}{\partial z^2} - \rho c \frac{\partial T(z,t)}{\partial t} = -\alpha F(t) e^{-\alpha z} \quad , \tag{6.1}$$

where z is the coordinate directing normally into the sample, t is the time, F is the laser power density incident on the sample (component reflected from the surface is subtracted), K is the thermal conductivity, ρ is the density, c is the specific heat, and α is the optical absorption coefficient of Si, respectively. These physical parameters are assumed to be independent of the temperature. If the sample surface is taken as $z = 0$ and the ambient temperature as $T = 0$, the initial and the boundary conditions are

$$T(z,0) = 0 \quad \text{at} \quad t = 0 \quad \text{and} \tag{6.2}$$

$$K \frac{\partial T}{\partial z}\bigg|_{z=0} = 0 \quad \text{at} \quad z = 0 \quad , \quad T(\infty,t) \to 0 \quad \text{at } z \to \infty \quad . \tag{6.3}$$

The first equation in (6.3) represents the assumption that the heat dissipation from the surface into the ambient due to the radiation and convection can be neglected compared to the conduction into the sample.

When the waveform of the laser pulse is approximated as a square with the width t_0 and the height F_0, the solution of (6.1) becomes, during the laser irradiation ($t \le t_0$),

$$T_1(z,t) = \frac{2F_0}{K} \sqrt{\frac{\kappa t}{\pi}} e^{-z^2/4\kappa t} - \frac{F_0 z}{K} \operatorname{erfc}\left(\frac{z}{2\sqrt{\kappa t}}\right) - \frac{F_0 e^{-\alpha z}}{\alpha K}$$

$$+ \frac{F_0}{2\alpha K} e^{\alpha^2 \kappa t} \left[e^{\alpha z} \operatorname{erfc}\left(\alpha\sqrt{\kappa t} + \frac{z}{2\sqrt{\kappa t}}\right) + e^{-\alpha z} \operatorname{erfc}\left(\alpha\sqrt{\kappa t} - \frac{z}{2\sqrt{\kappa t}}\right) \right] \tag{6.4}$$

and after the termination of the laser pulse ($t > t_0$),

$$T_2(z,t) = T_1(z,t) - \frac{\alpha \kappa F_0}{2K} \int_0^{t-t_0} e^{\alpha^2 \kappa t'} \left[\operatorname{erfc}\left(\alpha\sqrt{\kappa t'} + \frac{z}{2\sqrt{\kappa t'}}\right) \right.$$

$$\left. + e^{-\alpha z} \operatorname{erfc}\left(\alpha\sqrt{\kappa t'} - \frac{z}{2\sqrt{\kappa t'}}\right) \right] dt' \tag{6.5}$$

where $\kappa = K/(\rho c)$ is the thermal diffusion coefficient and erfc(x) is the complementary error function defined by

$$\text{erfc}(x) = \frac{2}{\sqrt{\pi}} \int_{x}^{\infty} e^{-\xi^2} d\xi. \tag{6.6}$$

When the absorption coefficient for the laser wavelength is sufficiently large ($\alpha\sqrt{\kappa t} \gg 1$), the surface temperature can be approximated as

$$T(0,t) = T_m(\sqrt{t/t_0} - \sqrt{t/t_0 - 1}) \quad , \tag{6.7}$$

(we take only the first term for $t \leq t_0$),

where

$$T_m = 2/\sqrt{\pi}[F_0 t_0/(\rho c \sqrt{\kappa t_0})] \quad . \tag{6.8}$$

T_m is the maximum temperature the surface reached at the termination of the laser pulse as shown in Fig.6.17b. Figure 6.17c shows the temperature distribution in the sample at $t = t_0$ for the laser wavelength 0.53 μm. In this figure, the temperature distribution when 0.1 μm thick oxide covers the Si surface is also shown by the broken line.

When the temperature dependence of the physical parameters or the exact pulse form of the laser is taken into account, a numerical calculation with

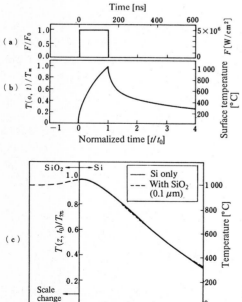

Fig. 6.17. Temporal change and spatial distribution of the temperature in Si crystal during laser annealing

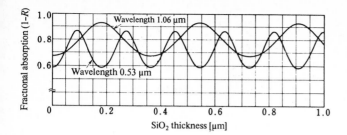

Fig. 6.18. Changes of the fractional absorption of laser light in Si due to
the variation of the oxide thickness

the computer must be carried out. For instance, when the decrease of the
thermal conductivity with the temperature rise and the increase of the free
carrier absorption coefficient due to the thermal generation of carriers are
taken into consideration, the temperature rises more rapidly than the approximated solution [6.40].

As shown by the broken line in Fig.6.17c, SiO_2 layers with the thickness
0.1 μm have little effect on the temperature distribution itself. The effect
of an SiO_2 layer appears in that it acts as an optical interference film and
the fraction of the absorbed laser power varies with the thickness of SiO_2
as shown in Fig.6.18.

b) Physical Process Under Laser Irradiation

Calculation of the temperature stated above assumes that the absorbed laser
energy is instantly converted to the thermal energy, i.e., the energy of the
excited electrons is transferred to the lattice system within several pico-
seconds. By this model, it is understood that ordinary melting of Si occurs
when the temperature calculated above exceeds the melting point.

Contrary to this, a new model has been presented in which the high-density
free electrons and holes excited by the laser form a plasma with a lifetime
0.1 - 1 μs and the apparent "melting" state is the result of shielding of the
bonding of Si valence electrons by this plasma, thus causing free movement
of Si atoms [6.41].

To evaluate the validity of these two models, an observation of dynamical
processes under laser irradiation is necessary. The measurement of surface
reflection was made by using a probing He-Ne laser beam and the increase of
reflection accompanying "melting" has been observed as shown in Fig.6.19
[6.42]. This dynamical measurement must be made more precisely not only for
the understanding of physical processes but also for the realization of practical, controllable LSI process equipments.

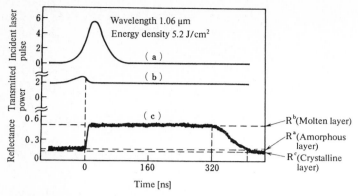

Fig. 6.19. Measurement of light reflected from Si surface during laser annealing. (a) Incident laser pulse, (b) transmitted laser light, (c) reflectance of He-Ne laser light (0.63 μm)

6.2.3 Other Beam Annealing Technologies

The laser annealing takes a comparatively long time to anneal the entire wafer because it scans the wafer by the beam with a diameter less than 100 μm. Contrary to this, the use of high-power flash lamps or arc lamps may have a practical merit because they can anneal a wider area in one shot [6.43]. The power density of these light sources, however, is smaller than that of lasers and the irradiation time per unit area on the wafer becomes longer. Therefore, the penetration depth of the heat does not become negligible compared to the wafer thickness and the assumption made to get the solution of (6.4) is no longer valid. The author solved (6.1) under the boundary condition of finite wafer thickness for various light sources.

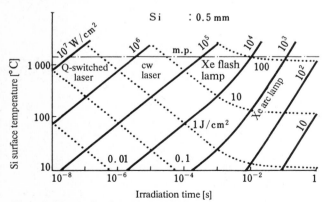

Fig. 6.20. Calculation of the temperature rise at the surface of 0.5 mm thick Si wafer for various light sources (in the case of strongly absorbing emission)

The temporal variation of the surface temperature $T(0,t)$ is shown in Fig.6.20. In this figure, each solid line represents the surface temperature rise under the light irradiation with specified power density and the dotted lines connect the points with equal energy densities. When the irradiation time is short, (6.4) is valid and $T(0,t) \propto \sqrt{t}$ as in the laser annealing. As t increases, $T(0,t)$ tends to increase as $T(0,t) \propto t$. In this region, the features of transient heating as shown for the laser annealing are comparatively weakened.

Large-diameter pulsed electron beams with an acceleration voltage of 100 kV, a current of 400 kA and a pulse width of 50 ns have been used to anneal the entire wafer in one shot [6.44]. A scanning electron beam, with an acceleration voltage of 20 kV and a beam current of 60 μA, can draw the pattern with higher speed than the laser scanning [6.45]. The difference between this electron beam and laser annealing is that the penetration depth of the electron beam is not so much different for various materials and is uniquely determined by the acceleration voltage. Damages created in the sample by an electron beam are under investigation [6.45].

Annealing has also been tried by using a pulsed proton beam with the energy 180 kV and the current $18 - 380$ A/cm^2 [6.46].

6.3 Thin-Film Deposition Techniques

Various metals and insulator films are applied in the production of integrated cirucits, for which film deposition techniques have been developed based on various principles. Since extremely fine lines are used for pattern generation, improvement on film quality, such as composition, defect density and thickness uniformity, is necessary for film deposition methods. For this purpose, reaction between films and substrates, temperature rise during deposition, and radiation damage must be studied in relation to the film and device characteristics. Moreover, the number of wafers that can be processed in a given time must be increased to lower production costs.

This section explains the chemical vapour deposition (CVD) in which a film is formed through a chemical reaction and the evaporation in which a film is formed through a physical reaction, together with their principles and characteristics; however, these two methods cannot be clearly distinguished. For example, both principles are used in the reactive sputter deposition method [6.47].

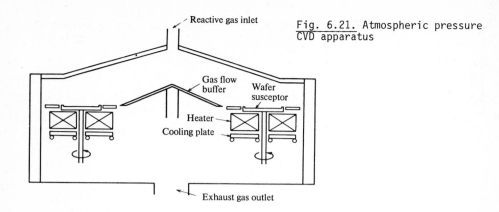

Fig. 6.21. Atmospheric pressure CVD apparatus

6.3.1 CVD Technique

The CVD technique is used to form a poly-silicon film or various insulator films on the substrate surface through thermal decomposition of source materials in the vapour phase. Figure 6.21 shows a typical atmospheric pressure CVD apparatus. The reactive gas is supplied via the gas inlet to form films on the surfaces of wafers arranged horizontally on the susceptor. To get films of a uniform thickness, the reactive gas must be homogeneously supplied and the temperature must be uniformly distributed over the wafer surfaces. For this purpose, the CVD apparatus has a mechanism to turn a wafer around its own axis and around an axis outside the wafer.

The susceptor shape must also be improved. Figure 6.22 shows how film-thickness uniformity is affected by the wafer susceptor shape in forming SiO_2 films. In this example, SiH_4 and O_2 are used as the reactive gas and N_2 is used as the carrier gas (wafers are heated indirectly via the susceptor). A cavity in the susceptor (Fig.6.22) enables uniform temperature distribution on wafers and improves film-thickness uniformity; however, since the wafer temperature changes slightly because of unsatisfactory snugness between wafers and the susceptor, a good reproductivity is not always obtained. Furthermore, in this method the number of wafers per batch is reduced with increased wafer diameter. Because of these defects the low-pressure CVD, which will be explained later, is mainly used today.

Nevertheless, the application of atmospheric-pressure CVD will continue due to the following reasons: the film deposition rate is higher than with other methods, it can be used to deposit films in various applications with some necessary modifications of component devices, and the CVD apparatus is relatively small and inexpensive.

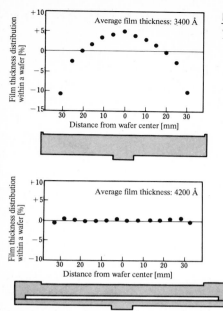

Fig. 6.22. Wafer susceptor cross section and 75 mm wafer film thickness distribution

Figure 6.23 shows a low pressure CVD apparatus [6.48]. An ordinary diffusion furnace is used as the reaction furnace in which wafers are placed
in a back-to-back configuration several millimeters apart. This means that
about 200 wafers can be processed regardless of the wafer diameter. Such an
apparatus is ideal for a high productivity. The wafer arrangement is realized
because the gas concentration in the reaction furnace becomes uniform due to
the high diffusion speed of the reaction gas whose molecules under low pressure (ordinarily 0.5 - 1.0 Torr) have a mean free path of 1000 - 1500 times that
under atmospheric pressure (760 Torr). The wafer heating system can heat the
wafers directly, realizing a uniform temperature distribution in each wafer,
as well as between the wafers, thereby providing good reproductivity. More-

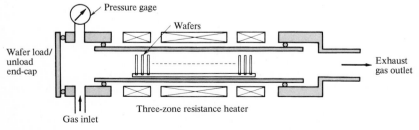

Fig. 6.23. Low pressure CVD apparatus

over, the film step coverage on the wafer surface is greatly improved by reaction gas molecules uniformly distributed on the steps. Because of its advantageous features, the low-pressure CVD technique is widely used to deposit
poly-Si and Si_3N_4 films, but there are problems to be solved for other films.
In forming PSG films for IC surface passivation, it is known that film-thickness uniformity is greatly affected by the relationships between the inner
furnace diameter and the wafer diameter as well as the distance between wafers
Attempts have been made to improve the film thickness uniformity by optimizing
these relationships, but the resultant technique still cannot be used in actua
production. In order to form a poly-silicon film containing impurities such as
P and As, a small volume of gases, such as PH_3 and AsH_3, is supplied in the reaction furnace, together with SiH_4 gas. These gases change the film growth rate
remarkably and degrade the film-thickness uniformity. Figure 6.24 shows the relationship between the deposition rate of poly-silicon film and PH_3 concentration. It can be considered that such a tendency occurs because the SiH_4 decomposition reaction on the wafer surface is greatly influenced by small volumes
of PH_3 [6.49] and because PH_3 gases are not sufficiently uniform in the reaction furnace. This means that the low-pressure CVD apparatus (Fig.6.23) is not
yet adequate when accurate reaction gas concentration uniformity is necessary
for the reaction furnace. In this case, improvement is necessary for geometric
relationships between the reaction furnace and wafers, reaction gas supplying
technique and so on.

Unlike an evaporation technique, since thermal decomposition of the reaction gas is used in the CVD technique, a high temperature is necessary for the
film deposition process (600 - 650°C for a poly-silicon film and 750 - 800°C for
an Si_3N_4 film). This causes a problem if there is a temperature limit for the
device fabrication process. For example, an Si_3N_4 film has good characteristics

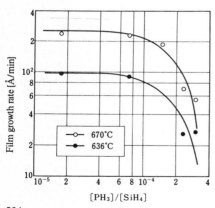

Fig. 6.24. Relationships between deposition rate of poly-silicon film and
PH_3 concentration

Table 6.6. Si$_3$N$_4$ film deposition conditions by various CVD techniques

	Atmospheric pressure CVD	Low pressure CVD	Plasma CVD
Reaction pressure	760 Torr	1 Torr	1 Torr
Reaction temperature	~800 °C	~800 °C	~300 °C
Throughput	10 wafers/lot	200 wafers/lot	10 wafers/lot
Reaction gas source	SiH$_4$ + NH$_3$	SiH$_4$ + NH$_3$ or SiH$_2$Cl$_2$ + NH$_3$	SiH$_4$ + N$_2$ or SiH$_4$ + NH$_3$
Deposition rate	~150 Å/min	~40 Å/min	~300 Å/min
Film thickness uniformity	± 10%	± 5%	± 10%
Film composition	stoichiometric (Si$_3$N$_4$)	stoichiometric (Si$_3$N$_4$)	nonstoichiometric (Si$_x$N$_y$)

as a surface passivation film, but the CVD technique cannot be used to form this film. This can be overcome by using the plasma CVD technique in which a film is formed promoting reaction gas decomposition reaction by plasma reaction at a low temperature. The Si$_3$N$_4$ film arranged by the plasma CVD technique has defects such as a chemical composition different from that of films made by other CVD techniques, and reduced stability and reproductivity; however, since the film can be deposited at a low temperature (300 - 350°C) and its film stress appears as compression, that is, the film is stable against cracks, this technique will be improved and used in the future. Table 6.6 gives comparisons of film deposition conditions for forming an Si$_3$N$_4$ film by atmospheric pressure, low pressure, and plasma CVD techniques.

6.3.2 Evaporation Technique

The evaporation technique is used to precipitate a film on a substrate surface through vaporization of materials, a physical reaction. This method is mainly applied to build up metal films. Vacuum evaporation [6.50] and sputter deposition [6.47] are widely used, and there is an ion plating technique [6.47] which is being developed for actual applications.

In vacuum evaporation, a metal target is heated to the melting point in a 10^{-7} to 10^{-6} Torr vacuum. The evaporation metal is sublimated or vaporized and resultant atoms are deposited on substrate surfaces held at a low temperature. The metal target is melted by various methods, such as resistive heating, rf inductive heating, and electron beam heating. The resistive and rf inductive heating systems can be installed at relatively low cost and applied to evaporation of metal films having relatively low melting points; however, impurities such as K and Na from the filament and crucible contaminate the deposited film. In the electron beam method, the evaporation source is directly heated by electron beams, thus the requirement to evaporate a highly pure metal film is fullfilled. Moreover, this method can be applied to the deposition of metal films consisting of Mo and W with high melting points. In this case, the substrate is damaged by electron beam and/or secondary X-ray, however, there exists no problem in practice because of the damage to be annealed after evaporation by a heat treatment at such a low temperature as 200 - 400°C [6.51]. An alloy film cannot be formed easily by vacuum evaporation because of the difference among metal vapour pressures. In flash evaporation, however, the film deposited on the substrate surface has the same chemical composition as the evaporation source [6.52].

Sputter evaporation is ordinarily applied for the production of alloy films. In this method, Ar plasma is generated by an electric field between an anode and a cathode (target) in an Ar gas at 10^{-3} to 10^{-2} Torr. The target material is sputtered by Ar ions, thereby growing an alloy film on the substrate surface. In this case, the alloy target and the film have nearly the same chemical composition; however, in this method, the film growth rate is low. Moreover, since the substrate is placed in the plasma, it is subjected to radiation damages. To remedy these problems, a magnetic field is applied orthogonal to the electric field to enclose the Ar plasma in the vicinity of the target, being used in high-speed sputter evaporation. Furthermore, the deposition rate is improved by the realization of high plasma density. In this method, the substrate is placed outside the plasma, and so radiation damages are almost negligible. High-speed sputter evaporation is applied today to deposit films of Al alloys such as Al-Si, Al-Cu, and Al-Si-Cu; metals such as Mo and W having high melting points; and silicides such as $MoSi_2$ and WSi_2. Target materials of various shapes are referred to as a planar magnetron, a sputter gun (S-gun), and a coaxial magnetron in this sputter process. Sputtered atoms or molecules are scattered under a vacuum of 10^{-3} to 10^{-2} Torr before they reach to the substrate surfaces, thus resulting in the good step coverages. Such a coverage is also based on the large areas of target materials. Unlike the sputter evaporation method, the vacuum evaporation method has a point evaporation source, emitting metal atoms or molecules. Furthermore, the sputter evaporation method allows the following: First, the substrate surfaces are slightly etched in a sputtering chamber applied with a reverse bias, then metal films can be deposited on substrate surfaces successively. This advantageous point can be used effectively, for example, in an Al multilayer interconnection in which a thin Al_2O_3 film

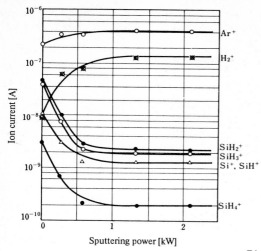

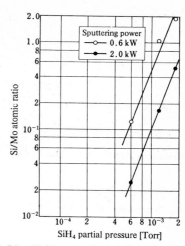

Fig. 6.25. Decomposed SiH_4 as a function of sputtering power. SiH_4 partial pressure is 6×10^{-4} Torr

Fig. 6.26. Si/Mo atomic ratio of the film as a function of SiH_4 partial pressure using sputtering power as a parameter

can be etched after deposition of the first Al layer, thus good electric contact is obtained between the first and the second Al layers.

Another interesting deposition technique is the reactive sputtering technique, in which gases such as O_2, N_2 and CH_4 are supplied in Ar plasma, then they are dissociated to deposit insulators such as oxides, nitrides, and carbides having high melting points. In the following example, Mo is applied as the target material and SiH_4 gas as the reaction gas to form a molybdenum silicide film [6.53] which has been developed as a material for the gate electrode of MOS LSI. Figure 6.25 shows the sputtering power dependence of product species dissociated by Ar plasma. The signals correspond to the ion currents detected by the mass spectrometer in the sputtering chamber. Below a sputtering power of 0.6 kW, the H_2 ion current increases by an order of magnitude along with the increase of the sputtering power. Conversely, the SiH_n ($n = 1$ to 4) ion current decreases by an order of magnitude. All ion currents saturate at a sputtering power of 0.6 kW. Figure 6.26 shows the relationships between the Si/Mo atomic ratio in a formed film and film deposition conditions. It is considered that Si which is created by the dissociation of SiH_4 in the Ar plasma is adsorbed on the surface of the target, then sputtered with Mo and deposited on the substrate [6.53-55]. Figure 6.27 shows the results of X-ray diffraction analysis ($28° \leq 2\theta \leq 51°$) for silicide films of various Si/Mo atomic ratios. Silicide films are found to have chemical compositions such as Mo_3Si, Mo_5Si_3,

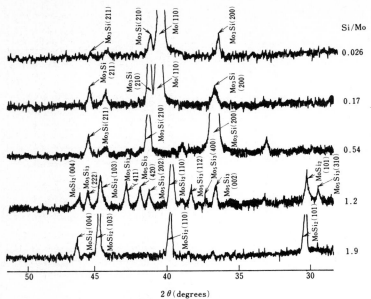

Fig. 6.27. X-ray diffractometer traces for the silicide films with various Si/Mo atomic ratios. The films are annealed for 30 min at 1200°C in N_2 atmosphere

and $MoSi_2$. With the reactive sputtering technique films having arbitrary chemical compositions, crystal structure, and electric resistance can be easily deposited by controlling partial pressures of reactive gases and sputtering conditions.

6.4 Metallization

Metallization technology becomes one of the most important technologies for fabricating LSI. It mainly means the process method for interconnecting the elements to each other on the silicon chip with a metal layer.

The following items must be considered for selecting materials from the point of fine patterning and electrical performance:

1) Electrical resistivity (to be low);
2) Adhesion to silicon or silicon dioxide (to be strong);
3) Ohmic contact resistance (to be low);
4) Reliability (against electromigration);

Especially for VLSI processes:

5) Refractory metal on the behalf of poly silicon;
6) Metal suited to submicron pattern formation.

Table 6.7. Properties of metals for metallization

Property \ Metal	Ag	Au	Al	W	Mo	Ni	Pt	Cr	Ti	Si	Poly Si	SiO$_2$	WSi$_2$ MoSi$_2$
Resistivity [$\cdot 10^{-6}$ $\Omega\cdot$cm]	1.59	2.3	2.7	5.5	5.7	6.8	9.8	12.9	5.5	-	1000		50 - 60
Melting point [°C]	960	1063	660	3410	2625	1455	1779	1890	1820	1430	1430	1710	
Thermal expansion [$\cdot 10^{-6}$/°C]	19.2	14.1	23.2	4.5	5.0	12.7	8.9	8.4	8.5	2.5	2.5	0.35	
Thermal conductivity [W/cm °C]	4.2	3.1	2.4	1.7	1.4	0.9	0.69	0.87	0.20	1.5	1.5	0.14	
Oxide formation energy [kc]	-0.59	+39.0	-376.7	-182.5	-161.0	-51.7		-250.0	-204.0	-192.4			
Adherence to SiO$_2$	weak	weak	very strong	strong	strong	strong	weak	very strong	strong		very strong		strong
Etching	easy	easy	easy	diffi-cult	easy	easy	diffi-cult	easy	diffi-cult	easy	easy		
Activation energy [eV]	1.98	1.96		1.48	4.9	2.4							
Electro migration	active	active	active	negli-gible	negli-gible	negli-gible	negli-gible	negli-gible	negli-gible		negli-gible		negli-gible

Table 6.7 shows several properties of metals related to the above conditions
Judging from the table, it is obvious that aluminum is the most preferable
metal. In fact, aluminum is widely used as a metallization material. However,
aluminum is apt to be weak for electromigration as the line width is becoming
narrower and the layer thickness is becoming thinner [6.56].

6.4.1 Fine Pattern Technology for Metallization

Two methods to fabricate the aluminum fine line pattern are discussed in the
following part. One is the method by plasma etching technology, and the other
one is the method by lift off technology.

a) Aluminum Pattern Formation by Parallel-Plate-Reactor-Type Plasma Etching

A plasma etching which is widely used for fabrication of LSI circuits is mainly
done by the barrel-type plasma etching apparatus. This method is now applied
to silicon nitride and poly-silicon film etching. However, this type of plasma
etching is not suitable for less than 2 μm line width patterning because of
the isotropic etching profile. Furthermore, the aluminum is not etched by this
type of plasma etcher, because aluminum oxide (Al_2O_3) on the surface layer cannot be removed.

Another method is the parallel-plate-reactor-type plasma etching with RF
diode sputtering. The etching reaction in a parallel plate reactor is due to
positive ion bombardment and chemical attacking [6.57]. Thus, the etching
profile is anisotropic without undercutting, which is desirable to fine patterning. Basic problems to be solved in this etching are damage due to bombarding silicon substrate, contamination and the method of end-point detection.

Especially contamination makes the aluminum etching rate change irregularly
because the aluminum surface is highly sensitive to oxidation or chemical reaction. Figure 6.28 shows an example of experimental aluminum etching data by
the parallel-plate-reactor-type plasma etching method [6.58]. In this case of
aluminum etching, there is an inhibition period which is related to remove the
surface oxide. In this experiment, it takes three minutes to remove the surface oxide.

For further discussions, Sect.6.1 explains the details of plasma etching
technology.

b) Pattern Formation by the Lift-Off Technology

As a method of obtaining excellent metal pattern line width control in micron
and submicron dimensions, a novel lift-off process using plasma deposition and
etching technology is described.

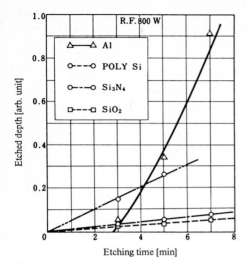

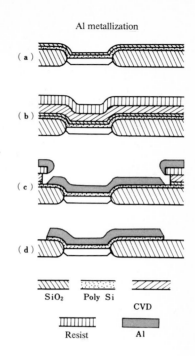

Al metallization

(a)

(b)

(c)

(d)

SiO₂	Poly Si	

CVD

Resist   Al

Fig. 6.28. Plasma etching rates for Al, poly Si etc., by parallel plate reactor type plasma etcher (800 W, 22°C 125 μHg. CCl₄ gas)

Fig. 6.29. Process flow step of the lift off for aluminium metallization

Generally, lift-off patterning [6.59] offers precise line width control, because line width itself of a metal pattern is determined by openings in a positive photo-resist layer rather than by a metal wet etching or plasma etching. In spite of its capability to form fine patterns, the process still lacks stability and reproducibility, but the process described in this section skillfully overcomes many problems. The outline of the process is shown in Fig.6.29 for final aluminum metallization [6.60,61].

Following poly-silicon deposition on a whole wafer after the contact hole opened, the poly-silicon layer is doped with impurity for low contact resistance by thermal diffusion. After this procedure, the surface layer of poly-silicon is oxidized thermally. As shown in Fig.6.29b, a silicon nitride film is deposited on the poly-silicon film by plasma CVD for a spacer of this lift-off process and the wafer is coated with a positive resist. At the next step, the resist is patterned by light or electron beam direct exposure. The silicon nitride film is then etched off by plasma etching using resist patterns as a mask and the oxide on the surface of the poly-silicon film is removed in an HF-buffered solution. Next, a thick aluminum film is immediately deposited by evaporation over the wafer. And then, resist and metal on the resist are removed by lift off. Finally, the residual silicon nitride film as a spacer and

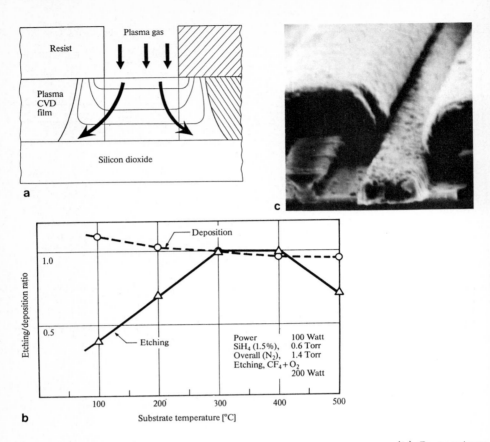

Fig. 6.30. (a) Formation of reverse trapezoidal shape spacer. (b) Temperature dependence of plasma etching and deposition for plasma silicon-nitride film. (c) SEM image of cross sectional view after Al evaporation

the poly-silicon layer are etched off by plasma etching. Through these steps, an aluminum pattern can be obtained.

The feature of this lift-off process to be understood best is as follows. The cross-sectional shape of the silicon nitride film is a reverse trapezoidal profile as shown in Fig.6.30a. When the shape forms a reverse taper, the discontinuity of metal between that on the substrate and that on the resist can be maintained easily, because the silicon nitride layer as a spacer is thicker than the evaporated aluminum layer.

The method forming a reverse trapezoidal shape of the spacer is as follows. If the plasma-etching rate of the plasma CVD silicon-nitride layer near the substrate is higher than that of the layer near the surface, the CVD layer is uniformly etched in both the vertical and lateral direction at the beginning

312

stage. At the middle stage, as the etching position goes deeper, the etching rate becomes higher and the amount of etching in the vertical direction is larger than in the lateral direction at the surface. At the final stage, since the etching rate of the CVD film near the substrate is still larger than that of the previous etching position, the amount of etching for the vertical direction is larger than that for the lateral direction. As a result, a spacer with a reverse trapezoidal cross section can be obtained.

In this case, the etching rate can be controlled by varying the deposition conditions as the film is growing. Figure 6.30b shows the dependence of the etching rate on the deposition temperature. The etching rate of the film deposited at 100°C is 40 percent of that of the film deposited at 300°C. Figure 6.30c indicates the reverse trapezoidal shape of the spacer after aluminum evaporation.

As an example, Figure 6.31 demonstrates the aluminum fine pattern in the submicron region. Direct electron beam exposure was used for very fine line resist (PMMA) patterning. Sequentially, through this lift-off process, a 0.5 μm line and space width aluminum pattern with 0.3 μm thickness was realized. As another example, Fig.6.32 shows an aluminum pattern for VLSI metallization by an SEM image.

In summary, this process is useful in developing 1 - 2 μm design rule LSI and submicron devices. Experimental results indicate that no unnecessary obstacles are encountered in LSI processing using this lift-off process.

6.4.2 Refractory Metal Metallization

When refractory metals are applied to the gate electrode like a poly-silicon, there are some merits as follows.

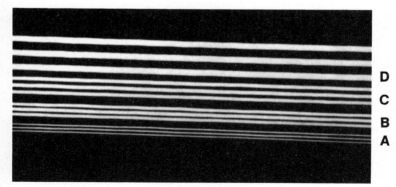

Fig. 6.31. Submicron Al patterning with the present lift off method

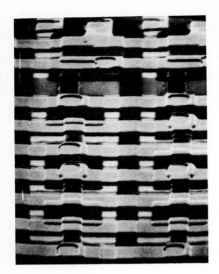

Fig. 6.32. Application to LSI metallization

1) Electrical resistivity of refractory metals is lower than that of a poly-silicon by two orders ($\rho = 5 \cdot 10^{-5}$ Ωcm).

2) Electrical migration does not occur contrary to the case of aluminum.

3) Fine line patterning can be realized by plasma etching as well as for a poly-silicon.

On the other hand, refractory metals have some demerits.

1) When molybdenum is exposed to high temperature and in an oxygen atmosphere, this surface is oxidized and the colour of the surface turns to black.

2) The ohmic contact resistivity between molybdenum and silicon after high-temperature treatment is extremely high because of the oxidation of molybdenum

3) Aluminum or gold cannot be bonded directly to molybdenum.

It is effective for improving the ohmic contact characteristics that a very thin platinum layer is inserted into molybdenum and silicon interface. By this method, contact resistance keeps as low as $2 \cdot 10^{-6}$ Ω·cm after high-temperature treatment. The reason of increasing the ohmic contact resistance is that silicide is formed between molybdenum and silicon even at temperatures lower than 600°C.

At the same time, mechanical stress appears between silicide and silicon substrates. Figure 6.33 shows the stress caused by various annealing temperatures between $MoSi_2$ and Si [6.62]. For relaxating the stress, a sandwich structure such as molybdenum, poly-silicon and silicon substrate is more suitable [6.63].

For further improvement, molybdenum or tungsten silicides are going to be applied in VLSI circuits.

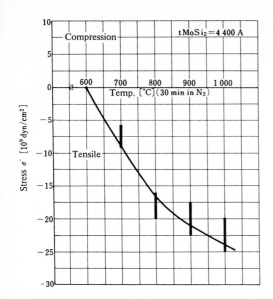

Fig. 6.33. The stress between $\overline{MoSi_2}$ and Si substrate by various annealing temperatures

6.5 Evaluation of Gate Oxide Film

With the development of oxide-coated semiconductor devices [6.64] such as the planar devices and the MOS LSI, a large amount of work has been done to study the surface charge density at the Si-SiO_2 interface.

The surface charge density, N_{FB}, and the surface state density, N_{SS}, are obtained from the capacitance-voltage (C-V) characteristics of MOS diodes. N_{FB} is strongly affected by the bias and temperature (BT) stress that causes the electrical instability of semiconductor devices. The instability of MOS structure has been attributed to the migration of positive ions (such as N_a^+ ion) within the SiO_2 film under electrically biased conditions at elevated temperatures.

The use of a phosphosilicate glass (PSG) film at the top of a thermally grown SiO_2 film provides an improvement in the stability and allows the fabrication of practical MOS LSI devices.

From about the middle of the 1970's, the significant instability due to the charge trapping in SiO_2 films has become an important problem with the development of short-channel devices, and there has been great interest in the Si-SiO_2 structure again because there are some basic problems:

1) interface state at the Si-SiO_2 interface,
2) fixed charge at the Si-SiO_2 interface,
3) transition region at the Si-SiO_2 interface,
4) impurity in SiO_2 film,

5) charge trapping center in SiO_2 film,

6) radiation damage in SiO_2 film,

7) defect in SiO_2 film,

8) conduction mechanism in SiO_2 film.

This section describes the corona charging and the avalanche injection techniques as the convenient evaluation method concerning 4) and 5).

6.5.1 Corona Charging Method [6.65]

The corona charging method is based on the deposition of ions (electrostatic charges) extracted from the corona discharge in air on the unmetallized SiO_2 surface. This method is a convenient technique for studying dielectric breakdown strength, mobile ionic contaminants, high field effects, and charge trapping centers in SiO_2. The potential at the SiO_2 surface increases up to the saturated level that the ionic charges arriving from the corona discharge are equal to the current (charge) through the SiO_2 film. The predominant ions from positive coronas are $(H_2O)_nH^+$ with $n = 1, 2, 3, \ldots$, depending on the relative humidity in air, and that from negative coronas are CO_3^- [6.66]. The kinetic energy of the ionic species are negligibly small. The current through the oxide is consistent with Fowler-Nordheim tunneling of electrons from Si into SiO_2 when the SiO_2 surface is positively charged, and with tunneling of holes from Si into SiO_2 when the SiO_2 surface is negatively charged.

The dielectric breakdown strength is evaluated by measuring the surface potential, the current, and the SiO_2 film thickness. For the case of the SiO_2 surface being negatively charged, the electric field needed to flow the same current through the SiO_2 is higher by more than a factor of 2 than for the case of the SiO_2 surface being positively charged.

The effects of ionic contaminants and the charge trapping in the SiO_2 are estimated by measuring the C-V curve shift after corona charging using an Hg or an In-Ga liquid metal for the MOS structure. Figure 6.34 shows the C-V curves before and after corona charging. Curve 1 is the C-V curve of the in-

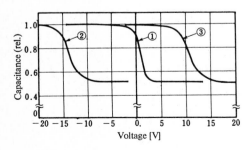

Fig. 6.34. Shift in the capacitance-voltage characteristics due to corona charging processes

itial and the after-positive corona. Curve 2 shows the C-V curve after-negative corona, indicating the buildup of positive charges (trapped holes) near the Si-SiO$_2$ interface. Curve 3 shows the effect of positive corona after negative corona. When the negative-corona charging is followed by positive-corona charging, the C-V curve is shifted to a positive value. This indicates that exposure to negative corona has generated electron traps in the SiO$_2$ [6.67].

It is found that the hole and electron trapping densities are reduced by phosphorus-ion implantation into the trapping area region (Si-SiO$_2$ interface for hole trapping and SiO$_2$ surface for electron trapping), and annealing in oxygen at 1050°C for 10 min as shown in Fig.6.35a and b [6.68].

The corona-charging method is a convenient technique for evaluating the insulator film in LSI process. This method, however, is not suitable for studying the effect of the metallization process on the Si-SiO$_2$ interface characteristics and the charge trapping centers in the oxide with the small capture cross section.

6.5.2 Avalanche Injection Method [6.69]

The avalanche injection technique is used to investigate the hot electron effect that restricts the scaling and the applied voltage for MOS devices in LSI

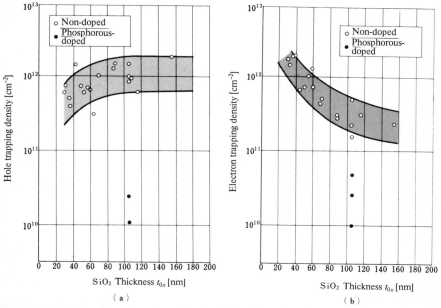

Fig. 6.35. a,b. Trapping density in SiO$_2$ of undoped and phosphorus-doped SiO$_2$. (a) Hole trapping density. (b) Electron trapping density

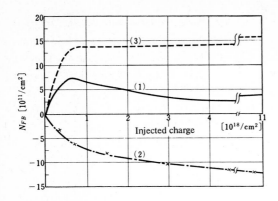

Fig. 6.36. N-shaped flat-band voltage shift in accordance with the injected charge density

and VLSI. The hot electrons are emitted from Si into SiO_2 in the vicinity of the highly biased drain junction of MOS transistors. Some of the emitted hot electrons are trapped in the SiO_2 and cause instabilities in the form of threshold voltage shift and transconductance degradation [6.70]. In this method, electrons are accelerated in the depletion layer of MOS structure by the 1 kHz to 1 MHz high voltage pulses and get a high energy. The high energy (hot) electrons are injected from Si into the SiO_2 film going over the potential barrier (3.1 eV) at the $Si-SiO_2$ interface.

The trapped electron density and the generated interface state density are obtained from the C-V curve before and after the MOS capacitor is avalanched for specified injected electronic charges. Figure 6.36 shows a typical result of the variation of the interface charge density, N_{FB}, as a function of injected charge density, N_{inj}, for a (100) MOS capacitor. Avalanche injection of hot electrons causes an N-shaped behaviour (Curve 1 in Fig.6.36) [6.71]. Curves denoted as 1, 2 and 3 in Fig.6.36 show the relationship of N_{FB} versus N_{inj}, the generated interface state density versus N_{inj}, and the trapped electron density versus N_{inj}, respectively, where Curve 1 is the sum of Curves 2 and 3.

The orientation of Si substrate strongly affects the generated interface state density. This density of the (100) capacitor has a value approximately half that of the (111) capacitor [6.71].

It is found that the chemical etching speed of the avalanched Si surface region by using the HNO_3-HF solution increases with increasing the N_{inj}. It is also observed that the avalanche injection current at a constant avalanche injection field increases in proportion to the number of injected electrons, contrary to the prediction by a field retarding caused by the charge trapping in the SiO_2.

Besides the avalanche injection technique many evaluating methods of hot electrons have been used such as the near avalanche injection, optical injecttion, photo-depopulation, photocurrent-voltage, high field current-voltage, and MOS transistor characteristics.

Thermally stimulated current, charge pumping, DLTS, cathode luminescence, Auger electron, and XPS methods are also used for evaluating the SiO_2 film or the $Si-SiO_2$ interface characteristics. Some of the physical and electrical properties mentioned above (1-8) are not fully understood yet.

6.6 Super Clean Environment

6.6.1 Cleanness and Its Monitoring

As the scale of integration will increase from LSI to VLSI and the pattern size to be used will become finer, the cleanroom as the production environment for the wafer fabrication should be improved in its grade. One of the major problems for the improvement in the cleanness concerns the size of particles to be controlled. Namely, although the cleanroom was controlled based on the number of dust particles larger than 0.3 μm up to the present, dust particles having smaller sizes should come into question in the clean space for VLSI. The relation between the size of dust particles and defects produced on the VLSI chip is multiple and very complicated. However, it would be general to consider that dust particles having the size of down to one tenth of that of the minimum pattern used in the LSI must be controlled, drawing the attention to defects which affect the reliability of LSI.

Figure 6.37 shows the minimum size of dust particles to be controlled corresponding to each step of the progress in the LSI technology. In this figure, the upper part shows the scale of integration of an MOS dynamic RAM and the minimum pattern size to be used as key numbers representing the progress in the LSI technology. The lower part shows the range of the size of dust particles to be controlled. As can be seen in this figure, the particle size of 0.3 μm is too large even for the 64 K bit dynamic RAM. It is also known that dust particles having the size of down to 0.09 to 0.1 μm should be controlled for the fabrication of 1 M bit dynamic RAM which is a target of VLSI for the time being.

A particle counter of the light scattering type which is generally used for monitoring the concentration of dust particles in the clean space can usually detect particles having the size of down to 0.3 μm. This detection limit results from the fact that the intensity of the scattered light from a particle of less than 0.3 μm in size is comparable to or less than that of the Rayleigh scattered light due to air molecules existing in the sensing volume of the

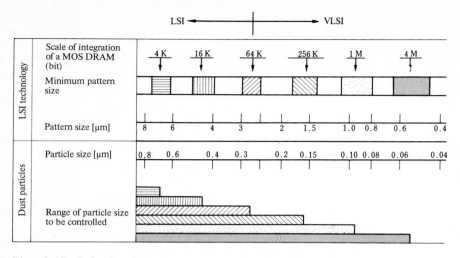

Fig. 6.37. Relation between LSI technology and size of dust particles

counter [6.72]. In a laser particle counter which was recently developed [6.73] the sensing volume has been decreased by about two orders of magnitude from that of the conventional one. This decrease causes the decrease in the density of the Rayleigh scattered light due to air molecules and enables the laser particle counter to detect particles having the size of down to 0.1 μm.

6.6.2 Super Clean Environment

a) Concentration of Dust Particles of Larger Than 0.1 μm

To investigate a superclean environment where dust particles having the size of more than 0.1 μm are controlled, first the knowledge of concentration of dust particles larger than 0.1 μm is necessary in a cleanroom which was designed based on the concentration of dust particles larger than 0.3 μm. Figure 6.38 shows some examples of the concentration of dust particles measured in various environments. The measurements were done by using the laser particle counter, and the vertical axis represents the integrated concentration of dust particles which is defined to be the total number of dust particles having the size of more than the value marked on the abscissa in the unit volume (liter). In this figure, curves indicated by (a) show the dust concentrations in conventional rooms which are almost the same as those outside a building. Curves (b), (c) and (d) show those measured in cleanrooms of class 5000 and 1000 and in a cleanbench of class 100, respectively. These classes are all defined with the 0.3 μm base. Curve (d) was given without working in the cleanbench which is

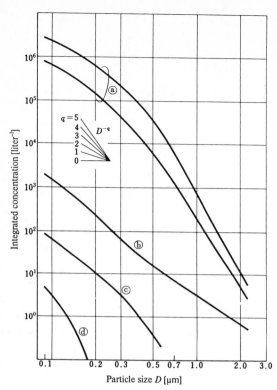

placed in the cleanroom where Curve (c) was measured. A group of lines indicated by D^{-q} gives the gradient of the negative q^{th} power of the particle size D. As can be seen in this figure, in conventional rooms, the integrated concentration at $D = 0.3$ µm, N (0.3), are in the range of $4 \cdot 10^4$ to $2 \cdot 10^5/1$, and that at $D = 0.1$ µm, N(0.1), from $7 \cdot 10^5$ to $2.5 \cdot 10^6/1$. The ratio of the integrated concentration at $D = 0.1$ µm to that at $D = 0.3$ µm, N(0.1)/N(0.3), is from 12 to 18. This means that the number of dust particles increases by a factor of around 15 if there is a shift in minimum size of dust particles to be controlled from 0.3 µm to 0.1 µm. Making the similar observations about cleanrooms, the ratio N(0.1)/N(0.3) is given to be about 21 for the room of class 5000 and about 23 for that of class 1000. As is understood from these facts, the ratio N(0.1)/N(0.3) is larger in cleanrooms than in conventional rooms, which suggests that the capture efficiency would decrease with decrease in the particle size from 0.3 µm.

In the cleanbench of class 100 represented by Curve (d), N(0.3) and N(0.1) are less than 0.1/1 and 3/1, respectively, and the concentration of $100/ft^3$

($\doteqdot 3.5/1$) is satisfied even for N(0.1). The ratio N(0.1)/N(0.3) is very large (more than 100). The concentration of dust particles during working depends on the work being done and cannot be said in general. In the case of the work usually performed in a cleanbench like the mask alignment work, however, the concentration increases by about a double over Curve (d), and N(0.1) is considered to be kept less than several hundreds/ft^3 even during working.

Figure 6.39 shows the particle size dependence of the dust generation rate in the range of more than 0.1 μm. These data were given by measuring the concentration of dust particles downstream in a cleanbench of class 100, while the work was being done upstream for generating dust particles. In this case, the work which generated a large number of dust particles was intentionally done so that the concentration was more than ten times as large as that measured with no work (Curve d in Fig.6.38). This means that most of the dust particles came from the dust generation due to the work. It is understood from these data that the gradient is generally small and, except for one case, q is in the range from 0.3 to 1.5, and the ratio N(0.1)/N(0.3) is 1.2 to 2.4.

Figure 6.40 shows some examples of the transmission rate of commercially available HEPA filters. Curves (a) and (b) are those for HEPA filters having the specification of capture efficiency of more than 99.97%, (c) and (d) of more than 99.99% and (e) and (f) of more than 99.999%. These transmission rates were obtained by calculating the ratio of two dust concentrations which were measured before and behind the HEPA filter using the room air stream as a sample air in a conventional room. It should be pointed out about samples (a) to (d) that the transmission rate increases as the particle size decreases from 0.3 μm and passes a peak value around 0.15 μm, then gradually decreases.

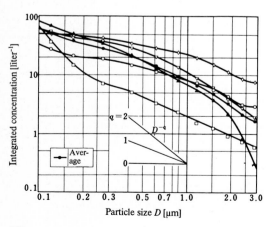

Fig. 6.39. Particle size dependence of dust generation

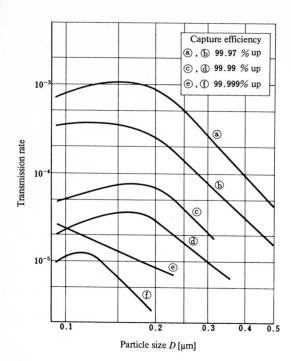

Fig. 6.40. Transmission rate
of commercially available
HEPA filters

All samples satisfy the specification of capture efficiency in the range of
the particle size of more than 0.3 µm, but the transmission rate exceeds the
specification below 0.3 µm in particle size except for samples (c) and (d).

c) 0.1 µm-Base Clean Environment

The average concentration of dust particles in a clean environment is expressed
in the following equation, if a HEPA filter is used as a final filter.

$$n \doteqdot n_M + n_G,$$

where

$$n_M = mS(1 - a) \quad , \quad \text{and} \quad n_G = g/K.$$

Here, n is the average concentration of dust particles in a clean environment,
$[1^{-1}]$, m is the concentration of dust particles in the outside air $[1^{-1}]$, S
is the rate of fresh air, a is the capture efficiency of HEPA filter, g is the
average generation rate of dust particles in clean environment $[1^{-1} \cdot min^{-1}]$,
and K is the air exchange rate $[min^{-1}]$.

The following part will describe the results of investigations about the
0.1 µm base clean environment, which were made by using the above equation

323

Table 6.8. Model calculation of 0.1 µm-base, class 1000 cleanroom

Particle size [µm]	m	$(1-a)$	n_M	N_M	n_G	N_G	n	N
$0.10 \sim 0.14$	$9.1 \cdot 10^5$	$2 \cdot 10^{-5}$	5.46	11.06	6.84	23.95	12.30	35.01
$0.14 \sim 0.20$	$8.0 \cdot 10^5$	$1.5 \cdot 10^{-5}$	3.60	5.60	5.13	17.11	8.73	22.71
$0.2 \sim 0.3$	$4.4 \cdot 10^5$	$1 \cdot 10^{-5}$	1.32	2.00	3.99	11.98	5.31	13.98
$0.3 \sim 0.5$	$1.9 \cdot 10^5$	"	0.57	0.68	3.19	7.99	3.76	8.67
$0.5 \sim 0.8$	$3.4 \cdot 10^4$	"	0.10	0.11	1.80	4.80	1.90	4.91
$0.8 \sim 1.2$	$2.4 \cdot 10^3$	"	0.01	0.01	1.00	3.00	1.01	3.01
$1.2 \sim 1.7$	200	"	0	0	0.59	2.00	0.59	2.00
$1.7 \sim 2.3$	26	"	"	"	0.37	1.41	0.37	1.41
$2.3 \sim 3.0$	4.2	"	"	"	0.24	1.04	0.24	1.04
> 3.0	1.0	"	"	"	0.80	0.80	0.80	0.80

$S = 0.3$ Unit: $(1-a)$ ... absolute number, others ... $[(\text{liter}^{-1})]$

and the data given before. Table 6.8 shows an example of model calculations for the 0.1 µm base, class 1000 cleanroom where the number of dust particles having the size of more than 0.1 µm is less than 1000 in a cubic foot. In these calculations, the larger one of two curves indicated by (a) in Fig.6.38 was used for the dust concentrations in the outdoor air, m, and the rate of fresh air, S, was fixed to be 0.3. For the transmission rate of HEPA filter, $(1-a)$, somewhat higher values than those of Curve (e) in Fig.6.40 were adopted. Values of n_G were calculated under the assumption that n_G is proportional to the negative square of particle size, D, which was based on the data given in Fig.6.39, and under the condition that $N(0.1) = N_M(0.1) + N_G(0.1) = 35 \ 1^{-1}$. (Here and in the following parts the capital letter with D in parentheses expresses the integrated value of the parameter expressed by the small letter at particle size D). As can be noticed in Table 6.8, $N_G(0.1)$ is equal to $3N_G(0.3)$, which means that G(0.1) is equal to 3G(0.3). To get N_G of the same level as that for the 0.3 µm base, class 1000 cleanroom, K is needed to be three times as large as one for the 0.3 µm base, class 1000 cleanroom. As K is usually designed to be one to two times per minute for the 0.3 µm base, class 1000 cleanroom, K is to be 3 to 6 times per minute for the 0.1 µm base, class 1000 cleanroom. This value corresponds to 0.125 to 0.25 m/s in the velocity of air flow, if a ceiling height of 2.5 m and a complete down-flow structure are assumed. The investigation described above indicates that the 0.1 µm base, class 1000 cleanroom, has no difficulty to realize.

Similar investigations were done for cleanbenches which are placed in the 0.1 μm base, class 1000 cleanroom described above. It is easily known from these calculations that $N_M(0.1)$ is small enough both for the 0.1 μm base, class 100 and class 10 cleanbenches, even if HEPA filters are applied having the capture efficiency of 99.97% on the 0.3 μm base. The air flow velocity should be 1.5 m/s for the 0.1 μm base, class 100 cleanbench, which could be realized. On the other hand, it should be 15 m/s for the 0.1 μm base, class 10 cleanbench. Because this value is not practical, the air flow velocity must be limited to less than 2 m/s, and the efforts should be made for suppressing the generation rate, g, to one fifth of that in the class 100 cleanbench.

7. Fundamentals of Test and Evaluation

With increasing integration level and circuit complexity toward VLSI, an advanced high technology is required in every part of the production of integrated circuits. Consequently, concerning test and evaluation technology there are several new problems, and it is required to establish new basic technologies for VLSI-testing.

Like design technology, the device integration is very complex in VLSI and, therefore, it is impossible to design a VLSI circuit manually. Today, Computer Aided Design (CAD) and Design Automation (DA) occupies a major part in VLSI design. But in the case of mask layout design where the ability of pattern recognition and the compact design are always required, the manual design is superior to CAD at the expense of longer design time. So even in VLSI it is probable to involve manual design in some parts. Thus, the automatic layout-design check system by computer means becomes essential for the VLSI design, especially for the one that includes manual designs or manual corrections.

If a VLSI circuit is designed fully automatically, a design check may not be needed. But actually such a design is rather a rare case; mostly it includes some manual design, and then the design check is needed. Thus, the automatic design check system is appreciated as a very important technology and many developing efforts have been paid to it.

Concerning device analysis, one of the big problems of VLSI devices is the heat-up of the chip including localized heat-up spots by the increased power dissipation as a result of enlarged circuit integration. To analyze this fact, the technology for measuring the temperature distribution of the VLSI chip is required precisely. Another big problem is the technology of probing fine patterns of VLSI chips, which is indispensable for failure analysis or the evaluation of VLSI chips. As the wiring pattern and the device size become very fine in VLSI, the conventional mechanical probing using a fine needle as a prober is impossible to be applied. For finer patterns, there is a necessity for non-contact probing methods including laser or electron beams.

With regard to technologies for VLSI testing the problems can be easily closed in by studying how to test the memory, the micro-processor and a future VLSI processor. The number of test patterns or test steps required for device testing increases extremely as the complexity increases. For example, in the "Walking Test" which is a method of memory function test, a 1 K bit RAM needs 2 million $(2 \cdot 10^6)$ test steps, but in the case of a 1 M bit VLSI RAM $2 \cdot 10^{12}$ test steps are required! These $2 \cdot 10^{12}$ test steps need about 50 h by a 10 MHz high-speed tester. This example indicates that, as the number of test steps increases drastically in highly integrated VLSI circuits, the time and the cost of testing expand enormously, too. In certain cases, the testing cost may become the limiting factor of the device cost. To realize VLSI testing within reasonable time and cost, high-speed operation and simplification of the test pattern must be realized. The high-speed testing technology is mainly studied here.

This chapter, showing the outline of the basic test and evaluation technology and examples of their advanced technology, makes clear the problems and the trends of VLSI test and evaluation technologies.

7.1 Testing and Evaluation of the Device Design

7.1.1 Testing of Device-Design Data

Recently, the elements of an IC increase more and more, and so any practical VLSI circuits with 200 k or more elements will be a reality in the near future. In the IC design process, designers change a logic circuit into a set of masks under given design rules, or they convert "software" into "hardware". Consequently, more elements on a chip are necessary, more working days for design, recognition of design errors and error detection are required. A design error means that the operation of a manufactured IC is different from the defined one [7.7,8].

On a display unit, the analog data for the visual check, or the mask patterns, and the digital data on a magnetic disk for processing by a computer, or mask data, are exchanged. Mask data are described by polygons with the coordinates of the vertices and the label corresponding to the IC production process. Under good conditions, mask data determine the planar dimensions of the mask and those of an IC. The steps of the process settle the sectional dimensions of an IC, the diffusion depth, the distance among layers, and the layer thickness. Thus, mask data and process steps form the IC structure.

These features direct the effective test strategy for the device-design
data, consisting of the following three steps:

1) design rule check;

2) topological circuit interconnection check;

3) IC functional simulation check.

Step 1: The minimum width of the polygons and the minimum distance between
two of them are compared with the values given by the design rules.

Step 2: The circuit $\mathbb{C}_M$ recognized from mask data is compared with the input
logic circuit $\mathbb{C}_L$, and the mismatching parts of $\mathbb{C}_M$ are detected from $\mathbb{C}_L$.

Step 3: The values of the most important circuit elements are calculated
from mask data, being the basis for the transition curves of any functional
circuit block for manual check.

Usually, Step 1 is processed by a minicomputer with an Applicon or Calma
display unit, and so on. But for Step 2 or Step 3, a commercial computer is
necessary instead of a minicomputer because of the required processing speed
and the design efficiency. Several methods for Step 2 or Step 3 have been
reported [7.1-8], but the best solution has not been found.

In one of the usual procedures for Step 2, mask data must contain a special
label relating to the fundamental circuit block (FB) being under consideration,
which is equivalent to a logic gate (OR, AND, NOT, and so on). A logic circuit
must be composed of only FB and a pseudo-FB, which is a transistor level circuit with an MOS load transistor labelled particularly with mask data. After
the equipotential patterns of the mask data are computed from which the FB
and pseudo-FB are subtracted the connections between FB and pseudo-FB are compared with those of a logic circuit.

Step 3 includes two simulation processes. The first one is a logical simulation for the operation by simple circuit elements, and the other one is the
transition and DC simulation using complicated circuit elements including stray
resistors, stray capacitors and accurate transistor parameters.

7.1.2 The Software System for Pattern Check

In the following a software system for mask data check after design is described. This check procedure is concerned with Steps 2 and 3 mentioned in
Sect.7.1.1.

The mask data in each integrated circuit are composed of polygons expressing
the aluminium layer, the diffusion layer, the first and second polysilicon
layers, and the contact holes.

The developed system is organized in relation to the independent six processes shown in Fig.7.1 [7.9]; all of them are in close correlation. They

328

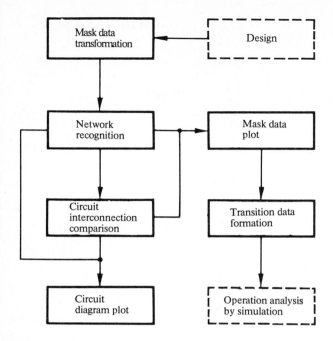

Fig. 7.1. Block chart of software system for pattern test

are as follows:
1) the mask data transformation process [7.35];
2) the network recognition process [7.36];
3) the network interconnection comparison process;
4) the transition data formation process;
5) the circuit diagram plot process;
6) the mask data plot process.

Starting with Process 1, all polygons are divided into trapezoids with parallel edges to the x coordinate axis using two algorithms of the top-bottom partition and the left-right partition.

Process 2 is made up of two steps based on two fundamental algorithms for trapezoids of chopping an overlap figure into them and detecting linkage between two of them. At a first step, mask patches of rectangels cut off from mask data are made, and then they are classified into fundamental ones and another one. At a second step, the circuit topology data are derived from each fundamental mask patch data based on the transformation shown in Fig.7.2 or copied from them.

In Process 3, the circuit interconnections $\mathbb{C}_M$ derived from Step 2 are compared with the logic circuit interconnection $\mathbb{C}_L$ with the input and output

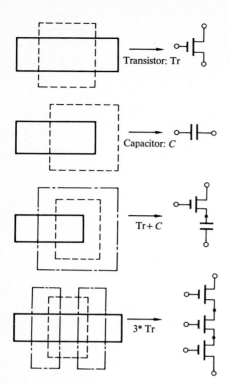

Fig. 7.2. Relationship between mask patterns and network elements

Transistor: Tr

Capacitor: C

Tr + C

3 * Tr

—— Diffusion — — — First-polysilicon-layer — · — Second-polysilicon-layer

labels to be homologous to those of $\mathbb{C}_M$ and n kinds of functional block circuit interconnections $\mathbb{C}_{Bi}$ (i = 1, 2, ..., n) used in $\mathbb{C}_L$. After $\mathbb{C}_{Bi}$ is mapped on $\mathbb{C}_M$ by an algorithm utilizing three number pairs, $\mathbb{C}_M$ is reduced to $\mathbb{C}_{M'}$. This cycle is repeated until $\mathbb{C}_{M'}$ is zero.

In Process 4, the capacitor values are calculated from the trapezoid areas, the values of the resistors and the transistor models from the length and the width in a group of the trapezoids between two given points and the peripheral capacitor values of the diffusion layer from the peripheral length of the polygon transformed from a group of trapezoids. Each element value is summed up to all vertices, and the processing data for transition or DC simulation are written on a disk.

This system has the following merits:

(i) The design is free of the test. Consequently, no label of any functional circuit block, any MOS load transistor and any terminal in mask data are necessary, and the design time for the logic simulation and the pattern layout decreases.

(ii) Arbitrary circuits can be treated.

(iii) The CPU time is short and the processing memory size required is small.

(iv) Each mask patch can apply to an updated process. Therefore, mask patches except those for design improvement or after machine trouble are available.

Experimental CPU time for 16 k bit dynamic RAM mask data with 608080 mask data and 36962 elements is 2.2 h on an ACOS/700 machine (2.3 h by theoretical equation) for Step 2, and that for a circuit with 39 elements and 10 stages about three seconds in Step 3.

7.2 Device Analysis and Evaluation

7.2.1 Methods of Device Analysis and Evaluation

Analysis and evaluation of a VLSI device should cope with the problem of miniaturization caused by high integration and high densification of the devices. A serious problem is the increase in temperature due to heat generation in the size-reduced device. This problem is related to the device reliability as well as the thermal design of the chip.

The temperature distribution in the IC pattern must be measured. It is impossible, however, to measure the three-dimensional distribution in the internal IC structure, and the heat distribution is examined by checking the temperature distribution on the IC's surface. This is normally done by using its infrared radiation. Since heat radiation differs according to the surface substance and the amount radiated off in the infrared naturally differs; a new method for measuring the surface temperature distribution is required. It is also impossible to detect correctly the fault in the microscopic elemental device out of the highly integrated VLSI pattern. Conventionally, a mechanical probe is used to detect faults in the IC pattern. Since the radius of the tip of the probe is at least about 10 μm, it is obvious that this kind of probe cannot be used for VLSI patterns, having a line width of 1 μm or less. There is a probability to damage the device to be measured by the mechanical force required with this method. Therefore, alternate methods using a laser beam, an electron beam, an ultrasonic wave, and so on are being under consideration. By these means, the probe does not contact the device directly but the interrelation is given by photon, electron, or sound-wave energy and its conversion to an electric signal. That is, the electric signal transmits relative energy changes and is converted to a visible picture.

For example, the laser-beam method is a non-contacting, non-damaging method
in which the laser beam is scanning the IC's surface as a microspot. The current generated by optical absorption in the semiconductor and the change in
the amount of light reflected from the surface are detected. The main characteristics are that the measurement can be performed in the atmosphere and
it can be easily carried out if the wavelength of the light used is smaller
than the object size to be detected. The electron beam method employs a scanning electron microscope and detects secondary electrons emitted from the
IC's surface. The generated current in the semiconductor takes the microspot
electron beam as a probe. Therefore, faults can be detected by observing
changes in voltage and current induced on the surface and inside the operating
IC, so that even very small areas can be observed.

There is also a method based on the reflection of focussed ultrasonic waves.
After considering the pros and cons of each method, the equipment shown below
was developed. If this basic equipment will be introduced to practical use,
faults in operating ICs will be easily detected and new causes of faults, unde-
tectable by the conventional electrical method, may be found [7.11,12,20,21].

7.2.2 Equipment so Far Developed

a) A System for Precisely Measuring the Temperature in the Infrared

In the system under discussion, although the surface temperature is measured
by the infrared radiation, a new emissivity calibration method has been developed by natural compensation. The correct temperature distribution can be
displayed on a color display.

In the past, the temperature distribution of the IC's surface was scanned
by infrared radiation for examination and analysis of the IC only [7.10].
Since the heat emissivity of the many-component materials such as Al and SiO_2
on the surface of the IC differs, however, a correct temperature distribution
could not be measured. Consequently, the relationship between the intensity
of the infrared radiation and the temperature in each part on the IC's surface
was measured by putting the sample in a constant-temperature environment. The
results were memorized. With these data, the intensity of the radiation under
IC-operation conditions is naturally correlated to the temperature, so that
a precise temperature profile can be determined. The electrical output signal
converted from the heat generated by the substance at a certain surface point
is obtained with respect to two known different temperatures. The slope of the
relationship can be obtained by taking the electric circuit for linearizing
the signal, whereby the output signal is proportional to the temperature. Then

the output signal at that point is measured and the product of this signal and the slope mentioned above is used to determine the temperature. With the output signal P_x being measured and the unknown temperature T_x in a small temperature range, we have

$$T_x = \frac{(P_x - P_1)(T_2 - T_1)}{P_2 - P_1} + T_1 \quad .$$

(7.1)

P_1 and P_2 are the output signals for the known temperatures T_1 and T_2. In this way the temperature distribution can be calculated for the entire surface of the IC. There are a few key points in the development of a system for doing the measurements and the determination of the precise temperature on the surface:

(i) The diameter of the infrared spot to be detected must be small. At the same time, it is necessary to micronize the width of the scanning step.

(ii) When the amount of heat radiation from the substance at a certain temperature is changed to an electric signal by a photoelectric conversion in the infrared detector, proportionality between temperature and electric output signal must be maintained.

(iii) The initial calibration must be correctly performed to divide the signal obtained from as few points as possible and to carry out calculation by (7.1) with respect to each point. For this purpose, it is necessary to utilize procedures of memory, calculation and display promptly.

The system accordingly developed consists of the following equipment: A minute dimensional scanning equipment which mechanically scans the surface whose temperature is measured in small steps (minimum pitch: 3.75 µm, maximum scanning width: 20 mm), an electric signal linearizer to maintain a proportional relationship between the magnitude of the infrared radiation and the temperature, equipment to keep the sample at a constant temperature, a memory, computational equipment for comparative calculations with respect to each point after dividing the surface into 256 bit points, and a color-display equipment to exhibit the output. The calculations and input/output control are performed by a ROM for programming and a keyboard. The superposition of the data actually measured on the calibration data, and the resulting temperature distribution are displayed by digitizing each level. The levels are displayed in color according to classified levels.

Figure 7.3 gives an overview of the system and Fig.7.4 indicates the block diagram.

The distribution of heat radiation before calibrating the emissivity of an appropriate IC (Fig.7.5) obtained by this system is as shown in Fig.7.6.

Fig. 7.3. External view of IC temperature distribution measurement system

Fig. 7.4. Block diagram of IC temperature distribution measurement system

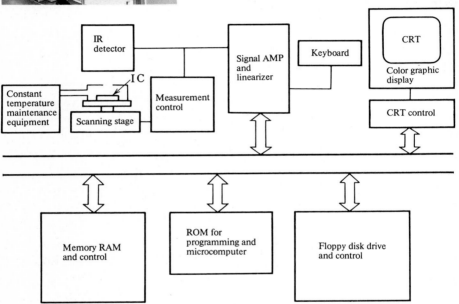

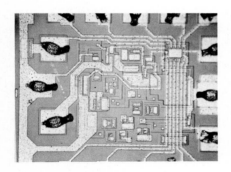

Fig. 7.5. IC pattern used for measurement

Fig. 7.6. Distribution before calibrating emissivity

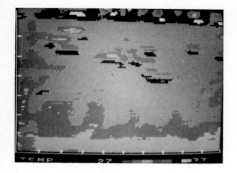

Fig. 7.7. Temperature distribution
of IC after calibrating emissivity

This indicates that the combination of the temperature distribution and the emissivity distribution, according to each substance, was obtained. By processing this combination, the correct temperature distribution can be obtained (Fig.7.7).

b) The Electron Micro-Probe Failure Analysis System

This system allows the contactless and non-destructive testing for a failure analysis or for a diagnosis of finely patterned devices. Especially, it can be used for the voltage measurement on the intereconnection of LSI chips [7.18-20]. The apparatus utilizes an accelerated and focussed electron beam as a probe and has the following characteristics:

(i) The fine probe can be easily realized (less than 0.1 μm diameter).

(ii) The probe does not interfere with the electric functions of the object to be measured and does not lead to errors during the measurements because it has very high (more than 10^6 MΩ) impedance and no stray capacitance.

(iii) Quantitative voltage measurements are possible through passivations.

(iv) The electron beam can be positioned quickly and precisely at every point on the LSI chips.

(v) Being noncontacting, it cannot cause mechanical destruction of the LSI-chip surface.

Using the dependence of the secondary electron emission on the surface potentials of the specimens, this apparatus measures the voltage at any point on the interconnection of an LSI circuit [7.12,13.16].

The specimen is irradiated by a finely focussed electron beam. This irradiation releases secondary electrons, backscattered electrons, Auger electrons, characteristic X-rays and several other types of radiation from a small part of the chip. Out of these signals secondary electrons being low in energy, are very sensitive to the surface potential. Usual SEM voltage contrast micrographs use this phenomenon and have been widely applied to the failure analysis of LSIs [7.14-17]. However, this voltage contrast gives only qualitative result to determine the local surface voltage because of the nonlinearity between signal intensity and surface potential, the retarding fields close to the surface of the specimen and the superposition of different contrasts, mainly topography and material contrasts. In the set-up described here the problem of nonlinearity has been solved by performing a calibration each time when a voltage measurement is to be made. It becomes possible by averaging the secondary electron signals to carry out noncontact voltage measurements with a voltage resolution less than 100 mV.

A schematic diagram of the system is given in Fig.7.8. The apparatus consists of a commercial SEM and the signal processing unit controlled by the microcomputer. The signal processing unit has several functions:

(i) Storage of the SEM image data. The secondary-electron image displayed on the SEM monitor is divided into 256 x 256 picture cells and then stored in a data memory. This image can also be displayed continuously at the TV rate on the monitor of this unit.

(ii) Accumulation of the secondary-electron signal. The secondary-electron signal from each picture-cell area is amplified and converted from analog to digital (8 bit), then added to the data memory corresponding to each repeated value for the specified number of times (256 times max).

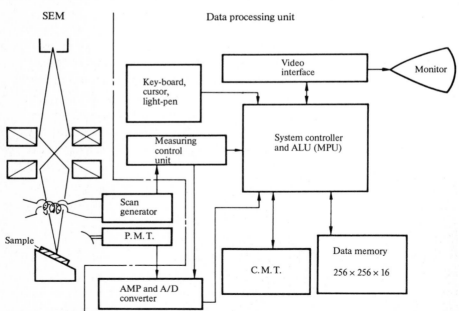

Fig. 7.8. The schematic block diagram of the electron micro-probe failure analysis system

(iii) Averaging. The secondary-electron signal of each picture cell within the area specified on the monitor of the signal processing unit by cursors is averaged.

(iv) Calculation of the difference between the signal mean values of the two specified areas.

(v) Providing a source for the calibration voltage. With a 10 bit DAC, the calibration voltage is generated in the predetermined range (10 V max).

(vi) Measuring the calibration data. The calibration voltage is applied to the area where the potential can be controlled precisely. After detecting the secondary-electron signal from this area, the mean value of the signal is calculated. Both the calibration voltage and this mean value of the signal corresponding to it are stored in the respective memory.

(vii) Voltage measurement. The mean value of the secondary electron signal is calculated within the measuring area. Then the voltage at that area will be provided with the calibration data and the linear interporation method. Finally, it is displayed on the monitor digitally.

(viii) Framing. Specifying the proper threshold value, the points where the signal is less or more than its value are selectively displayed on the monitor.

(ix) Eliminating the abnormal data. According to the specified threshold value, abnormal data are eliminated during the calculation of the mean value of the secondary-electron signal intensity.

(x) Data saving. The image data, the accumulation data of the secondary-electron signal, the calibration data and others are saved on the CMT.

Figures 7.9-12 represent some examples of the voltage measurement and calibration results combining these functions. Figure 7.9 shows an example of calibration data displayed on the monitor. Figure 7.10 gives an example of the measured voltage. Figure 7.11 demonstrates a relation between the secondary-electron signal and the calibration voltage. An example of the

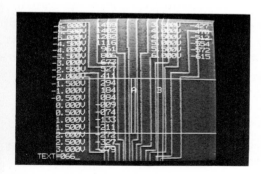

Fig. 7.9. An example of the measured calibration data

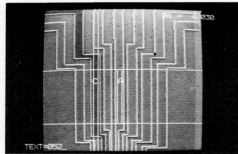

Fig. 7.10. An example of the voltage measurement result

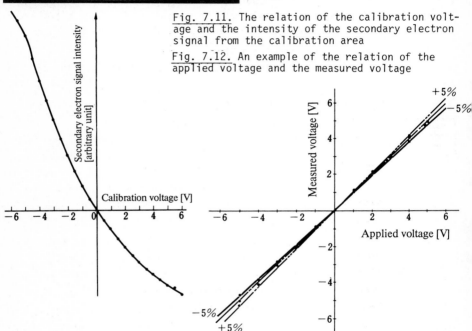

Fig. 7.11. The relation of the calibration voltage and the intensity of the secondary electron
signal from the calibration area

Fig. 7.12. An example of the relation of the
applied voltage and the measured voltage

measurement results for the relation of the detected voltage to the applied
voltage with a test sample is displayed in Fig.7.12. It can be recognized by
these examples that the electron micro-probe failure analysis system makes it
possible to measure the voltage at the interconnection of an LSI chip with the
electron beam by performing a calibration each time during the voltage measurement. The measurement error is less than ±5%.

c) A Device Analysis System Based on Laser Scanning

The principle of a device analysis system based on a laser scanning technique
makes use of the fact that semiconductor devices are photosensitive. In ad-

dition, by monitoring photogenerated currents while a focussed spot of a laser beam is being scanned over the specimen device, a contactless probing technique can be realized [7.21-31].

According to the incident radiation travelling over the device, electron-hole pairs are generated in the chip. The carriers can recombine locally or, if an electric field exists in the depletion region of reverse-biased pn-junctions, electrons and holes are separated causing a current in a closed current loop. By measuring the difference in the amount of the photogenerated current related to the DC currents, the logic level of the internal circuit can be detected.

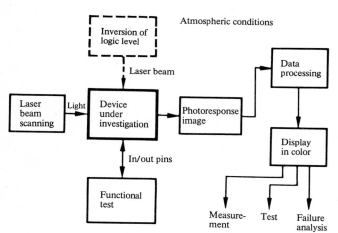

Fig. 7.13. Functional diagram of a device analysis system based on laser scanning techniques

The functional diagram of the system is shown in Fig.7.13. The device under test is scanned with a finely focussed laser beam while the device is being operated in dynamic (or in static) mode by applying the external signals. Consequently, the photogenerated currents induced by the radiation of the laser beam are measured and converted into a two-dimensional photoresponse image, stored in a memory in real time and then processed with a microcomputer (MPU). The results are displayed on a color monitor. If necessary, the logic level of the internal circuit can be changed without external contact, for example by the carrier injection into the specified circuit (e.g., memory cell) by the radiating laser beam.

The overall schematic and pictorial diagram of the system is shown in Fig.7.14. The system is composed of four basic units: laser scanning microscope (LSCM), LSI evaluation unit, signal detection unit, and image processing unit.

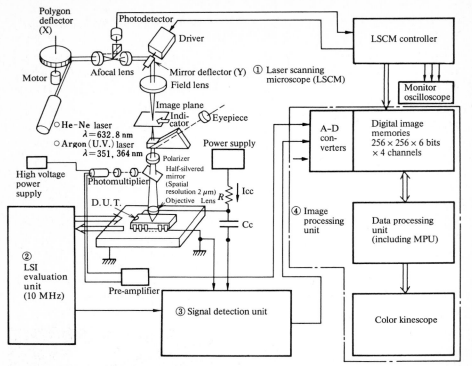

Fig. 7.14. Block diagram of a device analysis system based on laser scanning techniques

(i) In the laser scanning microscope, the laser beam from a radiation sourc is deflected in orthogonal directions to generate a scanning raster which is successively focussed to a fine spot (2 μm ∅) on the device.

(ii) In the LSI evaluation unit, functional test-pattern signals (maximum 10 MHz) for device operation are generated. At the same time, the output signals from the device are compared with the correct data and also, a sampling pulse is generated for the sample-and-hold amplifier.

(iii) In the signal detection unit based on a sample-and-hold technique, photogenerated currents superimposed on the power-supply currents are extracted and amplified.

(iv) In the image processing unit, the measured photogenerated currents are digitized and stored in memories in real time in the form of a two-dimensional photoresponse image and then processed with MPU. Consequently, the required information (e.g., logic level of the memory cell) are displayed on a color screen.

As a result, this system has some characteristic features. For example, in dynamic type LSI such as dynamic MOS RAM, the logic levels of internal circuits can be measured during the dynamic operation, without contacting, in the atmospheric surrounding and moreover, in a nondestructive manner and displayed in color. The depth resolution can be adjusted by selecting laser sources with different wavelengths (He-Ne laser and argon uv laser) as the penetration depth depends on the wavelength of the incident radiation. A

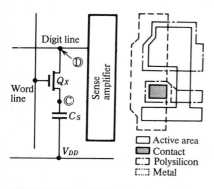

☐ Active area
▨ Contact
⊏⊐ Polysilicon
⠿ Metal

Fig. 7.15. Circuit diagram of memory cell and memory cell structure

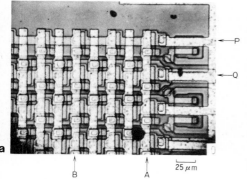

a

B   A   25 μm

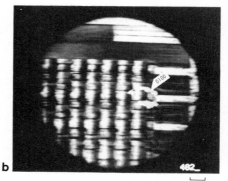

b

25 μm

Fig. 7.16. A typical example of measured results on 4 K dynamic MOS RAM (a) Optical photomicrograph

Location of memory cell	Address (octal)				
	MSB	B3	B2	B1	B0 LSB
P-A		0	0	0	0
P-B		0	0	0	1
Q-A		0	1	0	0
Q-B		0	1	0	1

(b) Logic level measurement of memory cell. The device is scanned with an argon u.v. laser beam (λ = 351, 364 nm. D $\approx$ 2 μm) while in its write cycle operation (f_c = 2 MHz.). Memory cell (0100) where "1" is being written is distinguished with red as indicated by an arrow and superimposed on He-Ne reflected image of green

341

maximum of four images for different measurement conditions can be stored in parallel and in real time as well and processed [7.28-31].

Figure 7.15 shows the circuit diagram of a memory cell of a 4 K dynamic MOS RAM together with the memory cell structure. Figure 7.16a represents the optical photomicrograph of the portion adjacent to the address 0000 bit (octal) in the device and Fig.7.16b a typical example of measured results. The device is scanned with a fine argon uv laser beam (wavelength $\lambda = 351$, 364 nm) focussed into about 2μm ∅ spot during its dynamic operation; that is, the logical data '1' is being scanned with write-cycle operation (2 MHz) between the address 0100 through 0107 in sequence and repeatedly as well. As the dynamic photoresponse image F is formed by sampling the power supply current when the logical data '1' is being written in the same memory cell (0100), the memory cell is distinguished with red color (indicated by an arrow) and superimposed on the He-Ne laser reflected image of green to indicate on a point-by-point basis. In the logical '1' state of memory cells, pn junctions between p-type substrate and n^+-diffused active areas of the cell transistor are deeply reverse-biased (about 15 V). Therefore, strong electric fields produced in the depletion region near the junctions enable an efficient collection of carriers (electron-hole pairs) generated by laser-beam radiation before they can recombine. These carriers therefore contribute to the photo-generated currents superimposed on the power-supply current path. Any inversion layer under the V_{DD} pattern of the cell capacitor has also significant influence; normally, it increases the photogenerated current.

This system can effectively be applied to test, evaluation and failure analysis of LSI and VLSI circuits. Furthermore, the ability of the signal injector with light interaction will be able to remarkably improve the capability of LSI and VLSI tester.

7.2.3 Microdevice Analysis and Evaluation

With device miniaturization, analysis and evaluation of minute areas, and fault analysis of such regions are required. There are two analysis methods: One in which microstructure can be directly observed and a substance can be detected; and one in which the analysis is performed indirectly from any change in electric characteristics. A scanning electron microscope (SEM), an X-ray micro-analyzer (XMA), a scanning Auger electron microscope (SAM), and so on are used for direct analysis methods. XMA and SAM are applied for composition analysis of the surface, and since both are provided with the function of SEM, the structure can be observed, too. SAM is the most effective method of

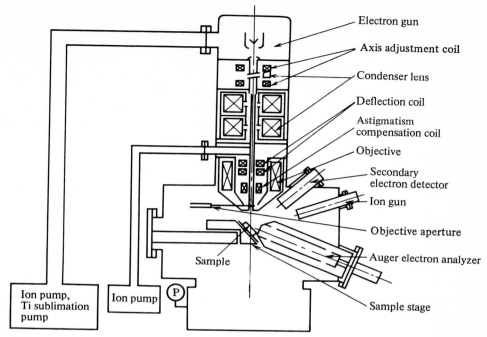

Fig. 7.17. Structure of microprobe electron spectroscope

analyzing micro-devices owing to its advanced micro-analysis feature. Therefore, an example of micro-device analysis and evaluation using SAM is demonstrated. A microbeam electron spectroscope has been used as SAM.

The results of an acceleration test carried out for a thin-film Al pattern
on an IC are analyzed. The spectroscope consists of an electron-beam irradiation system, an ion-irradiation system, a sample stage, and an electron-energy
analyzer, as shown in Fig.7.17. An Auger analysis has been performed using an
incident electron beam with a diamter of 1 μm or less. The spectrum is derived from an energy analysis of Auger electrons emitted by electron-beam
irradiation. Analysis in the depth direction is carried out by etching the
sample with the ion beam radiated from an ion gun using mainly Ar^+.

Then, the thin-film Al pattern on the IC was made to cause a fault through
an acceleration test and the fault was analyzed. Figure 7.18 shows an example
of an SEM picture of the failed thin-film Al stripe. The Al stripe consists of
aluminum deposited on SiO_2 to a thickness of 0.13 μm. After assembling the
Al stripe on a TO-5 stem, an electric-current acceleration test was carried
out at 200° C and $2 \cdot 10^6$ A/cm^2.

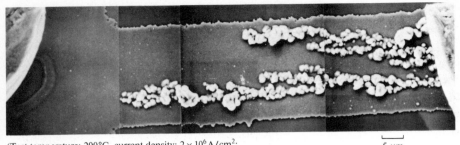

(Test temperature: 200°C, current density: $2 \times 10^6 \text{A/cm}^2$;
disconnector timer: 310 hours, film thickness: 1300 A°)

5 μm

Fig. 7.18. SEM picture of a failed Al stripe

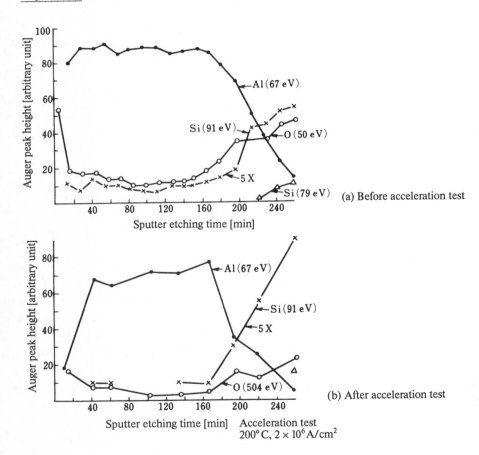

Fig. 7.19. Depth profiles of Auger peak in an Al stripe

344

The result is characterized in a hillock produced continually along the electron flow. It is assumed that this hillock occurs at the particular place where the resistance increases [7.32]. An analysis in the depth direction has been carried out before and after the acceleration test to examine the cause of increase in resistance. It has been found that, before the acceleration test, Si is distributed uniformly from the Al surface in the depth direction, as shown in Fig.7.19a. Assuming that Si was produced in Al by a heat reaction between SiO_2 and Al [7.33], even if pure Al was deposited at first, it is considered that a heat reaction occurs due to increase in temperature during the assembly process or the like.

It is thought that, after the acceleration test, Si was redistributed by the heat reaction during the test because it appears separately on the surface of Al and near the interface of $Al-SiO_2$ [7.34]. Consequently, the distribution of Si in Al could be determined from an internal reaction as shown in Fig.7.19b by means of an acceleration test in connection with the increase and decrease of resistance.

7.3 Device Testing

7.3.1 Test Methods

The test methods for VLSI circuits are basically those of conventional ICs. Generally, the following three items are tested:

    (i) DC characteristics test (or static characteristics test).

    (ii) Function test.

    (iii) AC characteristics test (or dynamic characteristics test).

As the level of integration increases to very high, a lot of functions are included in VLSI circuits. In this respect, VLSI becomes not only the integration of circuits but the integration of the system itself. Consequently, the test of a VLSI circuit becomes a test of a circuit and a system as well. According to the extended complexity, external pins are multiplied to more than 100 pins, for example, and the test steps are multiplied sharply in every test item. But the most distinctive point of VLSI testing is the extremely increased and complicated function test in comparison to that of conventional ICs because the VLSI circuit must be checked as a system.

In semiconductor memories, in the case of read-and-write operations of each bit cell, many kinds of memory-function tests such as 'walking' and 'galloping' have been done to detect such an error as the mis-writing to other bit cells. Take the 'walking' test, for example. At first, as the walking '1' test, write '0' to all bit cells of the memory, and after checking all the '0', write '1' to the first cell. Checking the '1' of the first cell and '0' of the remaining cells, write the first cell back to '0'. Repeating in the same way cell by cell, all bit cells are checked. The walking '0' is the converse operation of the walking '1'. And, finally, an N bit memory needs $2N^2$ steps of write-and-read operation in the walking test. As the number of bit cells is very large in VLSI memories, the volume of memory-function test patterns increases extremely. As a result of the test, time and cost expand. In order to overcome these problems, an advanced VLSI tester is required that can generate complex function test patterns and can execute the function tests in high speed. Two methods are usually applied to generate complex function test patterns. One is a software method that produces test patterns by logic simulation with a computer or processor unit in the tester, the other one is a hardware method. It compares the logic function pattern of the device under test with the pattern generated from another checked good device.

In the case of a VLSI circuit integrating a total system, the test must include that of the system, which normally requires a dynamic characteristic test in various kinds of the system function. So, in the VLSI tester, besides the function of the DC characteristics and the function test, the dynamic characteristics test becomes the essential function checking the accurate speed in various system functions.

7.3.2 The VLSI Tester

As described in the previous section, the VLSI tester is required to do the following functions:

(i) It must be able to handle different numbers of pins.

(ii) DC parameter testing and high-speed AC parameter testing at the various modes of the operations must be properly done.

(iii) High-speed function testing to various patterns must be possible.

Concerning these requirements, today's LSI circuits with more than 120 pins are to be tested. For the test pattern rate in the function test, a frequency of about 100 MHz is realized. Recently, the testers which can perform the high-speed AC parameter tests have been announced. An advance in the tester design may be expected to enter the VLSI area.

When considering the tester advances to be expected in the future, it becomes one of the most important problems to increase the operating speed of the hardware considerably. The main problem is to speed up the pattern generator which constitutes the heart of each tester and to make it more

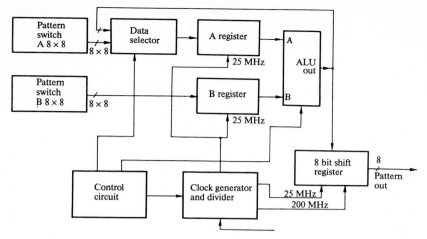

Fig. 7.20. The schematic block diagram of the very high speed address pattern generator built as a trial

flexible. Here, as an example, the pattern generator which uses the parallel data processing techniques as one of the most important speeding up methods is described. It is built as a trial, which has an output of eight channels in configuration, realizing a very-high-speed pattern generation at a rate of about 200 to 300 MHz. The 3 to 10 times speed improvement of the generating rate is achieved by parallel data processing techniques.

Figure 7.20 is a schematic block diagram of the equipment built as a trial. Figure 7.21 demonstrates the detailed diagram of the parallel data processing unit. A usual pattern generator forms every test pattern with processing the data step by step. To do this, the pattern generator using the parallel data processing method realizes a speed improvement by forming the test patterns as follows: The several-steps data (eight steps in this case) are processed in a lump. Then, the data are converted from parallel to serial. This can easily be done by programming the simple data not only as random patterns but also as algorithmic patterns, such as pingpong, galloping, or marching patterns. These types of pattern are often applied to address memories during testing. In this way, a very flexible and high-speed pattern generator can be realized.

Figure 7.22 shows an example of the data setting with a pingpong pattern simplified to four channels and four steps, in the parallel data processing methods. The following pattern can be created by these data setting alone:

$0 \rightarrow 1 \rightarrow 0 \rightarrow 2 \rightarrow - - - - - - - - \rightarrow 0 \rightarrow 14 \rightarrow 0 \rightarrow 15 \rightarrow 0 \rightarrow 0$.

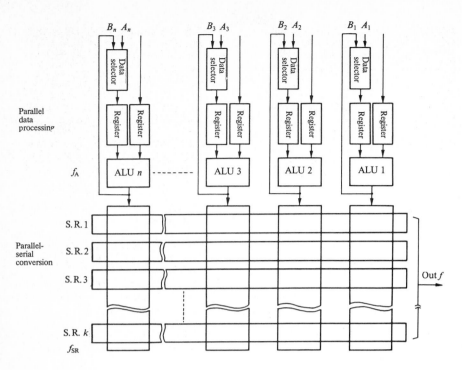

Parallel data processing

f_A

Parallel-serial conversion

Out f

f_{SR}

Fig. 7.21. The detailed diagram of the parallel data processing unit

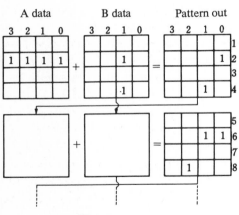

A data   B data   Pattern out

Ping pong pattern

$(0 \rightarrow 1 \rightarrow 0 \rightarrow 2 \rightarrow 0 \rightarrow 3 \rightarrow 0 \rightarrow 4 \cdots)$

Fig. 7.22. An example of the data setting for the pingpong pattern (the blank areas imply "0")

If the usual pattern generator should form this pattern under setting the A register data '0' and the ALU operating mode fixed 'ADD', the data of the B register may be dizzily changed as follows:

$0 \rightarrow 1 \rightarrow -1 \rightarrow 2 \rightarrow -2 \rightarrow \text{-------} \rightarrow 15 \rightarrow -15$.

348

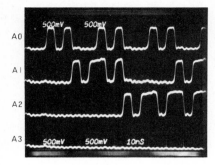

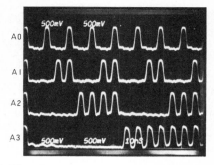

Fig. 7.23. The measured waveforms of the galloping pattern (the lower 4 bits only)

Fig. 7.24. The measured waveforms of the pingpong pattern (the lower 4 bits only)

From the above description it is clear that the algorithmic-pattern generating method for the equipment built as a trial is considerably simplified.

Figures 7.23 and 24 are examples of measured waveforms of these patterns usually applied to test memories, generated by the equipment under discussion. Both of these photographs show the waveforms of the lower 4 bits (A0---A4) in the patterns of an 8 bits configuration. Figure 7.23 exhibits the galloping pattern addressed as

$$0 \to 0 \to 1 \to 0 \to 1 \to 2 \to 0 \to 2 \to 3 \to \text{-------} \ .$$

The pattern rate is 200 MHz. Figure 7.24 gives the pingpong pattern with an address

$$0 \to 1 \to 0 \to 2 \to 0 \to 3 \to 0 \to 4 \to \text{-------} \ .$$

The pattern rate is 300 MHz.

As described above, the high-speed operation of the pattern generator is realized by the parallel-data processing method. By this means, very-high-speed function testing of devices becomes possible. This seems to be a step forward to realize a very-high-speed LSI test system which can deal with future VLSI circuits. The development of the peripheral units and the testing-system configuration which can make the most of such high speed of the pattern generator, are essential aspects to be realized by testers in the future.

8. Basic Device Technology

One of the most important features of VLSI technique, especially in view of VLSI devices, is the change from the conventional integration of circuits (Integrated Circuit: IC) to the integration of systems itself (Integrated System: IS) as the result of an increased level of integration.

To realize such VLSI systems, many efforts for high performance VLSI devices such as scaling down and higher integration methods have been undertaken. But many problems are based in the development of VLSI device technology, such as the kind of performance to be needed, the kind of device technology which must be developed to realize such a performance, and device limitations to be challenged.

This chapter is designed to clarify the outline of the basic VLSI technology, not only by listing examples of VLSI circuits, but also by illustrating relevant problems.

8.1 Background

The complexity of LSI has been estimated to increase by a factor of two every year nowadays. Even though the rate of increase seems to have slightly been reduced recently, the complexity of LSI is increasing every year, and actually started the VLSI days.

As the level of integration is growing into VLSI, such systems as small computers, microprocessors and memories up to mega-bits capacity, become possible to be realized on one chip. This means that VLSI is the technology changing the integrated circuit (IC) to the integrated system (IS). So in the discussion of VLSI device technology, an approach from the system technology side becomes very important. On the level of SSI-MSI logic ICs, every system can be constructed by collecting several kinds of ICs from selected standard logic IC families. But generally in the VLSI range, as the VLSI device is a part of the system itself, every different system requires another special VLSI circuit. Thus, the VLSI technique enhances the number of specialized highly integrated circuits, a process which is called customing.

On the other hand, trials to realize the standardization of VLSI circuits or to reduce the types of VLSI parts have been undertaken, mostly applying system technology. The very effective standardization of hardware can be achieved by using microprocessor, PLA (Programable Logic Array) and P-ROM. In all these IC's, the required system functions are realized by software programing. In general, memory is simple in function, but it is always required in large quantities. So it is a very suitable device for the VLSI technique, and the level of integration is indeed increasing most rapidly.

To extend the performance of a system, both an increasing function by a higher integration level and an increasing speed by advanced circuits are needed at the same time. Generally, to get a higher speed in logic circuits the power dissipation has to be increased. On this sight, the level of circuit integration seems to be limited by the total power dissipation. To satisfy both of these demands, new high-speed logic circuits operating on low-power levels or so-called "low-energy circuits" are coming up. Methods to develop low-energy circuits are reducing the power supply voltage, the logic swing voltage, the stray capacitances of devices, and the application of high-performance devices. In other words, the development of the VLSI device technology is normally directed toward realization of low-energy logic circuits and device structures, being comparable with those circuits.

To evaluate the performance of logic circuits the figure of merit and physical limitations of logic circuits have to be studied here. The performance of a VLSI logic device can be discussed by its ability of information processing when used in systems. Generally, in a logic circuit, as shown in Fig.8.1, to improve the logic speed, higher-power dissipation is required, if the energy or the power delay product of this circuit has to remain constant.

In a logic circuit operating at a constant logic energy E, the maximum logic speed or minimum delay time is defined as T_0. Then, $1/T_0$ ($= f$) means the maximum bits of information processed within a unit time. So, this factor

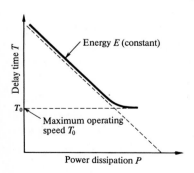

Fig.8.1. Energy E of a logic circuit and the maximum logic speed (minimum delay time) T_0

can be regarded just as the ability of information processing. When H is defined as the product of $E \cdot T_0$,

$$H = E \cdot T_0 = \frac{E}{1/T_0} = \frac{E}{f} \quad [\text{Joule} \cdot \text{s}] \quad , \tag{8.1}$$

it means the logic energy that is needed to realize a unit information processing ability. Thus, a smaller value of H indicates higher logic performance,
and H can be used as a useful figure of merit for logic circuits, especially
in VLSI circuits where system performance becomes the important aspect of
evaluation.

The value H with the dimension of Joule $\cdot$ s can be seen as a fundamental
value of each logic operation, but individually for every logic circuit.
Thus, the 'H' is introduced as 'logic quantum' here.

The physical lower limit of the logic quantum is Planck's constant h
($6.6 \cdot 10^{-34}$ J $\cdot$ s); it is determined by the Heisenberg's 'uncertainty principle' that a lower value than h cannot exist.

The lower limit of energy can be defined by the thermal energy kT $\cdot$ log2
at T degree K. As the delay time of typical VLSI devices is limited by the
carrier transit time for traversing the active volume of the device, its size,
such as the channel length and the base width, become the limiting dimensional
factor.

Energy and operating speed as limiting factors are illustrated in Fig.8.2.
This picture demonstrates that the physical limitations are far below from
today's logic circuits, with one exception: the Josephson junction logic approaches the physical limitations fairly closely.

To realize a better device performance, a new design concept called 'scaling
rule' has been introduced [8.1,2]. It indicates the proportionality of device
characteristics under the condition of constant field strength, when the size
of the device is scaled down to 1/s. The device parameters are changed in a
manner shown in Table 8.1. For example, if the propagation delay time t_{pd} and
power dissipation P are reduced to 1/s and $1/s^2$, respectively, the logic energy
P $\cdot t_{pd}$ is proportional to $1/s^3$, which means a remarkable improvement of device
characteristics. Through this method it becomes possible to predict the performance of future scaled-down devices by extrapolating from the performance
of today's devices to a certain degree.

It must be borne in mind that, among the parameters in Table 8.1, the CR
time constant corresponding to the signal delay of the wiring between devices
cannot be scaled down. This means that, as the device integration increases
much more and the wiring in it becomes more complex, wiring will become a main

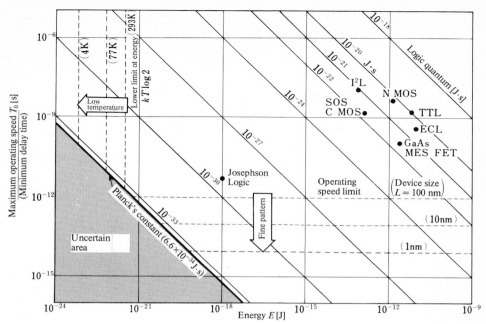

Fig.8.2. Logic quantums and physical limits of logic circuits

Table 8.1. Device scaling rule

Items	1/s
V_{DD}	$1/s$
t_{pd}	$1/s$
P	$1/s^2$
$P \cdot t_{pd}$	$1/s^3$
R	s
C	$1/s$
$C \cdot R$	1

factor of signal delay in future VLSI circuits. The question arises in complicated VLSI circuits how to realize low wiring time constants.

The short channel transistor in MOS technology satisfies as a result of
scaling rules both high integration and high performance requirements simultaneously. So, the microfabrication technology down to the micrometer or submicrometer range is the fundamental problem in VLSI development. But by reducing the device size as a result of application of the micro-fabrication
technology, several physical limitations which affect the performance are

showing up in devices. It is the most important question how to realize devices with reduced size as well as low-energy operation, avoiding some phyical limitations which must be recognized, by further reduction of device size. Some of those problems are studied in this chapter.

8.2 Limitations for Miniaturization

There are three important technological factors [8.3] to realize VLSI circuits: (i) improvement of the device and the circuit, (ii) increase of the silicon chip area, and (iii) miniaturization of device size. This section is addressed to the third factor, the miniaturization of device size.

It is obviously impossible to shrink the device size indefinitely; for instance, the gate length of an MOS transistor cannot be made zero. Generally, a physical quantity such as the potential or the current, normalized into two non-overlapping ranges, is used as a digital signal in digital circuits. Devices for these circuits must be controlled by this signal and must reproduce, in a good manner, the physical quantity as the digital signal. If this turns out to be impossible for a certain device size in the course of shrinkage of the device size, then that is a limitation.

There are several physical phenomena which are limiting the miniaturization in today's view. They can be divided into two categories: (i) those which affect device characteristics and (ii) those concerning the device fabrication. In the following section, firstly the physical phenomena of the first category, then that of the second category, and finally device size limits of a DSA MOS transistor and a memory switch transistor are discussed in sequence.

8.2.1 Physical Limiting Factors Related to Device Characteristics

a) Weakening of Insulation

When the device size is going to shrink, the electric field strength in materials, such as semiconductors and insulators being employed in the devices, becomes higher if the applied voltages are not made lower. But each of the applied voltages has a certain non-zero minimum value for the device to do a given function properly. On the other hand, each of the materials being considered has also a certain maximum for the applied electric field to avoid breakdown. Therefore, the device-size reduction also has a certain limitation.

One of the physical mechanisms for that is the avalanche breakdown. When the electric field strength exceeds a certain critical value E_c, the number of

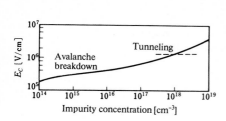

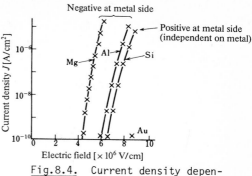

Fig.8.3. The critical electric
field strength in silicon

Fig.8.4. Current density dependence on electric field in metal-
SiO_2-Si systems

generated electron-hole pairs increases abruptly so that the ability of insulation breaks down. For semiconductors such as silicon, E_c depends on the
impurity concentration (Fig.8.3) [8.4]. The other mechanism is the tunnel effect. For silicon, E_c is determined by this mechanism for impurity concentration greater than 10^{18} cm^{-3} [8.5]. It is known for an SiO_2 layer that current
flow through it increases exponentially as its thickness decreases. Considerable current begins to flow when its thickness becomes less than about 5 nm.
If, on the other hand, the thickness is greater than 5 nm, the current flow
through it is determined mainly by the Fowler-Nordheim tunneling. Figure 8.4
shows the current density as a function of field strength for a metal-SiO_2-Si
system. A value E_c of about $6 \cdot 10^6$ V/cm [8.6] is obtained if the maximum current density is limited to about 10^{-10} A/cm^2. In addition, the impurity concentration in the lower doping side of a silicon pn junction is limited to
about $1.3 \cdot 10^{19}$ cm^{-3} [8.7].

As a further limiting factor, the so-called punchthrough phenomenon is
mentioned relevant to an isolation method using junctions such as pn junctions.
This is typically related to a system consisting of two junctions being series
connected and at least one of which is back biased. An example demonstrating
this effect is the system of source and drain junctions of an MOS transistor.
It may be assumed that a voltage is applied between source and drain with the
drain junction back biased. As the voltage is increased, a depletion layer
is spreading from the drain junction into the bulk. Normally, the electric
field from the drain junction is terminated at the edge of the depletion layer
in the bulk, and has no effect on the source junction. But when the voltage
becomes higher and the edge of the drain depletion layer comes close to the
edge of the source depletion layer, the drain field begins to influence the

355

potential distribution near the source junction. Finally, the potential barrier of the source junction is lowered, so that majority carriers in the source region are injected into the depletion layer of the drain. The result is a considerable drain current, which begins to flow. This is the so-called punch-through current. For a given drain voltage, the minimum channel length due to this phenomenon is approximately the sum of the depletion layer width of the drain junction and that of the source junction. In the case of a short channel length device, the maximum applied drain voltage is limited by the punchthrough effect rather than the breakdown due to electric field [8.8].

b) Impurity Fluctuation

The number of small-size characteristic regions in which impurity distribution can be considered uniform for a device-characteristic point of view (e.g., a cubic with a side of the depletion layer width) is going to decrease along with the device size. On the other hand, impurities being introduced into a semiconductor have a statistical fluctuation. Therefore, the number of impurity atoms in each of the small-size regions has also a certain fluctuation and this causes the remarkable fluctuation of the device characteristics as the number of the characteristic regions decreases. For example, if a pn junction of of area A is not to be influenced by both the avalanche breakdown and the punch-through at a design voltage of V_0, this voltage must have a certain margin due to the charge fluctuation in the characteristic regions associated with the pn junction. This design-voltage margin ΔV_0 is expressed as

$$\Delta V_0 \simeq \left(\frac{qE_c}{4\varepsilon_s}\right)^{1/2} \left(2 \ln \frac{E_c^2 A}{4V_0^2}\right)^{1/2} \tag{8.2}$$

where ε_s is the permittivity of the semiconductor and q is the electron charge [8.9]. It has been shown that ΔV_0 ranges from 0.5 to 1 V when E_c is assumed as $5 \cdot 10^5$ V/cm. Also in the case of MOS transistors, a threshold voltage fluctuation ΔV_{th} occurs due to a charge fluctuation ΔQ in the depletion layer under the channel. Let us denote S an area of the channel, n the number of devices, N the impurity concentration, t_{ox} the thickness of the gate oxide film and Δ_{ox} the permittivity of it. Then ΔQ and ΔV_{th} can be expressed as follows [8.9,10].

$$\left.\begin{aligned} \Delta Q &= q(2 \ln n)^{1/2} S^{1/4} N^{1/2} \\[2mm] \Delta V_{th} &= \Delta Q \frac{t_{ox}}{\varepsilon_{ox}} \end{aligned}\right\} . \tag{8.3}$$

c) Electromigration

When a DC current flows through a conducting material like metal, a motion of
atoms composing it may be observed. This phenomenon is called electromigration
[8.11,12]. Metallic films such as Al, Au, Mo, Cr and W, which are used in integrated circuits as electrical conductors, are sometimes disconnected due to the
electromigration and this is one of the most important causes of device fail-
ures. It has been suggested that MTF, the mean time between failure, is proportional to J^{-n}, where J is the current density, $n \approx 3$ for large J and $N \approx 1$ for
small J [8.13]. For aluminum films, the current density is generally limited to
about 10^5 A/cm^2 in practical applications, as MTF is very short (below $2 \sim 3$
months) if the current density exceeds 10^6 A/cm^2. Since the size of the metallic conductor films must be made smaller along with the device size, the limitation in the current density determines the maximum current which can flow or
the minimum size of the conductors to be used in the integrated circuits.

d) Heat Generation and Cooling

Heat energy generated by power dissipation inside a device must be removed to
the outside of it by some means; otherwise, the temperature in the device continues to rise until the device becomes impossible to operate or destruction,
the so-called burnout, results.

Cooling with air is one method of heat transfer. It provides a rate of heat
transfer of the order of 10^{-3} W/cm^2/deg in free convection and 10^{-2} to $8 \cdot 10^{-2}$
W/cm^2/deg by forced air-flow. If a temperature rise of 40°C is permissible,
then heat can be transferred to air at rates of 0.1 to 1 W/cm^2/deg. Cooling
with liquids provides substantially increased rates of heat transfer. Especially if the temperature difference between the device surface and the liquid
is high, the heat energy carried away by vapour bubbles due to boiling of
liquid at the surface is increased considerably. Beyond a certain point of temperature rise, however, the vapour is formed so rapidly that it insulates the
hot surface from the liquid and the rate of heat transfer passes through a maximum. Maximum heat-transfer rate of this evaporate cooling depends on the liquid used, but 20 W/cm^2 is considered to be a realistic maximum in electronics.

Since the heat energy per unit area which can be transferred from the device
surface to the outside has a maximum Q_m [W/cm^2], the integration density of the
devices is limited roughly to Q_m/P [cm^{-2}], P[W] being the power dissipated by
a unit device.

357

e) *Drift-Velocity Saturation of Carriers*

It is well known that the carrier drift velocity becomes a nonlinear function of the electric field strength at higher fields and tends to saturate to a cartain limiting value [8.14,15]. This is called the drift velocity saturation, and the limiting velocity is the saturation velocity. The saturation velocity of electrons in silicon was reported as 10^7 cm/s [8.16] in the surface inversion layer.

The operating speed of the device is limited by the finite carrier drift time due to this phenomenon. In an MOS transistor with a channel length L, for example, its cut-off frequency tends to be proportional rather to 1/L than to $1/L^2$. This effect also limits the current density J [A/cm^2], namely an inequality $J \leq qNv_s$ holds, N [cm^{-3}] being the carrier concentration and v_s [cm/s] the saturation velocity. This gives a minimum time which is taken to charge or discharge a stray capacitance associated with a device and becomes another factor in the limitation of operating speed.

f) *Impact Ionization*

Carriers accelerated by an electric field in a semiconductor can cause electron-hole pair generation through collision with the semiconductor lattice. The generation rate grows as the electric field strength increases and in an electric field, being high enough, generated carriers can repeat the generation process, so that carriers are rapidly produced in a high number. The avalanche breakdown treated in Sect.8.2.1a is the case that the rate of increase becomes infinity. Before coming into the avalanche breakdown regime, carriers so generated can stray in the semiconductor or in an insulator adjacent to it and cause a kind of faulty operation.

For example, in an n-channel MOS transistor biased in the current saturation region, electrons injected into the drain depletion layer are accelerated by the drain field and generate electron-hole pairs by impact ionization. A part of these electrons is injected into the gate oxide film and causes a variation of the threshold voltage (hot-electron effect) [8.17]. On the other hand, the generated holes are observed as the substrate current but they also have the possibility to be accelerated again on their travelling through the depletion layer to the substrate and to generate other electron-hole pairs by impact ionization. If this happens near the depletion layer edge, newly generated electrons may be injected into the substrate due to the small potential difference between the substrate and the depletion layer. These electrons can stray in the substrate and cause an error of the dynamic memory.

Table 8.2. Limiting values

Lower limit of thickness for SiO_2 films	5 nm
Upper limit of electric field strength in SiO_2 films	$6 \cdot 10^6$ V/cm
Upper limit of impurity concentration for lower-concentration portion in Si	$2 \cdot 10^{19}$ cm^{-3} (pn) $1.3 \cdot 10^{19}$ cm^{-3} (MOS)
Saturation velocity of electrons in Si	10^7 cm/s (bulk) $6.5 \cdot 10^6$ cm/s (inversion layer)
Upper limit of current density in Al thin films	10^6 A/cm^2
Upper limit of heat transfer	1 W/cm^2 (air cooling) 20 W/cm^2 (liquid cooling)

This section is summarized by showing, in Table 8.2, a list of limiting values which are related with the physical limiting factors.

8.2.2 Physical Limiting Factors Related to Device Fabrication

Many kinds of tools are used today in device fabrication. Among them, a basic tool, which transfers circuit design information to a circuit pattern arrangement with good resolution, is a particle beam using such as photons, electrons and ions. In the following, limitations related to this process are discussed briefly according to the arguments by WALLMARK [8.18].

It is well known that in the case of photon beams the minimum beam size is limited roughly to the wavelength and that is usually about 200 nm. This limitation is due to the uncertainty principle of Heisenberg. Let the uncertainty of position be ΔL, that of momentum be Δp, and h be the Planck's constant. Then the inequality

$$(\Delta L)(\Delta p) \geq h \tag{8.4}$$

holds. This leads to

$$\Delta L \geq \frac{hc}{2E} \tag{8.5}$$

for a photon beam and to

$$\Delta L \geq \frac{h}{2}(2mE)^{-1/2} \tag{8.6}$$

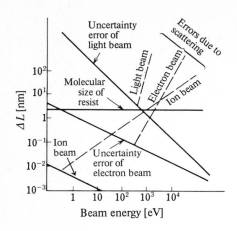

Fig.8.5. Uncertainty of position for particle beam

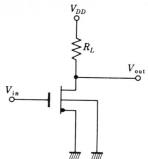

Fig.8.6. An inverter circuit using a DSA MOS transistor

for an electron or an ion beam, where c is the velocity of light, E is the ener by of the beam, and m is the mass of the particle. The inequality (8.6) shows that ΔL becomes much smaller than 1 Å if a $10 \sim 30$ keV electron beam is used. However, as a resist is used in the actual pattern transfer process, ΔL is limited by the size of molecules composing it ($1 \sim 10$ nm). In addition, ΔL is limited by scattering of the beam in the target (the resist here) and is increased with the beam energy. Taking account of these limiting factors, the dependence of ΔL on the beam energy is shown in Fig.8.5.

When a segment is drawn with a tool which has the position uncertainty ΔL, the length of it must have some uncertainty Δl. For the case that two segments are drawn near to each other, the minimum distance between them is limited by the proximity effect. Considering these points statistically, the uncertainty of length Δl is expressed as Δl = 4.6 (ΔL). Then, as ΔL = 2.5 nm is obtained from Fig.8.5, the uncertainty of Δl becomes about 10 nm.

8.2.3 Examples of Minimum Device Size

For an inverter circuit composed by MOS transistors with uniform channel doping the minimum channel length of its driver MOS transistor has already been discussed [8.7]. Here, the minimum size of a DSA MOS transistor used as a driver in an inverter circuit is firstly considered and then that of a conventional MOS transistor used as a switch transistor in a memory circuit is discussed by taking into account the impurity fluctuation.

a) Minimum Size of a DSA MOS Transistor

The minimum distance between source and drain of a DSA MOS transistor, which is used as a driver transistor in an inverter circuit (Fig.8.6) is to be ob-

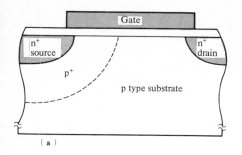
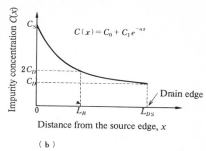

Fig.8.7. (a) Simplified cross section of the DSA MOS transistor. (b) An approximation of the impurity distribution in the base region

tained as a function of the power supply voltage V_{DD}. The punchthrough phenomenon and the gate oxide breakdown are considered to be the limiting factors for this purpose.

Figure 8.7a shows a simplified cross section of the DSA MOS transistor. The impurity distribution in the base region is approximated by using an exponential-like function, as shown in Fig.8.7b, where C_D is the impurity concentration at the drain edge and C_S is that at the source edge. The distance between the source and a point where the impurity concentration is $2C_D$ is defined as L_B. This is considered as the effective channel length of the DSA MOS transistor. Also, let us define $R = L_B/L_{DS}$, and C_D, C_S and R be given constants related to the impurity distribution.

The depletion-layer width x_s from the source edge and x_d from the drain edge are calculated by using the one-dimensional depletion approximation by taking the impurity distribution into account and assuming that charge neutrality holds outside the depletion layer. It is also assumed that when $x_s + x_d < L_{DS}$ holds, punchthrough does not occur, but when $x_s + x_d = L_{DS}$ is valid, it just begins. This L_{DS} value gives the minimum value.

Next, we consider the limitation due to the electric-field strength in the gate oxide film. Let F_{ox} be the maximum allowable electric field strength in the oxide film and t_{ox} be the thickness of it. As the maximum gate voltage of the driver in Fig.8.6 is equal to V_{DD}, the inequality

$$\frac{V_{DD} - V_{FB} - 2\phi_F}{t_{ox}} \leq F_{ox} \qquad (8.7)$$

must hold. If the threshold voltage V_{th} of the dirver is chosen as $(1/2)V_{DD}$, the relation between the impurity concentration N_A and t_{ox}

$$\frac{1}{2}V_{DD} = V_{FB} + 2\phi_F + \frac{t_{ox}}{\varepsilon_{ox}}\sqrt{4q\varepsilon_s N_A \phi_F} \qquad (8.8)$$

can be calculated, ϕ_F being the Fermi potential corresponding to N_A. If V_{th} is determined by the impurity concentration at the source depletion edge x_s, then ϕ_F and N_A in (8.7 and 8) can be expressed as

$$\left.\begin{aligned} N_A &= C(x_s) = C_0 + C_1 e^{-ax_s} \\ \phi_F &= \frac{kT}{q}\ln\frac{N_A}{n_i} \end{aligned}\right\} \qquad (8.9)$$

Let us define $Z = N_A/n_i$ and substitute (8.8 and 9) into (8.7). After some manipulation, we obtain

$$V_{DD} - 2V_{FB} - \frac{4kT}{q}\ln z - \frac{2\sqrt{4q\varepsilon_s n_i kT}}{\varepsilon_{ox}F_{ox}}\sqrt{z\ln z}\left(V_{DD} - V_{FB} - \frac{2kT}{q}\ln z\right) \geq 0 \quad .(8.10)$$

Here V_{FB} is a flat band voltage, k is the Boltzmann constant, T is the absolute temperature in degree Kelvin, ε_{ox} is the permittivity of the gate oxide film, and n_i is the intrinsic carrier density.

If C_S, C_D and R are given, then the upper limit of L_{DS} is determined from (8.10). Figure 8.8 shows the result when R = 0.2, $C_D = 10^{16}$ cm^{-3} and $C_S = 2 \cdot 10^{18}$ cm^{-3}. F_{ox} is chosen as $3 \cdot 10^6$ V/cm to have some margin and V_{FB} as -1 V. Curve A represents the upper limit of L_{DS} due to the gate oxide breakdown and Curve B the lower limit due to the punchthrough. For example, when $V_{DD} = 2$ V, the lower limit of L_{DS} is 0.53 µm and the upper limit is 1.13 µm.

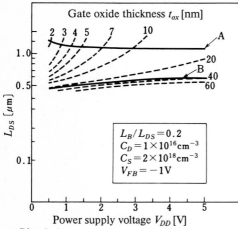

Fig.8.8. Distance between source and drain, L_{DS}, of the DSA MOS transistor

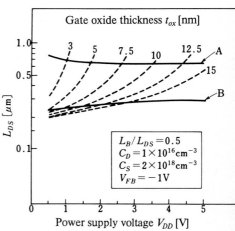

Fig.8.9. Distance between source and drain, L_{DS}, of the DSA MOS transistor

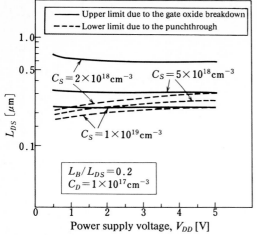

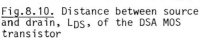

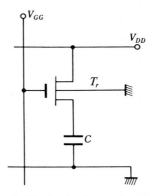

Fig.8.10. Distance between source and drain, L_{DS}, of the DSA MOS transistor

Fig.8.11. Equivalent circuit of the one-transistor memory cell

Any L_{DS} between the Curve A and B can be realized by choosing the thickness of the gate oxide film appropriately. Figure 8.9 shows the result when $R = 0.5$ and C_S and C_D are chosen the same as in Fig.8.8. If $V_{DD} = 2$ V and $t_{ox} = 10$ nm, then L_{DS} is 0.3 µm and the effective channel length is 0.15 µm. Figure 8.10 shows the result when $R = 0.2$, $C_D = 10^{17}$ cm^{-3} and C_S is taken as a parameter. In this case, if $V_{DD} = 1$ V and $C_S = 10^{19}$ cm^{-3}, then the minimum L_{DS} is 0.18 µm and the effective channel length is about 0.04 µm.

b) Minimum Size of a Switch Transistor

One of the limiting sizes of the switch transistor used in a one-transistor memory cell given in Fig.8.11 is discussed next. The impurity distribution in the channel region is assumed uniform and denoted as N_A. In Fig.8.11, when the switch transistor T_r turns on, then a capacitor begins to be charged. The terminal voltage is assumed to become equal to the power supply voltage V_{DD} after the charging is finished. In this case the gate voltage of T_r must be greater than $V_{thH} + V_{DD}$, and it is assumed that $V_{GG} = V_{thH} + V_{DD} + V_{thL}$ by taking into account a margin via V_{thL}, where V_{thL} and V_{thH} are the threshold voltages of T_r when the capacitor terminal voltage is 0 V and V_{DD}, respectively. Therefore, the maximum field strength in the gate oxide film becomes $(V_{GG} - V_{FB} - 2\phi_F)/t_{ox}$ and this must be less than F_{ox}. That is, letting $\phi_F = (kT/q) \ln (N_A/n_i)$,

$$\frac{V_{GG} - V_{FB} - 2\phi_F}{t_{ox}} \leq F_{ox} \tag{8.11}$$

must hold. Now let us choose V_{thL} equal to $(1/2)V_{DD}$, to yield

$$\frac{1}{2}V_{DD} = V_{thL} = V_{FB} + 2\phi_F + \frac{\sqrt{4q\epsilon_s N_A'\phi_F}}{C_{ox}} \quad , \tag{8.12}$$

$$V_{thH} = V_{FB} + 2\phi_F + \frac{\sqrt{2q\epsilon_s N_A(2\phi_F + V_{DD})}}{C_{ox}} \quad , \tag{8.13}$$

where $C_{ox} = \epsilon_{ox}/t_{ox}$. From (8.11-13), C_{oxmax}, the maximum allowable value of C_{ox}, can be obtained as a function of V_{DD}.

Now there is a fluctuation of the threshold voltage due to the impurity fluctuation as expressed in (8.3). This must be, of course, in a range around a designed value of the threshold voltage. Here it is assumed to be expressed as a ratio to V_{thL}, that is, $r = \Delta V_{th}/V_{thL}$ is assumed to be given as a constant. Then, using C_{oxmax} so obtained above and (8.3), the relation

$$\frac{q(2\ln n)^{1/2}S^{-1/4}N_A^{1/2}}{C_{oxmax}} \leq rV_{thL} \tag{8.14}$$

is obtained. Moreover, if L denotes the channel length and W the channel width, the area S can be considered equal to WL. Then, from (8.14) it follows that

$$\frac{q^4(2\ln n)^2 N_A^2}{(C_{oxmax}rV_{thL})^4} \leq WL \quad . \tag{8.15}$$

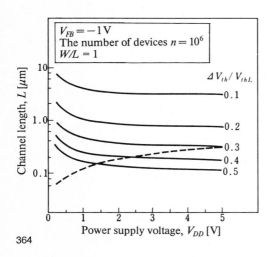

Fig.8.12. Limiting size of the switch transistor in the one-transistor memory cell due to the impurity fluctuation

Figure 8.12 indicates the minimum channel length calculated from (8.15) as a function of V_{DD}, where $n = 10^6$ and $W/L = 1$ are assumed and r is given as a parameter. The dashed curve shows the minimum channel length due to the punchthrough, which is assumed to occur at 2 V_{DD}. When ΔV_{th} is expected to be small, Fig.8.12 explains that it is difficult to use submicrometer channel-length devices, and it becomes more difficult at smaller power supply voltage. If the ratio W/L is greater than 1, then a shorter channel length than that shown in Fig.8.12 can be used. For example, if $W/L = 4$, then the minimum channel length is half of that in Fig.8.12.

8.3 Prediction of Device Performance Advancements

8.3.1 Requirements

It is widely understood that a driving force behind pushing the integration density higher and higher lies in our desire to realize high performance as well as high reliability of a system at the lowest possible cost. It is needless to say that quite a variety of basic technologies is necessary to realize high integration density, such as device design, circuit design, processing, packaging, testing and quality assurance technologies and so on.

The first thing obtainable and noticeable through the use of microlithography by realizing scaled-down elements as well as high integration density is higher speed. A few years ago, it was typical to have 8 - 10 ns/stage delay for MOS transistor circuitry with 5 - 6 μm channel length. Nowadays, however, one attains 2 ns/stage delay with 2 - 3 μm channel length MOS transistor circuitry and even anticipates to reach subnanosecond speed with the shrinkage of transistor sizes down to the submicrometer region. The situation is very similar to bipolar technology and very fast speed (0.2 ns/stage) is anticipated to be realized with the application of such advanced technologies as submicrometer emitter width, oxide isolation and so on. Scientists working on real-time signal processing, super high-speed computers are not content with bipolar technology and are further pursuing the development of even faster device technology such as GaAs, InP III-V intermetallic compounds, which feature higher mobility than that of silicon, and Josephson junction devices.

BRANCOMB of IBM once predicted [8.19] that the key to the realization of high-speed systems is its miniaturization to contain the increase of signal propagation delays as much as possible and there will be a computing machine 150 billion times more powerful than the present one by 2078, which best describes our dreams.

The second demand for higher integration density is the lowering of power consumption. We must decrease the power consumption per function to realize higher integration density and to change the circuit mode from static operation to dynamic operation and, further, it is advantageous to introduce the CMOS technology. Miniaturization of components on a chip helps to reduce parasitic capacitance as well as to improve the power-delay product which is, without doubt, another driving force.

The third condition is the lowering of costs. If various technological development efforts for a higher integration level lead to lowering the processing costs, reducing problems in such areas as device design and testing, and maintaining reasonably high production yield by introducing a redundancy technique, then the cost per function can be reduced by pushing the integration level. A manifestation becomes apparent if one compares the price of a 16 Kbit RAM, which can be purchased today at about the same price of 1 Kbit chip available in the early 1970s.

The fourth challenge is higher reliability. It has been mentioned earlier that, when the integration density is enhanced, the number of items to form a system is reduced as well as the number of connecting points, leading to an enhancement of reliability and the miniaturization of the system. As a result of this tendency, the costs are lowered by reduction of the number of processing steps, the most important effect of all. Recently, soft-errors due to α-particles have been detected and it became apparent that miniaturization of components does not necessarily lead, in each case, to an enhancement of system reliability; the similarity with hot-electron injection into oxide is evident and requires further studies.

The fifth need is related to the nonvolatility of memory. Although semiconductor memory defeats magnetic-type memories cost-wise, it is still far away from the ideal nonvolatile memory. Nonvolatility is only partially realized as EPROM, E^{2}PROM, or a battery backed-up CRAM fashion. In this relation there is a challenge to further improvements and new ideas.

8.3.2 Forecast of Integration-Density and Speed-Performance Trends

As far as the memory development is concerned, the integration density increased twice per year in the past, but volume production-wise it seems like density quadrupled every third year (Fig.8.13). However, as it is experienced with the 64 Kbit dynamic RAM today, it seems to take much more time for bringing the product from development to production with the increase of memory density because of various difficulties involved. The 1 Mbit dynamic RAM is expected not to be put into practical use until the end of the 1980s.

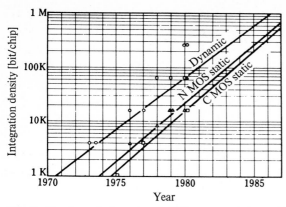

Fig.8.13. Annual trends of MOS memory integration density

Static memory-density increase is also visibly significant. At the present time, its density has reached the 64 Kbit level and will reach the 1 Mbit level by the end of this decade. So far as the bipolar memory is concerned, the 64 Kbit density level memory has been already announced at the research and development level [8.20], however, it is unlikely that bipolar technology will pursue the high density direction because of power and cost problems coming up more and more. Instead, bipolar memory development will be directed toward the ultra-high speed area with relatively small integration density. However, even in the ultra-high speed area, limitations will be felt with silicon bipolar technology and those ultra-high speed memories will be gradually replaced by GaAs ones with an access time of a few nanoseconds.

Figure 8.14 shows the annual trends of access time in static memories. It indicates that the MOS static memory will increase its integration density while maintaining 10 - 50 ns access time, and the MOS dynamic memory will increase its density more and more, while maintaining its access time around 100 ns.

When looking to logic devices, Fig.8.15 reveals the trend of integration level. When simply extrapolating that trend, there may be expected logic chips with 1 M elements on it by the end of the 1980s. However, the problems are how to define, design and test such a complex system with several hundreds of thousands K gates on a chip. Hence, integration density will gradually saturate for several reasons.

Figure 8.16 shows development trends of bipolar logic devices [8.21]. Standard logic devices seem to have a saturation of integration density at around 500 gates per chip. But array logic devices such as gate arrays, PLA etc. have a remarkable increasing tendency and could be the answer to the demand for

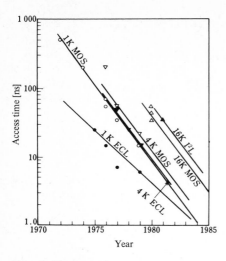

Fig.8.14. Annual trends of static
memory access time

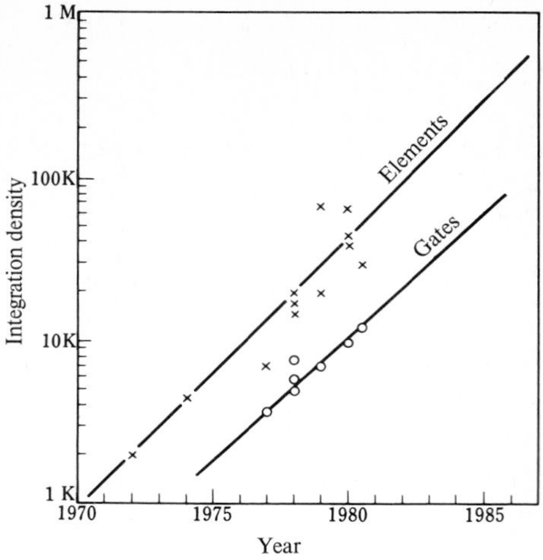

Fig.8.15. Annual trends of MOS locic
device integration density

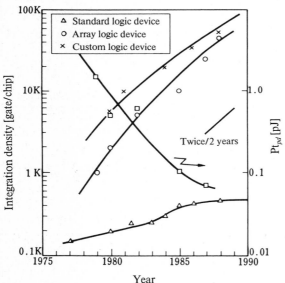

Fig.8.16. Forecast of bipolar
logic device integration
density

shortening of development time. On the other hand, full custom logic devices
will be reaching, from a 10 K-gate integration level in 1980, to a 100 K-gate
level in 1990. As already pointed out when considering MOS logic devices, the
integration density will be limited by power-dissipation and system-definition
problems today and more serious in the next future.

So far as the speed goes, in the latter part of the 1980s, there will be devices with subnanosecond propagation delays using practical 1 μm dimensions or less in GaAs technolgy. It has potential capability of attaining an order of magnitude shorter propagation delay.

8.4 Examples of Device Structure

G.E. Moore, the former President of Intel showed in his by now well-known lecture [8.22] that the integration density increased twice per year in the past based on various technical contributions. According to this investigation the integration density showed about 64000 times increase during the period from 1959 to 1975, contents of which are divided into such factors as 20 times contributions from chip size increase, 32 times increase attributable to component density increase based on lithographic progresses, about 100 times contributions from improvements and innovations in device structures and circuitry. Above all, device structure and circuit innovations have been predominant in the past and will be the stimulating factors in the future.

Some examples related to device structure improvements are such as brought into bipolar ICs the change of pn junction isolation to oxide isolation, invention of I^2L/MTL [8.23] (integrated injection logic/merged transistor logic) and putting it into practical use, introduction of coplanar self-isolation structure [8.24] in the MOS technology, the replacement of the Al gate by self-aligned polysilicon gate structure [8.25], introduction of double polysilicon [8.26] and multi-layer metal interconnect structures, and so on.

The best example of circuit innovation is the early 8 transistor per cell [8.27] to nowadays 1-transistor/1-capacitor dynamic memory cell [8.28]. As shown in the last example related to the memory cell innovation, circuit improvements seem to be reaching a limit in some fields. Hence, integration density improvements will owe greatly its driving force to the progress of lithography. As a matter of fact, the rate of increase of integration level reflects this situation and slows down from quadrupling every third year to quadrupling every fourth year.

As described previously, raising of the integration level will largely depend on the progress of lithographic technology, although down to 1 μm patterning techniques are now being established through the use of 10:1 reduction-type projection aligner as well as the reactive ion etching technology. One of the remaining problems is the various pattern alignment accuracy, which requires usually 1/4 to 1/10 of the minimum dimension.

Based on this background and recognizing the importance for the future, various self-aligning structures were studied and such structures as DSA, MSA, and QSA have been developed at the VLSI Cooperative Laboratory.

Another problem related to device construction is the isolation of integrated components. The technology currently used is mainly the so-called co-planar technique, which shows relatively large dimensional difference between the patterned and the finished dimensions due to the bird-beak phenomenon [8.29], being one of the main obstacles for a higher integration density. Dielectric isolation such as SOS (Silicon On Sapphire) could be a good approach to overcome this barrier.

When the dimensions of the MOS transistor are scaled down, so-called short channel effects are coming into play, and the I-V curve becomes triode-like from the pentode-like characteristics. There are ideas to utilize that characteristic as well as the three-dimensional structure of the so-called SIT (Static Induction Transistor) [8.30] for a further improvement of the integration technique.

Concerning I^2L, this device solved most of the problems related to isolation in an ingenious way, but there remained relatively-slow-speed problems. This area has been recently challenged and seems to have been solved with the introduction of a new structure having less than 1 ns propagation delay [8.31]. So, this technology could add a multitude of potentialities to bipolar ICs such as A-D, D-A converters, analog-digital hybrid type of circuitry as well as ECL-I^2L hybrid type of circuitry.

When the miniaturization of a silicon device is pushed to its extremes, limitations of the material will be felt in the application area where the ultra-high speed is required. In this field, intermetallic compound devices such as GaAs and InP could play a role in the future. In addition, it cannot be excluded that Josephson junction devices might follow as a next generation of devices.

8.4.1 The DSA MOS Transistor

The DSA (Diffusion Self-Align) MOS transistor was initially proposed to improve the high-frequency characteristics of an MOS transistor. From the viewpoint of an integrated circuit component, it is difficult for a conventional MOS transistor to make its channel length less than a certain value which is mainly determined by the necessity to limit the variation of its characteristics within a proper range and accuracy in the photolithography process. On the other hand, shorter channel length is realized effectively for the DSA

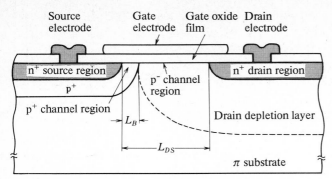

Source electrode Gate electrode Gate oxide film Drain electrode

n⁺ source region

p⁺ channel region

p⁺

p⁻ channel region

L_B

n⁺ drain region

Drain depletion layer

L_{DS}

π substrate

Fig.8.17. Cross sectional structure of an original DSA MOS transistor

MOS transistor by using impurity diffusion processes at the same positioning and therefore results in the operation with higher speed.

Figure 8.17 shows a simplified cross sectional structure of the original DSA MOS transistor. A high resistivity p-type substrate is used and a high impurity concentration p^+-channel region is formed around the source region. A p-type impurity in this case is introduced through the same window as used for an n-type impurity diffusion to form the source region and length of the p^+-region LB is determined accurately by the difference of the diffusion length between the p-type and the n-type impurities from the same window edge.

Next, the operating characteristic of the DSA MOS transistor is discussed briefly in comparison with that of a conventional MOS transistor. As the threshold voltage is generally determined at the highest potential part in the channel, it is determined by the impurity distribution in the p^+-channel region for the DSA MOS transistor. The impurity concentration is highest near the source region and decreases horizontally towards the drain and vertically toward the depth direction. Therefore, if the same threshold voltage and the thickness of the gate oxide film are assumed for the DSA MOS transistor and a conventional one, the impurity concentration near the source region can be made higher for the former than for the latter. So, the dependence of the threshold voltage on the distance between the source and drain regions becomes less and the punchthrough current can be limited in the DSA MOS transistor.

Channel resistance of the p^--channel region is less than that of the p^+-channel region since the effective gate voltage in the former becomes greater than that in the latter. The electron mobility also does so due to impurity difference between them. Therefore, the DSA MOS transistor can be approximated by a series-connection of two MOS transistors. One of them has the channel length of L_B and the other one has that of $L_{DS} - L_B$ and with a low resistance, too. Therefore, it has lower on-resistance and larger transconductance g_m.

The drain field is almost terminated in the p^--channel region when a high drain bias is applied, so that its effect on the p^+-channel region is small. Hence, the drain conductance is small as well as the degradation of output conductance under high drain bias.

As a high resistivity substrate is used, a depletion-type MOS transistor can be easily formed on the same chip and then by using this as a load in the so-called ED-configuration, high-speed, high-density and low-power DSA-ED-MOS integrated circuits can be realized.

One of the drawbacks of this structure is to show asymmetrical character-istics when source and drain are interchanged with each other. In spite of this, it is sometimes used as a pass transistor by utilizing the fact that the threshold voltage in inverse operation becomes less than that of normal operation, and the capacitance between the drain region and the substrate is small. The source-junction capacitance is larger in the original structure, and such an improved device has been proposed in which only a small portion of the source region is in contact with the p^+-channel region.

Next, an example of fabrication of the DSA MOS transistors will be dis-cussed. N channel MOS transistors are also fabricated for comparison. A con-ventional silicon gate process was used except that all etching processes were replaced by dry etching with a reactive ion etching method to realize fine patterning. In doing these processes, special attention has been drawn to growing the gate oxide film which determined transistor characteristics so that interface characteristics did not degrade. Namely, to remove carbon [8.32] which was introduced into a silicon substrate through C_3F_8 gas dry etching of a pad silicon dioxide film in the LOCOS structure, the silicon substrate was exposed in O_2 plasma after the etching to remove a surface carbon pile [8.33, 34]. Using this O_2 plasma processing instability for the BT treatment of an oxide film growing after the dry etching was avoided and 1 μm channel length DSA MOS transistors were realized [8.35].

Both the DSA and N-channel MOS transistors were fabricated on p-type, (100), 100~150 ohm·cm silicon substrates. A double channel doping was done for the latter. Figure 8.18 shows the dependence of the threshold voltage and the drain breakdown voltage on the distance L_{DS} between the source and drain for the DSA MOS transistor and the effective channel length L_E for the N channel one (the gate length is the same for $L_{DS} = L_E$). For the short channel case, especially for the case that L_{DS} or L_E is nearly equal to 1 μm, the threshold voltage stability of the DSA MOS transistor is remarkable and also BV_{DS} is greater by more than 4 V (dotted line in the figure shows BV_{DS} at $I_D = 10$ μA). The double-channel doping in the N-channel MOS transistor is not enough in this case to limit the short-channel effect and the punchthrough.

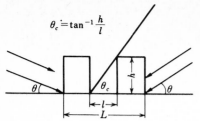

$$\theta_c \doteq \tan^{-1} \frac{h}{l}$$

Fig.8.24. Illustration of shadowing effect on MSA process

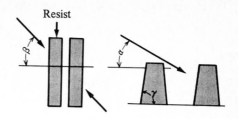

Resist

Fig.8.25. Incident beam angle to resist walls

As illustrated in Fig.8.24, a pair of walls are lithographed, separated by the distance l with wall height h. When the parallel ion beam enters at the angle of incidence θ_i smaller than $\theta_c = \tan^{-1} h/l$, the area between the walls is not irradiated by the beams entering from both directions, thereby providing a process for pattern selection.

Furthermore, as shown in Fig.8.25, parallel beams irradiate with azimuthal angle β and incident angle α. If $\cot\alpha \cdot \cos\beta \geq l/h$, a shadow region will exist between the walls.

Assuming that the inclination angle of resist wall is γ, the condition satisfying the shadowing effect is given by

$$\cot \alpha \cdot \cos \beta \geq l/h \quad , \tag{8.22}$$

$$\cot \alpha \cdot \cos \beta \geq \cot \gamma + l/h \quad , \tag{8.23}$$

where α is the angle of incidance, β is the azimuthal angle, γ is the inclination angle of resist wall, l is the space between resist wall, and h is the resist height.

In the following some experimental results using the MSA technique are given. The MSA MOS process steps are shown in Fig.8.26.

a) Field oxidation and gate oxidation.
b) Channel doping for threshold voltage adjustment.
c) Polysilicon deposition by CVD and impurity doping for lowering the sheet resistivity.
d) Resist walls formation by lithography.
e) Oblique ion etching of polysilicon and oxide film on source and drain contact region except intervened regions between resist walls.
f) Ion implantation for forming the N^+ region of source and drain and then the deposition of refractory metal such as molybdenum for metallization.

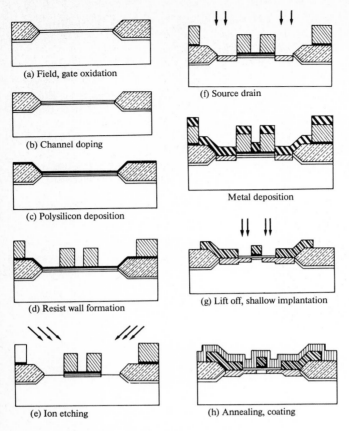

(a) Field, gate oxidation

(b) Channel doping

(c) Polysilicon deposition

(d) Resist wall formation

(e) Ion etching

(f) Source drain

Metal deposition

(g) Lift off, shallow implantation

(h) Annealing, coating

Fig.8.26. Process steps of MSA MOS transistor

g) Then, residual metal on resist is removed by lift off process, and neighbouring regions of gate are implanted for connecting source, gate and
   drain electrically.

h) Finally, wafer is oxide coated for passivation and annealed for acti-
   vating implanted ions by thermal treatment.

By these process steps, an MSA MOS FET can be fabricated. The MSA MOS FET
fabrication process has the following advantages:

Since the position of the source and drain contact region and the gate can
be determined by a single glass mask or one electron beam exposure, this process does not require the severe dimensional tolerance necessary for the conventional process requiring the alignment of three of four glass masks. This
method is efficient for forming a deep source/drain junction in the electrode

378

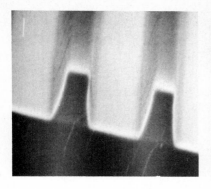

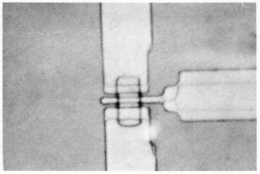

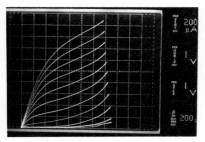

Gate length (L)	1.0 μm
Gate width (W)	6.0 μm
Gate oxide thickness	460 Å
P: 1×10^{16}/cm^2, 25 keV	
As: 2×10^{15}/cm^2, 170 keV	

Fig.8.27. Cross sectional view of resist pattern

Fig.8.28. Top view of MSA MOS FET with 1 μm gate length

Fig.8.29. V-I characteristics of 1 μm gate length MSA MOS FET

contact regions and a shallow source/drain junction in the vicinity of the gate of a short channel MOS FET. It is effective for protecting from the short channel effect, and molybdenum wiring is optimal for fine processing because of its small grain and also provides low resistance.

Figure 8.27 shows the resist pattern with a resist height 3.2 μm and a resist width 2.0 μm. Figure 8.28 gives also the top view of MSA MOS FET with a gate length of 1 μm using the MSA process.

The V-I characteristics of this MOS FET are demonstrated in Fig.8.29. As an application of MSA process, the process for a dynamic RAM with one transistor and one capacitor, as shown in Fig.8.30a, is described in the following.

A wafer surface is oxidized thermally for field isolation, then the silicon dioxide and nitride films are formed on it for gate and capacitor insulators. Further, polysilicon films for gate and capacitor electrodes are deposited on the wafer. Next, a resist pattern is formed lithographically as shown in Fig.8.30b, which is the top view of this pattern. This resist thickness is about 3 μm for making resist walls. Then, the bare polysilicon film which

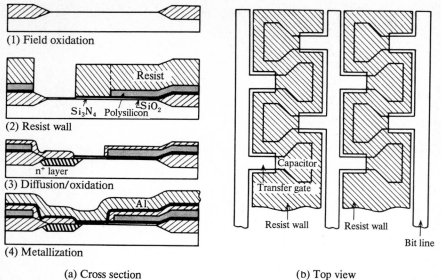

(1) Field oxidation

(2) Resist wall

(3) Diffusion/oxidation

(4) Metallization

(a) Cross section   (b) Top view

Fig.8.30. Process flow of contact holeless type D-RAM using MSA process

is not covered with resist, is etched off by plasma etching. After that, by
a diagonal ion beam etching from the direction of arrow mark shown in
Fig.8.30b, which is parallel to a bit line, the SiO_2 and Si_3N_4 films on the bit
bit line region are removed. However, these films on the cavity region corresponding the transfer gate can remain because of shadowing effect of resist walls.

After the removal of the resist, the bare silicon for a bit line region
and residual polysilicon for capacitor electrodes are doped by thermal diffusion with phosphorous for lowering sheet resistance. Next, regions of bit
line and polysilicon layer are covered with thick oxide by a high pressure
oxidation method for passivation shown in Fig.8.30a (1-3). At that time, the
surface above the transfer gate region cannot be covered with an oxide layer
because the silicon nitride layer of this region prevents the surface from
oxidation.

After the silicon nitride film is removed, a new oxide layer for a gate
insulator is formed on the region. Finally, an aluminum film is deposited on
the wafer and an aluminum wiring pattern can be fabricated as a word line by
the conventional photolithography which is shown in Fig.8.30a (4). Through
this process, the diffusion N^+ layer, metal line and polysilicon layer serve
for a bit line, a word line and a capacitor electrode, respectively.

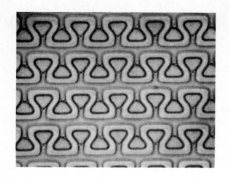

Fig.8.31. Top view of MSA D-RAM cell

In this experiment, we obtained the following results. Oxide thickness was about 9000 Å on the bit line with 6 atm, 900°C, 60 min pyrogenic oxidation. Also, the sheet resistance of N^+ layer and polysilicon layer with 5000 Å thickness after phosphorous doping, was 35 $\Omega/\square$ ($x_j = 1$ μm) and 30 $\Omega/\square$, respectively.

One of the most important advantages of this process is that the position of a bit line, transfer gate and capacitor can be determined by only a thick resist pattern. A single alignment suffices for the whole fabrication. A significant reduction from the three or four mask alignments required for conventional method is realized.

Figure 8.31 shows the top view of MSA dynamic RAM cell pattern. As mentioned in the paper [8.40], this technology is also applicable to the bipolar transistor.

8.4.3 Quadruply Self-Aligned Stacked High Capacitor RAM [8.76]

The one million bit dynamic random access memory (d-RAM) is the major target in high density VLSI memories in the 1980s. Here, the basic technology to realize a one million bit memory by QSA SHC (Quadruply Self-Aligned Stacked High Capacitor) RAM will be described. In implementing high density memory, the reduction in cell area and power dissipation is necessary. Therefore, one transistor one capacitor 1 Tr-1 C dynamic memory is utilized because of its small device number per cell and small power dissipation required to sustain the memory data.

To achieve high packing density by a one transistor-one capacitor cell, the reduction in device dimension and the improvement of the cell structure is the most important demand. According to the device scaling theory, when the device dimension is reduced by a factor of n, then the packing density increases by a factor of n^2. By application of the 2～3 μm design rule, 64 k bit dynamic memories have been realized while by using the $1 \sim 1.25$ μm design rule, 256 k bit memories have been produced. By a simple extension of these techno--logies, a design rule of 1 μm or less will be necessary to realize the one million bit memory. However, two limiting factors must be considered reducing the device dimension in a dynamic memory. The one is related with the supply voltage. Due to the subthreshold current of MOS transistors, a threshold voltage V_T of the order of $V_T \approx 7 \times 2.3 \, \alpha kT/q$ ($\alpha \approx 2$) is necessary. This value depends only on temperature, and at room temperature $V_T \approx 0.8$ V has to be realized independently of device dimensions. Therefore, the supply voltage cannot be lowered down to less than 3 V, in contradiction to the requirement of the scaling theory. If the device dimension is down scaled under constant voltages, source-drain punchthrough or hot carrier injection into the gate oxide will be the most serious problems. The lower limit of channel length is considered to be around 0.8 μm. The second problem is related with the soft errors caused by the penetration of high energy α particles through semiconductors. The amount of the charges stored in the capacitor of memory cell is lowering as the area of memory cell decreases. The amount of charges generated in the cell by an α particle is around 0.03 pC. If the voltage variation due to these charges are to be kept within 1 V, the capacity of the cell cannot be reduced less than 0.03 pF, again in contradiction to the scaling theory. Thus, the solution of these problems is the key point to achieve high density dynamic memory.

Considering the limitations of straightforward down scaling, new device structures must be searched. So far, Si gate memory cell and double polysilicon cell were developed (Sect.8.5.2d). An Si gate memory cell with a cell area of 130 μm^2 in 1.25 μm design rule was reported. This cell area corresponds to the number of squares 83.2 ($= 130$ μm$^2/(1.25$ μm$)^2$). A double level polysilicon cell with a cell area of 45.6 squares ($=71.25$ μm$^2/(1.25$ μm$)^2$) was also reported. However, an ultimate cell area of 4 squares $(4F)^2$ should be possible, in principle, by some planer technology (Sect.8.5.2d). In this sense, Si gate and double polysilicon cell still leave room for further improvements.

QSA SHC RAM achieves the ultimate cell area 6F^2 within the framework of 1 Tr-1 C-cell. Figure 8.32 shows the structure of a 2 bit cell. As switches, QSA MOS transistors are used. The self-registering nature of QSA MOS transistors eliminates the area for registration tolerance. Furthermore, the storage capacitors are stacked over the switching transistors and, thus, the cell area is further reduced. Utilizing a Ta_2O_5 film with dielectric constant 5 times as high as that of SiO_2, a small capacitor area was achieved. By the combination of these techniques, the cell area of $3F \times 2F$ can be realized by the QSA-SHC cell.

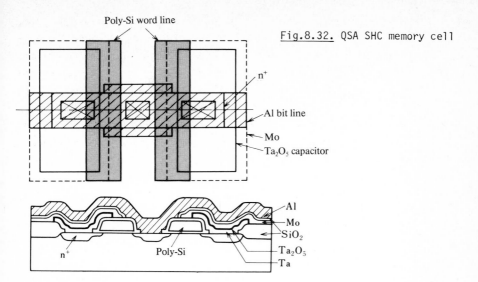

Fig.8.32. QSA SHC memory cell

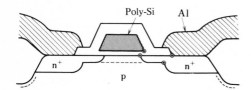

Fig.8.33. QSA MOS transistor

Figure 8.33 shows the QSA MOS transistor. As shown by the marks ⊛ , the four areas, gate electrode, shallow diffusion area, deep diffusion area, and contact area are formed in a mutually self-aligned manner. Thus, this transistor is named as QSA (Quadruply Self-Aligned) MOS. The main features are as follows.

1) Short channel effect is eliminated by shallow n^+ source drain area adjacent to the channel area.

2) Interconnect resistance of diffused area is reduced by deep n^+ diffused area.

3) Contacts between Aℓ interconnections and source drain area are formed in self-align technique with the gate electrode and the edge of deep junction area. Therefore, density of interconnection to the transistor is high and resistance of the circuit can be reduced.

4) Overlapping between gate electrode and source drain area is minimized, and therefore the Miller effect can be reduced.

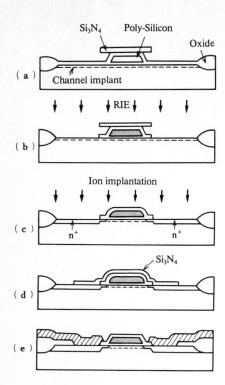

Fig.8.34. QSA MOS process

Figure 8.34 shows the outline of the QSA MOS process. The polysilicon gate is undercut by wet chemical etching or plasma etching, and the side wall of the polysilicon gate and the Si substrate are oxidized by thermal oxidation (a). The wafer is etched by anisotropic sputter etching using the shadowing effect of Si_3N_4 ledge (b). After ion implantation into source drain area (c), selective oxidation is made except the contact area by covering the contact area with an Si_3N_4 mask (d). Finally, the electrodes to the self-align contact are formed by metalization process. Thus, gate electrode, shallow n^+ area, deep n^+ area and contact area are formed in mutually self-aligned manner.

The features of QSA SHC RAM cell are now described. The dielectric constant of Ta_2O_5 film (about 22) is 5.5 times as large as that of SiO_2 (3.9). Therefore, the capacitor of the memory cell can be increased by the same factor with the same capacitor area and the memory cell has enough charge storage capability to reduce the probability of soft errors by α particles. The minimum cell area of double polysilicon cell is determined by the C_B/C_S ratio (Sect.8.5.2d), which in turn determines the amplitude of output signals from the cell to the bit line. In contrast, this limitation is removed and the cell area can be reduced down to the cell area the layout limitation allows.

In this cell, the capacitor is stacked over the switching transistor and field oxide area. Then, the Si active area to place the capacitor is not necessary, and the cell area is determined by the area necessary for the switching transistor. By using the QSA MOS as a switching transistor, source, gate and drain are self aligned and n^+ drain and bit line are common for each two bits. This makes the length of the cell along the bit line be reduced down to 3F, while the width of the cell along the word line can be reduced to 2F. Thus, the cell area $3F \times 2F$ can be realized.

In addition to large storage cell capacity, the small n^+ regions in source and drain and the use of Aℓ bit line facilitate the reduction of probability of soft error due to α particles. Since the C_B of Aℓ bit line is small, the C_B/C_S ratio is also increased.

A 10 mask process is required for QSA SHC process, including Ta_2O_5 formation process.

Using $3F \times 2F$ cell and $F = 1$ µm design rule, the 1 M bit memory array can be packed into 3×2 mm^2 area. With 2 µm design rule, it can be put into 6×4 mm^2 area.

$3F \times 2F$ cell and $3F \times 4F$ cell were designed with 2 µm minimum feature size and 0.5 µm alignment tolerance. The actual cell areas were 6.5×4 µm^2 and 6.5×8 µm^2, respectively, due to layout margin. Reduction of cell area of a factor of $1/8 \sim 1/4$, and the same amount or the double amount of storage charge were achieved as compared to double polysilicon cell of 64 K bit memory. Refresh cycles of 256 were assumed. The lowering of high level of the cell due to α particle can be kept within 0.7 V for the $3F \times 2F$ cell and 0.4 V for the $3F \times 4F$ cell.

Using a 6.5×8 µm^2 cell, a test vehicle for 1 M bit memory was designed. One sense amplifier was shared by 256 cells and, therefore, 4096 bit sense amplifiers were necessary. To reduce the power dissipation of so many sense amplifiers, a dynamic sense amplifier was used (Sect.8.5.1b). This sense amplifier was successfully layed out with 4F pitch, the same as that of the memory cell, by the clever use of QSA MOS. The chip area of 1 M bit memory including memory cells, sense amplifiers X decoders, Y decoders and output buffer amplifiers was packed into 9.2×9.5 mm^2.

Figure 8.35 shows the comparison between the cell area of a QSA SHC memory and those of other dynamic memories. The QSA SHC cell exceeds in density and can be a most powerful approach to the 1 M bit memory. Problems to be solved are the establishment of Ta and Ta_2O_5 process and their reliability, and the reduction of word line resistance for example by using $MOSi_2$ gate.

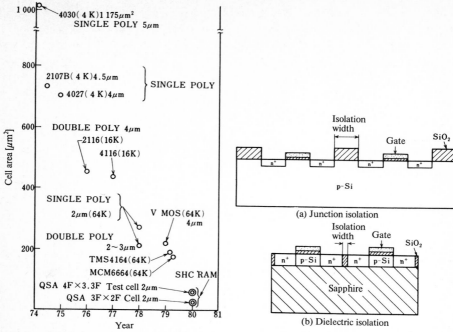

Fig.8.35. Comparison of cell area of dynamic memories

Fig.8.36. Comparison of isolation technologies

8.4.4 Dielectric Isolation

The dielectric isolation technology separating the devices in semiconductor-integrated circuits with dielectric material from each other, has been indicated to have some merits. Those are:

1) The electrical capacitance of interconnection is small and it is possible to get high speed operation.

2) The devices are absolutely dielectrically isolated and it is possible to eliminate the overhear in electrical crossbar exchanger which depends on the leakage current between each of the devices.

3) In principle, the dielectric isolation width can be smaller than the junction isolation width and it is possible to get higher packing density.

Especially item 3) is a strong point for VLSI technology.

The comparison of MOS LSI isolation technologies are shown in Fig.8.36. In the case of conventional junction isolation technology, the isolation width is determined by the punchthrough in the depletion layer between devices as shown in Fig.8.36a. In the case of dielectric isolation technology, the isola-

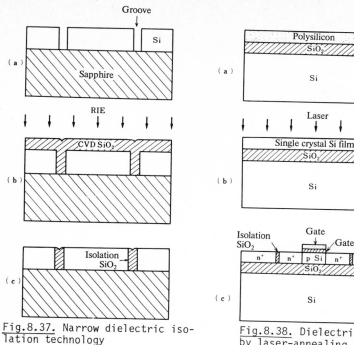

Fig.8.37. Narrow dielectric isolation technology

Fig.8.38. Dielectric isolation device
by laser-annealing technology

tion width is determined by dielectric strength of dielectric material. When a
voltage of 10 V is applied, dielectric isolation by SiO_2 is sufficient only
with the width of 10 nm and the packing density becomes higher than shown in
Fig.8.36b. In this way we can produce VLSI with submicron dielectric isolation width, using direct writing electron beam lithography or X-ray lithography [8.41].

In the case of narrow dielectric isolation shown above, it is needed to
bury the dielectric material like SiO_2 into the silicon groove [8.42]. The
process is shown in Fig.8.37. The processing steps are (a) formation of narrow isolation groove by reactive ion etching (RIE), (b) burying the isolation
groove with chemically vapour-deposited (CVD) SiO_2, and (c) removal of the
surface SiO_2 and remaining the SiO_2 in the groove.

The representative dielectric isolation device is the SOS (Silicon On
Sapphire) MOS FET. A shortcoming of SOS is the sapphire substrate cost. To
solve the problem, laser annealing technology is developed to get a single
crystal silicon film from polycrystal silicon film on amorphous dielectric
substrate [8.43]. In this technology, as shown in Fig.8.38a, a polycrystalline silicon film is grown on SiO_2 film by CVD technology, the polycrystalline silicon film is changed to single crystal film by laser annealing

387

(Fig.8.38b), and dielectrically isolated devices are formed in the single crystal silicon film (Fig.8.38c). The cost of the substrate comes to nearly the same as bulk silicon substrates. Another weak point of SOS MOS FET are the problems of leakage current at low voltage level and mobility reduction with the reduction of silicon film thickness. The reasons of the leakage current are back channeling at silicon-dielectric substrate interface due to high-density interface states, and stacking faults of silicon film. To prevent the back channeling, channel stopper formation at silicon-dielectric substrate interface by ion implantation is effective for leakage current reduction [8.44]. Besides, silicon ion implantation at silicon-dielectric substrate interface and annealing after that are effective for reduction of leakage current, too, because of stacking faults reduction in silicon film by solid phase epitaxial growth [8.45]. Using these technologies, leakage current at low voltage level of thin silicon film MOS FETs comes to the same level as bulk silicon MOS FETs [8.46].

Concerning the mobilities, the reasons of mobility reduction with the reduction of silicon film thickness has been attributed to the increase of stacking faults at silicon-dielectric substrate interface [8.47] and the increase of compressive stress at the interface [8.48]. But recently, the author proposed an opinion that series resistances involved in the channel of thin film MOS FET, i.e., contact resistance and diffusion layer resistance increase with the decrease of film thickness, and so apparent mobilities seem to be reduced. In this case, the silicon film thickness is limited physically and cannot be reduced so thin, and so the narrow isolation technology becomes more important.

Application of submicron technology for dielectric isolation will be a realistic way to make high switching speed and high density VLSIs.

8.4.5 I^2L

I^2L (Integrated Injection Logic) or MTL (Merged Transistor Logic) is a high density and low delay power product bipolar device. Its speed is also being improved down to the propagation delay time less than 1 ns.

Figure 8.39 shows the basic structure of an I^2L gate [8.49]. The basic inverter consists of lateral pnp transistor and npn transistor. The base of the pnp transistor and emitter of the npn transistor are grounded to the common n^+ substrate. An isolation between the pnp transistor and npn transistor is not necessary. The ground line is not necessary, either. Thus, the layout of I^2L gates can be made very compact. The role of emitter and collector in the npn transistor is reversed in the normal sense. The low cut off frequency of the

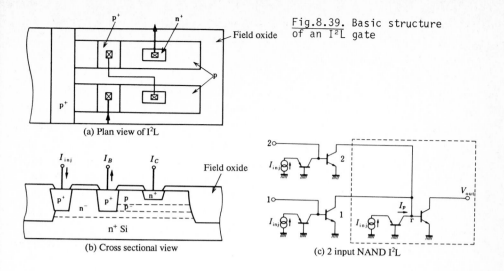

Fig.8.39. Basic structure of an I^2L gate

(a) Plan view of I^2L

(b) Cross sectional view

(c) 2 input NAND I^2L

inverse npn transistor was the main reason for the slow speed of I^2L, as compared to other bipolar circuits, but today much improvement has been done.

Figure 8.39c shows the two input NAND gates. When the two input levels 1 and 2 are low (L), the level of node r is low (L) and the npn transistor is cut off and thus the output level is high (H). When either one of the input levels is H, the output level is L. In this case, the time during which the npn transistor switches on t_{on} is determined by the current I_p supplied by the pnp transistor which is in turn controlled by the injector current I_{inj}. The time during which the npn transistor switches off t_{off} is determined by the time required for the npn transistor of the previous stages (say transistor 1) to sink the current I_p. This sink current is given by $\beta_{eff} I_p$, where β_{eff} is the inverse current amplification factor of the npn transistor.

In the low current regime, and when $\beta_{eff} \gg 2$ (as is usually the case), t_{off} is smaller than t_{on} because the discharging current $(\beta_{eff}-1)I_p$ is larger than the charging current I_p with respect to the node r. Thus, the propagation delay time is given by

$$t_D \approx \frac{t_{on}}{2} \approx \frac{C_T V_{BE}}{2I_p} \quad . \tag{8.24}$$

Here, C_T is the capacitance of the node r including wiring and junction capacitance, and V_{BE} is the voltage swing between base and emitter.

To reduce C_T, oxide isolation as shown in Fig.8.39 and reduction in the devise dimension are necessary. In this region, as current increases, t_D decreases.

389

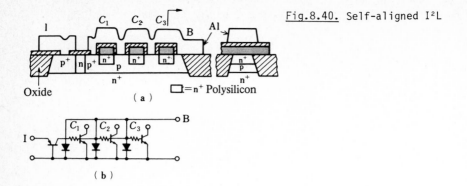

Fig.8.40. Self-aligned I²L

$\square$ = n⁺ Polysilicon

(a)

(b)

At high current regime t_{off} is larger than t_{on} because of minority carrier storage. Then, t_D is given by

$$t_D \approx \frac{t_{off}}{2} \approx \frac{1}{2}\frac{Q_{storage}}{\beta_{off}I_p} \quad . \tag{8.25}$$

Here, Q_{store} is the electronic charge stored in the base region. As the current increases, Q_{store} also increases and then the decrease in t_D gradually slows down and reaches a constant value.

Recently, using self-aligned contact-technique, and reducing the volume ratio of base to collector to 2 : 1, as shown in Fig.8.40, the charge storage is reduced and thus $\tau_p \simeq 1$ ns is achieved with 2.5 μm design rule. Here, the reduction of base resistance by Aℓ interconnection is also contributing to this high speed [8.50].

Figure 8.41 shows a result of simulation when the device dimension is scaled down. The minimum t_D is predicted to be around 150 ps. The device parameters are shown in Table 8.3.

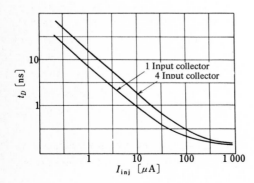

Fig.8.41. Simulation results for devices with minimum line width 0.5 μm. Other device parameters are listed in Table 8.3

Table 8.3. Device parameters and performance of down scaled I^2L

Junction depth	extrinsic base	$2000 \, \overset{\circ}{A}, \, 1 \cdot 10^{20} \, cm^{-3}$
	npn collector	$1000 \, \overset{\circ}{A}, \, 2.3 \cdot 10^{20} \, cm^{-3}$
Thickness of epitaxial layer		$2000 \, \overset{\circ}{A}$
Substrate doping		$3 \quad 10^{20} \, cm^{-3}$
base width,	pnp	$1000 \, \overset{\circ}{A}, \, 2 \cdot 10^{18} \, cm^{-3}$
impurity concentration	npn	$1000 \, \overset{\circ}{A}, \, 2 \cdot 10^{18} \, cm^{-3}$
	npn emitter base	23 ps
Time constant	npn collector base	31 ps
	pnp emitter base	165 ps
Base substrate diode		180 ps
Area of the cell (one output collector)		$7 \, \mu m^2$

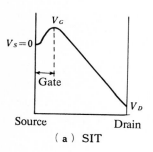

(a) SIT

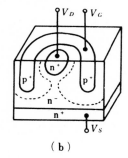

(b)

Fig.8.42. Potential distribution and structure of SIT

8.4.6 SIT (Static Induction Transistor)

SIT (Static Induction Transistor) is a voltage-controlled FET device exhibiting triode-like non-saturating I-V characteristics [8.30]. The device operates in a punch-through mode. The majority of the carriers that flow from source to drain are controlled by the gate potential barrier as shown in Fig.8.42a. The drain current I_D is given as a function of gate voltage V_G and drain voltage V_D by the following exponential form

$$I_D = I_0 \exp\left[-\frac{q}{kT} \eta \left(V_G - \frac{V_D}{\mu}\right)\right] \quad . \tag{8.26}$$

Here, k, T, q, μ, I_0, η are Boltzmann's constant, absolute temperature, electronic charge, voltage amplification factor, a constant current value determined by the device structure and a constant factor less than 1, respectively. Because of exponential dependence of I_D on V_G and V_D, the SIT has a large current driving capability and large conductance and, therefore, high speed.

Figure 8.42 shows the vertical SIT structure. Like conventional junction FET, there are two operational modes, namely normally off type and normally on type. In the normally off type SIT, the gate electrode is forward biased and, therefore, maximum operating voltage is restricted within about 0.8 V. This results in small delay power product, although this also means that tight control of threshold voltage is necessary. Due to high transconductance, high speed operation is possible.

Normally on type SIT has large current driving capability because its gate electrode is reverse biased and large logic swing is available, as implied by the current equation.

Features of SIT are summarized in Table 8.4.

Normally on type SITs are used in low noise high power audio amplifiers.

Table 8.4.   Characteristics of SIT

Operating mode	Gate bias	I-V characteristics	Threshold voltage	Gate capacitance	Time constant
Normally off	forward	$V_G=0$	positive small	storage small	small
Normally on	back	$V_G=0$	negative, absolute value is large	no storage	minimum

Normally off type SITs are used in LSI. Figure 8.43 illustrates an SIT Logic (SITL) of I^2L type circuit configuration. The inverse npn transistor of conventional I^2L is replaced by the vertical SIT. By this structure and drain channel of 6×6 μm^2, the propagation delay time of 5.2 ns/gate and

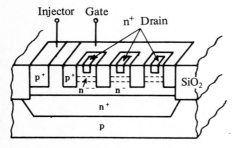

Fig.8.43. I^2L type SITL

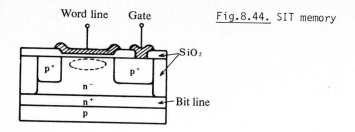

Fig.8.44. SIT memory

power dissipation of 100 μW/gate are obtained with 3 μm design rule [8.51].

Figure 8.44 shows the SIT dynamic memory. A p^+ gate is negatively biased and the gate electrode of MOS diode is used as a word line, the buried n^+ line is used as a bit line. Storage charge is stored in the potential well of the MOS diode. Three levels, write in voltage (V_H), holding voltage ($\sim \frac{1}{2} V_H$) and read out voltage (0) are applied to the word line. In writing, the bit voltage of V_H is applied both to the cells into which '0' is written and to the unselected cells.

8.4.7 GaAs IC Technology

GaAs ICs have high speed for two reasons, i) high electron mobility and ii) use of semi-insulating substrates. Figure 8.45 shows the planar Schottky barrier FET and its drain current I_D versus gate voltage (V_G) relation [8.52]. As in other FET structures, there are normally on (D type) and normally off (E type) type devices. In the normally off type device, the gate bias is confined to about 1 V (Schottky barrier height). In general, D type devices have high current driving capability and therefore high speed.

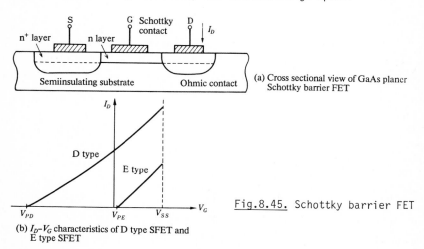

(a) Cross sectional view of GaAs planar Schottky barrier FET

Fig.8.45. Schottky barrier FET

(b) I_D-V_G characteristics of D type SFET and E type SFET

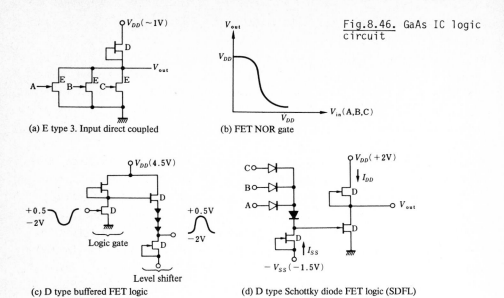

(a) E type 3. Input direct coupled

(b) FET NOR gate

Fig.8.46. GaAs IC logic circuit

(c) D type buffered FET logic

(d) D type Schottky diode FET logic (SDFL)

Figure 8.46 shows representative circuit configurations of GaAs FET ICs. Figure 8.46a demonstrates a 3 input direct coupled ED circuit. Output voltage V_{out} is shown as a function of input voltage V_{in} (A or B or C) in Fig.8.46b. Since the range of input voltage and output voltage is identical, this circuit can be directly coupled to the next stage of the circuit. The logic swing is kept within 1 V. This circuit is very simple and requires a minimum number of transistors. However, tight control of threshold voltages of FETs are necessary (Table 8.5). The D-type buffered FET logic needs a negative logic swing

Table 8.5. Comparison of GaAs IC ring oscillator

Circuit type	Gate width [μm]	Gate length [μm]	Power dissipation [mV/gate]	Delay [ps/gate]	Power delay product	Fan in	Fan out
D-buffered logic	20	1	40	86	3.9	2	2
	10	0.5	5.6	83	0.46	1	1
	50	0.5	41	34	1.4	1	1
D-Schottky diode-FET logic	5	1.0	0.17	156	0.027	2	1
			0.48	107	0.051		
	10	1.0	0.34	120	0.040	2	1
			0.73	95	0.069		
	20	1.0	1.10	99	0.087	2	1
			2.26	75	0.170		
E-direct coupled FET logic	–	0.6	1.2	30	0.036	1	1
	20	1.2	0.1	300	0.03	1	1

(0.5~-2 V) for the input gate while the output voltage of the inverter stage is positive. Therefore, level shifting buffer circuit is necessary. Because of high driving capability of D-type MES FET, this type of circuits has high speed but high power dissipation. Due to large area of level shifting diodes, the packing density is rather low. Schottky diode FET logic as shown in Fig.8.46c employs the D-type FET but level shifting is made in the low current input circuit. The level shift diodes operate as diode logic as well. This type of circuit can realize high speed and medium power and density. By using an ion-implanted planar process, 8×8 multiplier with about 1000 gates, 5.3 ns multiplication time and power dissipation of 2 W were developed.

The ion-implanted planar GaAs FET as shown in Fig.8.45a is fabricated as follows. n and n^+ layers are formed by the successive selective ion implantation of Se and S into Cr doped or non-doped semi-insulating GaAs substrate. Photoresist is used as the mask for the selective ion implantation. The depth and impurity doping level of n layer is critical for the control of the threshold voltage of FET. Reproducibility of ion implantation technique is better than that of epitaxial technique. This structure is planar and, therefore, high packing density can be realized. The key processes are annealing the ion implanted layer, CVD passivation, Schottky contact and Ohmic contact formation.

Figure 8.47a shows a novel FET structure called HEMT (High Electron Mobility Transistor) [8.53]. As shown in the band diagram of Fig.8.47b, electrons generated in the depletion layer of n type $Al_xGa_{1-x}As$ are transferred into the interface region of non-doped GaAs to form an electron layer. At liquid nitrogen temperature, electron mobility as high as 32500 cm^2/Vs (10 times that of GaAs at room temperature) has been realized. By using 2 µm gate length, cut off frequency of 8.2 GHz and values 20% higher than that of conventional GaAs FET have been obtained at room temperature.

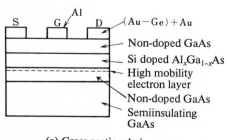

(a) Cross sectional view

Fig.8.47. HEMT (High Electron Mobility Transistor)

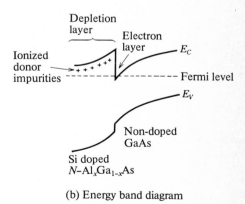

(b) Energy band diagram

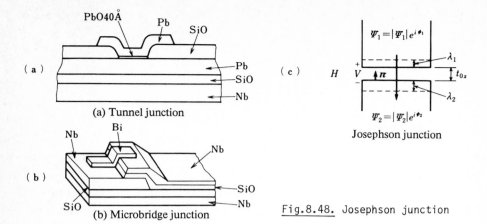

(a) Tunnel junction

(b) Microbridge junction

Josephson junction

Fig.8.48. Josephson junction

Ballistic electron transport (BET) is expected as the gate dimension of GaAs FET is reduced down to less than the electron mean free path [8.54]. The operation of the FET in this region is just like that of a vacuum tube of very small dimension. According to a theoretical prediction, switching speeds of $0.5 \sim 1$ ps should be possible.

Integration of digital circuits and optical devices such as injection laser on one GaAs chip is becoming important as the fibre optical communication progresses.

Without calling in question the high performance of GaAs devices, the production technology is still premature as compared to Si technology, due to lack in the knowledge of how to control GaAs crystals. The establishment of fabrication technology will be the key factor for the future development of GaAs ICs.

8.4.8 Josephson Junction Devices

A Josephson junction is a weak link between two superconductors. There are two types of weak link, namely tunnel junction and microbridge. Figures 8.48a and b illustrate these types of junctions respectively.

According to the microscopic theory of superconductivity developed by BARDEEN, COOPER and SCHRIEFFER (BCS), the wave function of the ground state of a superconductor can be written as $\Psi = |\Psi| e^{i\phi}$, where $|\Psi|$ is its amplitude and ϕ is the phase. Two superconductors have respective wave functions $\Psi_1 = |\Psi_1| e^{i\phi_1}$ and $\Psi_1 = |\Psi_1| e^{i\phi_2}$ as shown in Fig.8.48c. According to Josephson's prediction, when there is a weak link between the two superconductors, there will be a supercurrent between them, which is described by [8.55],

$$I_j = Aj_1 \sin \phi \quad , \quad \phi = \phi_1 - \phi_2 \tag{8.29}$$

$$\frac{\partial \phi}{\partial Y} = \frac{2e}{h} V \tag{8.30}$$

$$\nabla^{(2)} \phi = \frac{2ed}{h} (H \times n) \quad , \quad d = \lambda_1 + \lambda_2 + t_{ox} \tag{8.31}$$

$$j_1 = \frac{\pi \Delta G_{nn}}{2eA} \tan h\left(\frac{\Delta}{2kT}\right) \quad . \tag{8.32}$$

Here, G_{nn} is the tunneling conductance and $2e$ in these equations implies that the supercurrent consists of an electron pair.

From (8.29,30), when $V = 0$, then $\phi = $ constant and, therefore, I_j is constant. This means that a constant supercurrent flows through the junction at zero bias voltage (D.C. Josephson effect). Maximum supercurrent is $I_m = Aj_1$. When $V \neq 0$, then the A.C. supercurrent $I_j = Aj_1 \sin[(2e/h)Vt + \text{const.}]$ will flow through it (A.C. Josephson effect).

In the case of tunneling junction, besides the supercurrent, there are displacement current I_c and single electron (quasi particle) tunneling current I_{gs} that are expressed as

$$I_c = C \frac{dv}{dt} \tag{8.33}$$

$$I_{qp} = \left(\frac{G_{nn}}{e}\right) \int N(E - ev)N(E)[f(E - ev) - f(E)]dE \quad . \tag{8.34}$$

Here, $N(E)$ and $f(E)$ are the electronic density of states and the electronic distribution function. The junction current consists of three components, namely

$$I = I_j + I_{qp} + I_c \quad . \tag{8.35}$$

There is a gap energy Δ for a single electron pair in the ground state to break into two single electrons in the excited states. Therefore, the single electron tunneling current I_{qp} will be zero below the bias voltage Δ/e. Thus, the I-V characteristics of the tunneling junction will exhibit a gap voltage such as shown in Fig.8.49a.

In the case of microbridge type junctions, the junction can be thought of as the composite of supercurrent, inductive current and resistive current. The I-V characteristics will be like that of Fig.8.49b.

As digital devices, tunneling junction is more widely studied than microbridges. So, the following discussion will be focussed on it.

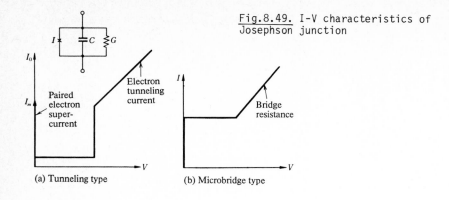

Fig.8.49. I-V characteristics of Josephson junction

(a) Tunneling type

(b) Microbridge type

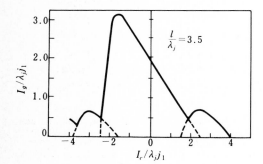

$\frac{l}{\lambda_j} = 3.5$

Fig.8.50. Threshold character-istics of inline gate (1: width of junction; λ_j: Josephson penetration depth; I_c: control current; J_1: tunneling current density)

When a control gate is put on a junction, the maximum supercurrent will be controlled by the magnetic field generated by the control current as implied by (8.31). Figure 8.50 illustrates the maximum supercurrent (denoted by gate current I_g) versus control current (denoted by I_c) relation. These curves are called characteristic relation. In the area below these curves, the junction is in a superconducting state, while above these curves it is in a voltage state where d.c. current is composed of only normal electron current carried by a finite bias voltage.

Figure 8.51a shows the equivalent circuit of what is called a 3 input in-line gate. Three control gates are overlapped on a single junction. There are two stable points in this circuit as shown by the points a and c, in the I_G versus V_G curve in Fig.8.51b. Transition between these points can be controlled by I_C and I_G. Figure 8.51c shows the characteristic curves of the 3 input in-line gate. In the positive sense of I_c, the self magnetic field generated by I_g is additive to that of the control gate, and small I_c will suffice to make the junction undergo a transition from superconducting state to a voltage state. In the negative sign of I_c, the opposite is the case. If any of the currents of

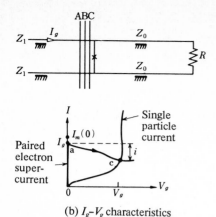

(b) I_g-V_g characteristics

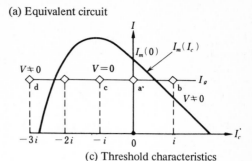

Fig.8.51. Input inline gate

(a) Equivalent circuit

(c) Threshold characteristics

the control gate A, B or C is +i, then the junction changes from the initial
superconducting state a to the voltage state b. In this case, the gate acts
as a 3 input OR gate. When all of the currents of the input gate A, B, C are
-i, then the junction changes from the initial state a to the voltage stated.
In this case, the gate acts as AND gate.

When each cycle of logic operation is completed, the junction must be returned to the superconducting state a. There are three modes of resetting,
namely self-resetting, non-latching and latching, depending on the circuit
loading. In integrated circuit, the circuit is generally latching and, therefore, resetting operation is necessary.

In the in-line gate, to obtain a current gain $I_g/I_c > 1$, the condition
$\ell > 3\lambda_j$ is necessary, where ℓ is the width of the junction and λ_j is the
Josephson penetration depth. By this condition, the minimum feature size
of the in-line gate has a lower limit.

Figure 8.52 shows the equivalent circuits and the threshold characteristics of 2 junction and 3 junction SQUID (Superconducting Quantum Inter-

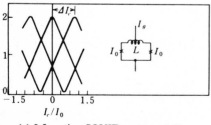

(a) 2 Junction SQUID $\Phi_0/LI_0 = 0.94$

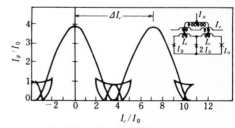

(b) 3 Junction SQUID $\Phi_0/LI_0 = 7$

Fig.8.52. Threshold characteristics of multijunction devices

ference Device). These circuits employ loops including two or three junctions. In this case, the magnetic fluxes are applied to the loops rather than to junctions themselves and, therefore, larger current gains are obtained with small junction areas.

As a logic device, the 3 junction SQUID is used. As shown in the threshold characteristics of Fig.8.52, current gain of about 3 is possible.

The third logic gate is CIL (Current Injection Logic) as shown in Fig.8.53 [8.56]. In this circuit, current I_A and I_B are injected into 2 junction SQUID. As shown in the characteristic curve (Fig.8.53b), this circuit functions as AND gate.

Comparison of these circuits is made in Table 8.6.

For a SQUID to make a transition to a voltage state, the product of loop inductance and loop current must be at leat larger than the magnetic quantum flux Φ_0. This condition limits the density of SQUID-type circuits. So, current

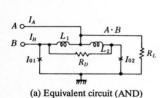

(a) Equivalent circuit (AND)

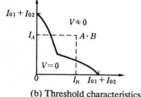

(b) Threshold characteristics

Fig.8.53. 2 junction current injection logic

Table 8.6. Comparison of Josephson logic gates

Circuit configuration	Minimum dimension	Circuit	Speed	Power
Inline gate	25 µm	4 + 4 adder	125 ps/gate	25 µW/gate[1]
		4 × 4 multiplier	236 ~ 275 ps/gate	
		Add + shift multiplication	6.6 ~ 3.0 ns 27 ~ 12 ns	
3 junction SQUID	5 µm	2 input OR	40 ps	
		2 input AND	105, 73 ps/gate	
		300 circuits[2]	5 ns/cycle	1.5 mW/chip
CIL	2.5 µm	4.5F/I 3F/0	36 ps/gate	
		2 input OR	13 ps/gate	
	1 µm		2 ~ 3 ps/gate (estimated)	1 µW/gate

[1] F/0 = 1.4   [2] Chip area (6.35 mm)^2

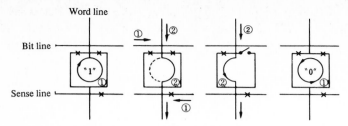

<u>Fig.8.54.</u> 3 junction NDRO memory cell

injection-type circuits without resorting to magnetic coupling are being looked for.

Figure 8.54 shows a 3 junction NDRO (Non Destructive Read Out) memory [8.57]. Permanent circular current along the loop is used for memory retention.

Clockwise current is defined '1', and anticlockwise current '0'. Word line is bidirectional and bit line and sense line are monodirectional. Reading is made by the coincidence of sense and bit line. Writing is made by the coincidence of word and bit line.

<u>Table 8.7.</u> Comparison of memory cells using Josephson junction devices

		Minimum feature size	Access time	Cycle time	Chip area
NDRO memory	64 Kbit full decoded	25 µm	4~23 ns	5~35 ns	$(6.35 \text{ mm})^2$
	2 Kbit	5 µm	1.8 ns	–	$(6.35 \text{ mm})^2$
	4 Kbit (1 kW × 4 bit)	2.5 µm	500 ps (estimated)	–	$(6.35 \text{ mm})^2$
DRO memory	16 Kbit	2.5 µm	15 ns		

<u>Table 8.8.</u> Conceptual design of superfast Josephson computer with 3033 architecture

Single CPU	1 ns cycle time	300 K circuits
Cashe memory	2 ns access time	256 K byte
Main memory	10 ns access time	64 M byte
Volume vessel of liqu. He systen		$10 \times 8 \times 8 \text{ cm}^3$ $6 \times 4 \times 4 \text{ cm}^3$
Power dissipation Josephson Semiconductor		10 W 20 kW

Characteristics of high speed 3 junction NDRO memory and high density 2 junction DRO (Destructive Read Out) memory are compared in Table 8.7.

Since the logic swing of Josephson junction device is of the order of gap voltage ~3 mV, nearly 3 orders of magnitude smaller than that of semiconductor devices, the delay power product is also three orders of magnitude smaller. Therefore, very compact system design is possible. It is estimated that 1 ns cycle time and the throughput of 250 Mips (million instructions per second) can be realized by packing an IBM 3033 mainframe computer into the volume $6 \times 4 \times 4$ cm^3. Table 8.8 summarizes the results.

A fundamental problem in Josephson tunneling junction device is the reliability. Upon repeated thermal cycling between liquid He temperature and room temperature, tunneling junctions become shortened in the rate, say $5 \cdot 10^{-6}$ failure/devices/200 cycles, because of grain growth of the electrode metal. Microbridges are comparatively free from it [8.58]. This problem to be solved before a computer system is constructed is now under an active investigation.

8.5 Device Structure

8.5.1 Logic Circuit

The directions of Si VLSI logic can be divided into two categories, namely high speed bipolar logic LSI and high density MOS LSI.

a) Ultrafast Logic Circuits

TTL (Transistor Transistor Logic) and its improved version, the LSTTL (Low Power Schottky TTL) are most widely used as low cost logic SSI and MSIs. Figure 8.55 shows an LSTTL implementation of a 2 input NAND gate. In this circuit, clamping the collector low level by a Schottky diode connecting collector and base prevents to some degree collector saturation responsible for lowering the circuit speed.

CML (Current Mode Logic) or ECL (Emitter Coupled Logic) are used as ultrafast logic devices. Figure 8.56a shows an ECL 2 input NOR/OR circuit. This circuit takes a form of the differential amplifier and operates in a constant current mode. Namely, the resistor connected to the emitter of all transistors plays the role of constant current source. This constant current switches to the branch performing the logical operation or to the reference transistor depending upon whether the input level A or B is higher than the reference voltage V_{ref} or not. Here, the logic swing can be designed to be very small so that 'on' transistors never saturate. The output emitter follower circuits

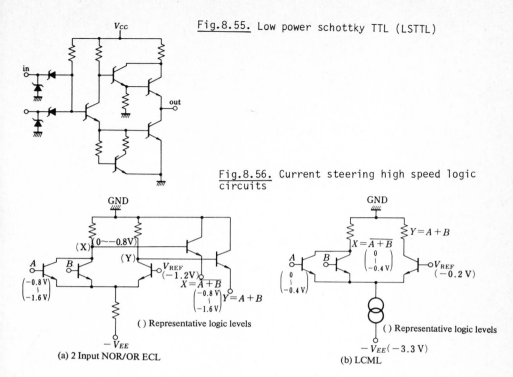

Fig.8.55. Low power schottky TTL (LSTTL)

Fig.8.56. Current steering high speed logic circuits

(a) 2 Input NOR/OR ECL

(b) LCML

( ) Representative logic levels

are the level shifter and current driver. Figure 8.56b shows the LCML (Low Level CML) circuit. Here, by choosing the logic levels 0~-0.4 V and reference level -0.2 V, the level shifting emitter follower circuits are eliminated. In this circuit device number and power dissipation are reduced and high speed is achieved by small logic swing.

To realize a high speed bipolar LSI, high speed transistors are necessary. For this purpose, reduction of base resistance or stray capacitance is attempted by self-aligning techniques or local oxidation technique. Figure 8.57 illustrates the isoplanar transistor (a) and the PSA (Polysilicon Self-Aligned)

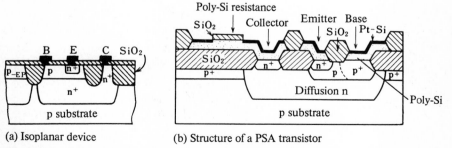

(a) Isoplanar device   (b) Structure of a PSA transistor

Fig.8.57. Self alignment techniques for a bipolar transistor

transistor (b) [8.59]. In the PSA process shown here, collector n diffusion, thick oxide formation, selective boron ion implantation into base region and subsequent drive-in process are made. Then, after polysilicon deposition, the polysilicon interconnection pattern is formed by local oxidation using Si_3N_4 masking. Then, phosphorous is diffused into the polysilicon area connecting to emitter and collector, while boron is diffused into the polysilicon area connecting to base region. Thus, self-aligned contacts to base emitter and collector are formed. Resistivity of polysilicon is reduced by a Pt-Si layer. The PSA technology can be applied to ECL or TTL, as well. For example, an 18 bit RALU (Register Arithmetic Logic Unit) by 400 ps ECL was developed.

In the following, simulation results about bipolar speed limitations are described [8.60]. Here, gain band width f_t and frequency of unity gain f_{max} are defined as

$$f_t = \frac{1}{2\pi}\left(\frac{\Delta I_c}{\Delta Q_h}\right) \qquad f_{max} = \sqrt{\frac{f_t}{2\pi R_B C_{BC}}}$$

where ΔI_c is the collector current variation, ΔQ_h is the variation of stored hole charge by base emitter voltage variation, R_B is the base resistance and C_{BC} is the base collector capacitance. Device parameters and predicted performance are shown in Table 8.9. Maximum value of f_{max} is 9.9 GHz.

In scaled down bipolar transistors, the bias voltage cannot be reduced less than about 1 V, because it must be larger than the minimum forward bias voltage of a p-n junction in the range of 0.7 V ($\simeq$ band gap of Si). Therefore, a tradeoff between speed an power dissipation will be very critical, and the technique of cooling the devices will be very important.

Table 8.9. Predicted performance of bipolar transistor and device parameter ($V_{BC} = 0$)

Parameter	Device 1	Device 2	Device 3	Device 4	Device 5
Emitter size [μm^2]	3×7	1×2	1×1	1×1	1×1
Emitter depth [μm]	0.4	0.2	0.1	0.05	0.03
Base width	0.25	0.2	0.1	0.05	0.03
f_t [GHz]	3.3	6.4	16.8	45.8	89.2
Collector current [mA]	2.11	0.31	0.38	0.62	0.73
Diffusion capacitance [pF]	2.5	0.24	9.6×10^{-2}	2.0×10^{-2}	9.7×10^{-3}
Current gain	75	96	53	63	105
f_{max} [GHz]	2.1	4.2	9.9	7.6	6.1
Base collector breakdown voltage	>2.0	>2.0	>2.0	2.0	2.0

b) High-Density MOS VLSI Logic

There are many reasons that spur high density in MOS VLSI, namely small areas of isolation, the multilevel interconnection of Si gate processes, the small device number per function or the small power dissipation of ED and CMOS circuits. Some interesting examples of these features are given here.

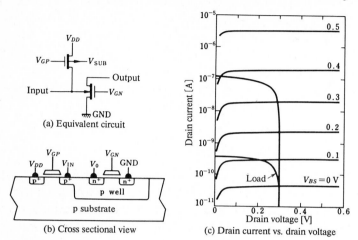

(a) Equivalent circuit

(b) Cross sectional view

(c) Drain current vs. drain voltage

Fig.8.58. Backgate input MOS (B MOS)

Figure 8.58 shows the B (back-gate input) MOS principle [8.61]. The structure of this circuit looks like that of a CMOS circuit, but the input terminal is a p^+ region shared by a drain of P-MOS and a p-well of N-MOS. The gate of N-MOS is biased in the subthreshold region. If the logic swing is selected within $0 \sim 0.5$ V, the input current to a p-well is never lost into the n-type substrate and the drain current of N-MOS is given as an exponential function of the input bias voltage (Fig.8.58c). By an appropriate choice of the threshold voltage and the gate voltage, the P-MOS transistor plays the role of constant current source, as shown in the same figure. The loading characteristics look more like an EDMOS behaviour. Due to a small voltage swing, the power delay product is very low, although the operating speed is not so high due to the input capacitance of the p-well.

Figure 8.59 shows a novel C-MOS inverter [8.62]. The single gate electrode is shared by a surface channel N-MOS transistor and a buried channel P-MOS transistor. Their source drain areas are arranged so that the current direction of each channel is mutually perpendicular. This circuit can achieve high density, as compared to the conventional C-MOS technique. But when the P-MOS

Fig.8.59. Single gate C MOS inverter

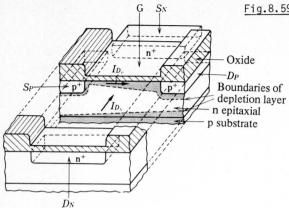

Fig.8.59. Single gate C MOS inverter

Table 8.10. Characteristics of EDMOS VLSI circuits

L (channel length) [μm]	Circuit	Delay	L/V sat [ps]	Comment	
				Interconnection delay [ps]	Capacitive delay [ps]
0.25 μm	R.O.	65 ps/gate	3.8 ps		40 ps
0.5 μm	R.O	100 ps/gate	7.4 ps		80 ps
1.0 μm	R.O	125 ps/gate	15 ps		160 ps
1.3 μm	R.O. F/O = 1	350 ps/gate		0.48 mW/gate	
	R.O. F/O = 3	1.9 ns/gate		0.13 mW/gate	
	2000 circuit (1 level metaliz.)	∼ 3 ns/inter-connect			
	8000 circuit (1 level metaliz.)	∼ 10 ns/inter-connect			
	8000 circuit (2 level metaliz.)	∼ 3 ns/inter-connect			
	Dynamic PLA	56 ns/cycle			
	Static PLA	13 - 21 ns/cycle			

transistor is on, and the N-MOS transistor is not off yet, there is some DC current loss.

Next, we describe perspectives of the scaled-down EDMOS-VLSI circuits. Table 8.10 shows some experimental results and theoretical predictions. A ring oscillator with 0.25 μm gate length has been achieved with a propagation delay

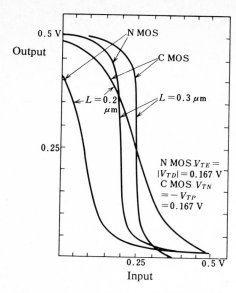

Fig.8.60. Transfer characteristics of C MOS, E/D N MOS inverter

of 65 ps/gate [8.63]. According to a simulation [8.65], a 8000 circuits LSI with 3 ns delay per intereconnection is obtained with a 2 level metal inter-connection [8.64]. In a 20 input - 9 output PLA (Programable Logic Array) with 105 product term, 56 ns and 13~21 ns cycle times, respectively, were obtained with dynamic 4-phase circuits and ED static circuits. Dynamic PLA has low power dissipation but it is difficult to achieve cycle times less than 50 ns due to the difficulty of a high slew rate of an internal clocking circuit, say 0.28×10^9 A/s.

By cooling to liquid N_2 temperature, the performance of N-MOS circuits is greatly enhanced, because of high carrier mobility, low interconnection re-sistivity, etc.

Some limitation of the ED-MOS technique is shown in Fig.8.60 [8.66]. At 0.2 μm gate length, the output high level of an ED-MOS inverter is less than the bias voltage of 0.5 V, because the E-MOS transistor is not sharply cut off due to subthreshold leak current even when the input level is zero. On the con-trary, a C-MOS inverter exhibits satisfactory transfer characteristics. As a result, combined with small power dissipation, this technique implies that MOS VLSI will eventually be dominated by C-MOS technology, as the device dimension is scaled down.

c) The VLSI System

As an example of VLSI during the 1980s, the target of the American VHSIC pro-ject is referenced here [8.67]. It is reported that with 0.5 μm design rule and

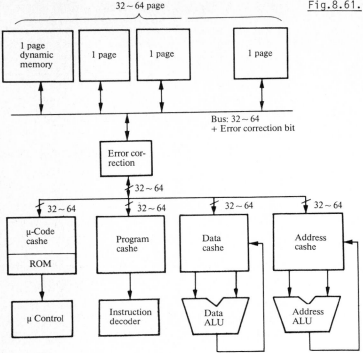

Fig.8.61. One-chip system

400×400 mil^2 chip size, the 2 M bit static RAM and 1 chip microprocessor with a $3 \cdot 10^{23}$ clock rate gate (bipolar microprocessor with 30 K gate $\times$ 200 MHz and MOS microprocessor with 150 K gate $\times$ 30 MHz) will be developed by 1986.

The architecture of VLSI-systems will be very different depending on their applications. Here, the merit of a one-chip system is illustrated by an example shown in Fig.8.61. In this system, the CPU (Central Processing Unit) cash memory and the main memory are integrated on one chip.

Figure 8.61 shows a conceptional design of a 32~64 bit MOS one-chip microprocessor with 1 million devices [8.68]. In this system, the CPU, the cash memory and the main memory are integrated on one chip. Thus, the system cylce time is expected to be drastically reduced, as compared to multichip systems, where the system cycle time is determined by the time required for data transfer between the CPI chip and the memory chip. Microcode design is adopted for the purpose of regularity, to simplify the design automation, the testability, etc. The main features of the system are summarized in Table 8.11.

Present-day 16 or 32 bit commercial microprocessors are multi-chip systems. However, by utilizing the maximum device number, many main frame features such

Table 8.11. One million device MOS microprocessor to be developed in the latter half of the 1980s

Block	Bit/word	Words	Number of words	Kilobyte	Device number ($\cdot 10^3$)	Area [%]
Memory						
Dynamic RAM	32~64	125~256	32~64	16~128	128~1024	
Static RAM	16~64	976~1744	8~9	2.563~12.625	82~404	
ROM microcode	32~64	1024~2048	1	4~16	32~128	
Total memory					242~1556	45~65
ALU	16~32		2		20~40	
Control					20~80	15
Communication I/0					10	10
Total circuit					292~1686	

as pipeline control, virtual memory storage, etc. are adopted and their performance is approaching those of high-end minicomputers or even mainframe computers.

Design of VLSI systems is very complex, including architectural design, logic design, circuit design and pattern design. Therefore, design automation is imperatively necessary, and the regularity of the design is very important. For this purpose, many approaches such as a building block, PLA, gate array, etc. are adopted. Regularity of the design is also essential for the testability, the verification of layout patterns, etc. For testability, self-testing circuits will also be included on the chip. All these features will strongly influence the architecture of future VLSI systems.

d) Gigabit Logic Systems

It is expected that gigabit logic systems with 1~10 ns cycle time will be possible by ultrahigh speed CML devices, GaAs-IC or Josephson junction devices. These systems will be employed in instrumentation, satellite or fibre-optics communication, supercomputers with a throughput as high as 1000 MIPS, or pattern recognition systems such as the image processor.

8.5.2 Memory Circuits

When memory circuits are grouped into static and dynamic types, the criteria are as to whether it is necessary to refresh the content of a memory cell periodically or not. Furthermore, the choice of the type greatly affects speed, power consumption as well as the cost aspect of the system.

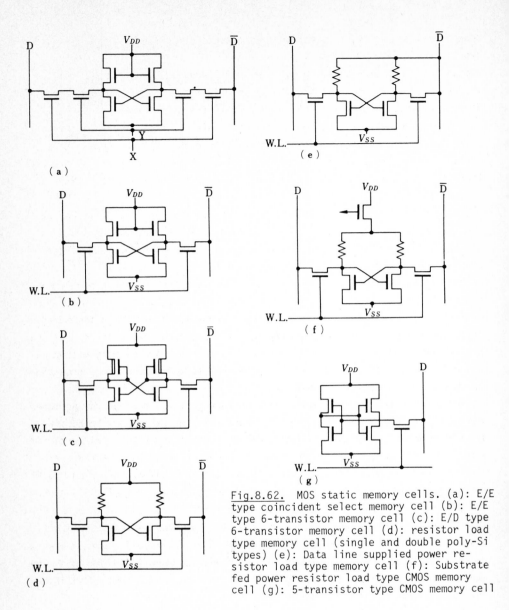

Fig.8.62. MOS static memory cells. (a): E/E type coincident select memory cell (b): E/E type 6-transistor memory cell (c): E/D type 6-transistor memory cell (d): resistor load type memory cell (single and double poly-Si types) (e): Data line supplied power resistor load type memory cell (f): Substrate fed power resistor load type CMOS memory cell (g): 5-transistor type CMOS memory cell

The static memory, specifically the RAM, has flip-flops as basic memory elements. The information content can be read out nondestructively, and it is maintained unless the power is cut off. This situation simplifies the system structure. However, the static memory requires multiples of elements for the cell. Taking the MOS memory as an example, it requires eight elements for the coincident type cell, as demonstrated in Fig.8.62a. Figures 8.62b and c show some transistor cells, Figs.8.62d and e exhibit 4 transistor- plus 2 resistor-type of cells, Figs.8.62f and g display the cells for C-MOS RAM in which 5 transistors or 5 transistors plus 2 resistors are employed, contrasting the 1 transistor- plus 1 capacitor-type dynamic cell and occupying large areas. Hence, the chip density is about 1/4 of that of a dynamic RAM.

Now dynamic memories are facing the difficulty of soft errors caused by α-particle incidence as well as the difficulty caused by the complexities of circuits. Thus, the development of the cell types is experiencing delays and there is a tendency that the static memory may catch up the dynamic RAMs in the future.

The H-MOS technology established with the advances in circuit technology and the practical application of microlithographic technology has improved the speed performance of MOS memories greatly. However, before entering the bipolar memory market, a couple of problems have to be solved to further increase the integration density of static memories. Some of these problems are explained now.

The first requirement is the lowering of power dissipation in the memory cell. The main approach is currently the utilization of polysilicon as a load resistance to decrease the cell area, but controllability of polysilicon resistors is poor and further study is necessary to establish that technology to increase the integration density.

The second condition is to minimize the power dissipation of the peripheral circuits such as decoders and others. There are some examples like MOSTEK's MK-4104 which is utilizing a dynamic circuit technique at the peripheral circuits, having cycle times longer than the access time because of the pre-charge time necessary for a dynamic type of operation. There are also MOSTEK's 4801-type memories which feature address transition detector circuits generating various internal clocks, by detecting address changes, which helps to lower the power dissipation, although the standby power dissipation is not low enough. On the other hand, there are Intel's 2147-type circuits which feature a $\overline{CS}$ signal controlled internal clock generator helping to maintain the power dissipation ratio between active and inactive modes high, in addition to equal access and cycle times.

An ideal technology to realize low-power dissipation would be to employ C-MOS technology, because the technology not only lowers the power dissipation but also shows a resistance to α-particles. If C-MOS/SOS process technology ever becomes cost-wise competitive to other technologies, this technology could be an idal one from the standpoint of power dissipation, high integration density because of favourable isolation area, α-particle resistance and latch-up free characteristics.

It is impossible to realize memory cells with the enhancement MOS transistor as a load to attain high impedance for decreased power dissipation as well as to minimize the cell area at the same time. Therefore, the depletion MOS transistor as a load has been widely used. However, limitations are being felt to realize a bit memory larger than a couple of tens of thousands from the standpoint of power dissipation as well as cell area.

Static memories with polysilicon load cells are being developed industry-wise from these points. Looking for further simplification, the enhancement transistor load cell including a positive application of one of the short channel effect, the weak inversion-region current as a load-bias operation has been investigated. Figure 8.63 shows theoretical as well as experimental values of the weak-inversion region current I_w. Figure 8.64 gives a static memory circuit, employing the weak-inversion MOS transistor (WIMT) as a load. The high-level voltage V_H in the flip-flop is plotted in Fig.8.65. Good cor-

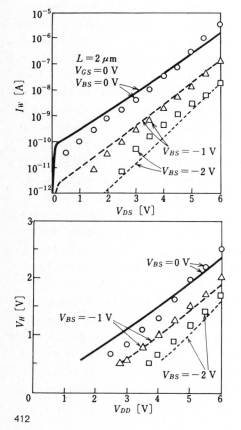

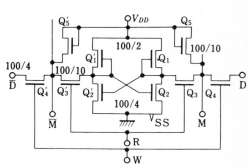

Fig.8.63. Theoretical and experimental curves of weak inversion current
Fig.8.64. Static memory circuit of WIMT coad cell

Fig.8.65. Theoretical and experimental values of high level voltage FF circuit

relation between theoretical and experimental values indicates that V_H, which is the necessary parameter to design the static memory cell, is predictable. A static memory circuit shown in Fig.8.64 has been investigated concerning its dynamic operation. The static memory cell with the WIMT as load element could be designed with a very small area by utilizing buried contact; additionally, low-power dissipation characteristics have been confirmed.

b) The CML Compatible High-Speed Static MOS Memory

In the VLSI area, high-speed, large-capacity cache memories, buffer memories, control-storage as well as main memories are highly demanded, and the trend is going to continue into the figure. Cycle time of these memories has been estimated to be around 50 ns based on the analysis of computer performance which should be available in the latter half of the 1980s.

To realize a 50 ns cycle time, it seems that not a dynamic-memory but a static-memory technolgy is necessary. We carried out a study using 1 μm design rule required for realization of a 1 Mbit dynamic memory. Such a system will probably correspond to the technology of 256 Kbit static RAM. When a 1 μm design rule is employed, lowering of the supply voltage down to around 3 V is necessary to prevent the short-channel effect as well as a hot-electron injection phenomenon, which will cause electrical instability. Also from the application point of view, it was decided to employ a CML compatible circuit at the interface, because future systems are high-speed ones and the probability is high that CML-like circuits will be used in the system. Besides, it will be very desirable that the MOS memory and logic circuits have to be CML compatible in that environment.

In this study, an MOS memory was designed so as to be L-CML compatible, where V_{DD} = 0.0 V, V_{SS} = -3.3 V, V_{REF} = -0.26 V have been chosen. Although the power supply is essentially, only one generator was built onto the chip to enhance performance of the memory. In addition, a power-down capability was added to the feature of the memory to save power. The block diagram of this static memory is shown in Fig.8.66.

Basic clocks, CS, $\overline{CS}$ ai ? the mutually complementary signals whose amplitude swings between -0.45 V and -0.05 V in the worst case. Therefore, the logic swing is in the range of 0.4 V; it is supposed to be supplied from the same CML gate. All the addresses, data input and R/$\overline{W}$ signals are driven from the CML gate with proper termination resistors. One terminal of the inputs is connected to an -0.26 V reference voltage so that the interface circuits should be operative with a signal amplitude of 0.2 V. The output of the memory is supposed to drive a differential-type receiver with proper termination.

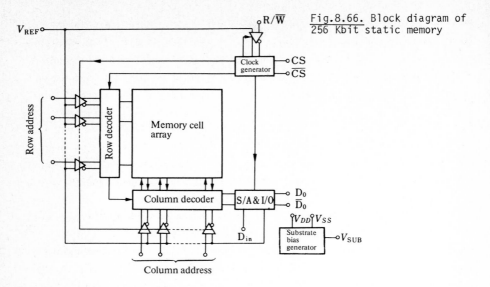

Fig.8.66. Block diagram of $\overline{256}$ Kbit static memory

The most important part of the memory is the cell, which is shown in Fig.8.67. The cell is of the FLIP-FLOP type with polysilicon load resistors. In addition, double polysilicon structures are employed to save space, where load resistors are layed out over the cell area without requiring extra areas. A second polysilicon layer is also used for the interconnection to other peripheral circuits to save areas. Additionally, in the high-density memory-cell layout, where the percentage of contact hole areas is significant, an under-

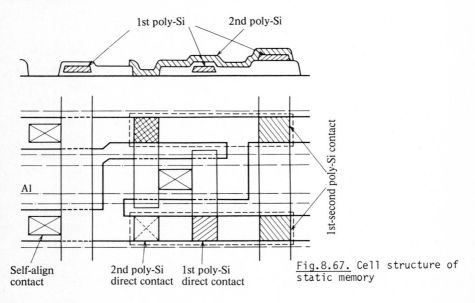

Fig.8.67. Cell structure of static memory

cut technique and a self-aligned gate as well as a self-aligned contact technique combined with anisotropic etching are utilized to solve the problem. The cell size is 9.2 μm × 12.8 μm, and the cycle time of the memory was estimated to be 37 ns by simulation.

c) The DSA MOS High-Speed Static Memory

A DSA MOS (Diffusion Self-Aligned MOS) transistor has the ability of a high speed VLSI element because of its higher transconductance g_m, higher source-drain breakdown voltage (BV_{DS}), lower junction capacitance resulting from a p-substrate ($100 \sim 150$ Ω·cm) and higher resistance to short-channel effect than a conventional N-MOS, as discussed in Sect.8.3.1.

In this section, we give a description of design and fabrication of a 1 Kbit static RAM. Its circuit and function block are shown in Figs.8.68 and 69, respectively. The main characteristics of the circuit are the connection of the lower capacitance drain of the DSA with a supply voltage line and the higher capacitance source with the ground line. So, high speed operation of the 1 Kbit RAM can be obtained easily. Furthermore, fine design rules are required for high-speed operation with such values; the source-drain distance is 1 μm, the

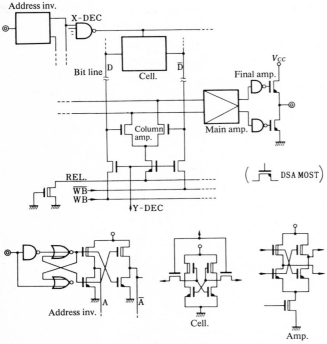

Fig.8.68. DSA MOS 1 Kbit static RAM circuit

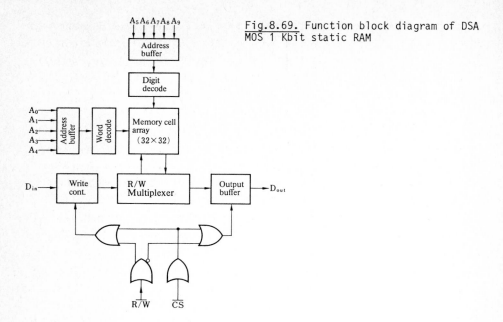

Fig.8.69. Function block diagram of DSA
MOS 1 Kbit static RAM

base width is 0.3 μm, and the line/space relation of the interconnection is
1.5~2.0 μm/2.0 μm, respectively. These design rules make possible a g_m of DSA
about three times higher than that of a conventional N-MOS and only a tenth of
the interconnection capacitance comparing to that of N-MOS.

The structural parameters of the DSA constructing the element of a 1 Kbit
static RAM are shown in Table 8.12. The gate oxide thickness of 300 Å is a
lower limit considering hot carrier injection into thin gate oxide under 5 V
of supply voltage.

Source and drain shallow junctions are obtained by As-ion implantation, and
also phosphorous reflow is applied to fabricate a fine-pattern VLSI.

Table 8.12. Structure of DSA-MOS 1 Kbit static RAM

Source drain distance L_{SD}	1.0 μm
Gate oxide thickness t_{OX}	300 Å
Source, drain junction depth X_j	0.30 μm
Process	n-channel Si gate
Memory cell	6 transistor E/D type
Cell size	34 × 18 μm
Chip size	1.4 2.1 mm
Power supply	5 V (single)

416

The operational characteristics of the 1 Kbit DSA RAM are as follows. The address access time is smaller than 10 ns, the power consumption is 400 mW and the figure of merit of DSA RAM is 1/3~1/2 smaller than that of a conventional N-MOS RAM. A very-high-speed 1 Kbit static RAM is obtained by using 1 µm DSA MOS transistors.

d) Dynamic Memories

During the 1970s, the density of semiconductor memories doubled every year. The main reason for this trend were the reduction in the number of devices per cell, the reduction in device dimensions, and an increase in chip area. The device number has decreased from six transistors required in a static memory to four, three and one transistor per cell by the development of the dynamic memory.

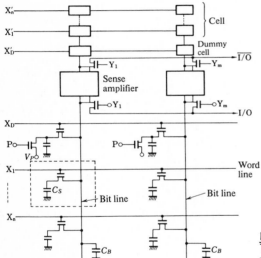

Fig.8.70. Equivalent circuit of 1 Transistor 1 capacitor dynamic memory

Figure 8.70 shows the one-transistor one-capacitor memory. The cell encircled by a dotted line can be selected by word line X_1 and bit line B_1. In the read cycle, the bit line is set floating to a level and then the word line X_1 and the dummy word line X_D' are opened. Then, a voltage difference ΔV will appear between the two bit lines depending upon the data of the cell. ΔV is given by

$$V = \frac{|V_{cell} - V_p|}{1 + C_B/C_S} \quad ,$$

417

where C_B and C_S are the capacitances of bit line and cell, V_{cell} and V_p are the levels of the cell and dummy cell. The difference signal is latched into a sense amplifier which, in turn, is connected to the I/O line by the Y select line.

To obtain a large difference signal, a minimum C_B/C_S ratio is desirable with minimum cell area. For this purpose, various cell structures have been developed. Figure 8.71 shows the double polysilicon cell. The QSA SHC-RAM cell using the Ta_2O_5 dielectric were also described earlier. The triple polysilicon cell or the V-MOS cell have also been proposed.

Figure 8.72 shows a balanced flip-flop sense amplifier. This amplifier is of the differential type and rather insensitive to fluctuations of device parameters. High sensitivity and low-power dissipation are realized by multiphase clocking.

Figure 8.73 illustrates the architecture of a 64 Kbit RAM. 65536 cells are organized into 256 rows by 256 columns and partitioned into two 128×256 sub-

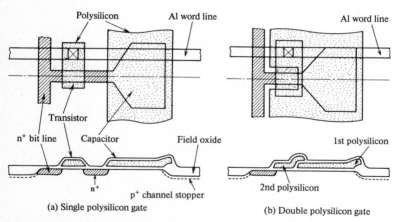

(a) Single polysilicon gate   (b) Double polysilicon gate

Fig.8.71. Structure of 1 transistor dynamic memory cell

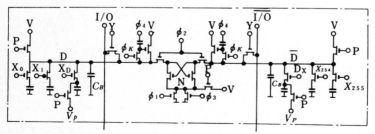

Fig.8.72. Balanced flip-flop sense amplifier

blocks. These subblocks are separated by 256 sense amplifiers. A cell is selected by a 8 bit X (row) address and a Y (column) address. These X and Y addresses are given to the memory via common 8-address pins in a time-multiplexed manner. By this scheme called address multiplexing, the number of address pins can be halved, and so a 64 Kbit and even 256 Kbit RAM can be packed into the 16 pin package [8.69].

In the future, high-density memories, adoption of redundancy and the self-testing function will be inevitable.

64 Kbit RAMs are now in production and 256 Kbit RAMs are being developed [8.69,70]. The One-Mbit RAM is probably under active research now.

During the development of 1 transistor dynamic memory, its limitations are also becoming clear, namely soft errors due to α-particles, hot electron injection, subthreshold leakage current problems etc. Therefore, new types of memory cells are being looked for (Table 8.13).

Table 8.13. New types of dynamic memory cell

	1 T/cell	MCM	VCCM	I^2L	SCM	TI
Charge	Electron	Electron	Electron	p-n junction	hole	hole
Device number	2	1	3	3	2	1
Number of control lines	2.5	2	3	3	3.5	3
Area	$6F^2$	$4F^2$	$8F^2$	$8F^2$	$6F^2 - 8F^2$	$4F^2$
Voltage level	2 level	3	3	3	3 ($\pm$)	3 ($\pm$)
Output	C_B/C_S	C_B/C_S	C_B/C_S	C_B/C_S	const. current	const. current
Levels of interconnection	2~3 levels	2	2 + epi	2 + epi	2	2~3
Contact/cell	0.5~1.5	0	0	2~3	0	2
Read out	Destructive	Destructive	Destructive	Nondestructive	Nondestructive	Nondestructive

Device structure and internal wave form of the MCM (Merged Charge Memory) are shown in Fig.8.74 [8.71]. It has two control lines, a word line and a bit sense (BSS) line. Electronic charges are dynamically stored under the overlapping area of the control lines. In the store mode, the BSS line is held on V_H (supply voltage) while the word line is set to zero. In the read mode, the BSS line is set floating to the level V_H, and the word line level is set to $V_H/2$. Then, each cell on the selected word line is connected to the n^+ charge injection (C.1) line and thereby filling and spilling operations are made.

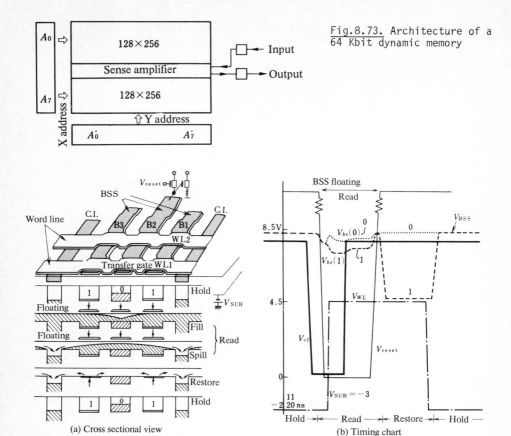

Fig.8.73. Architecture of a 64 Kbit dynamic memory

Fig.8.74. Structure and timing chart of MCM

Thus, the initial charge stored in each cell is compared to the charge after fill and spill operations are made. The difference between them is detected by capacitive coupling as a small-level shift of each floating bit line. This signal is latched into a sense amplifier provided with each column. When data are read out, the stored data are destructed (destructive read out), and they must be written back into cells from the respective column sense amplifiers. For this purpose, the word line is set to $V_H/2$ and the BSS lines are set to V_H or $V_H/2$ depending on the data to be written back; thereby a charge injection into each cell is made, and then the BSS line and the word line are returned to store mode. This cell can realize the theoretically minimum area with a plan capacitor 1C, $4F^2$. Here, F is the minimum feature size. But, three control levels are necessary and the C_B/C_S ratio cannot be reduced independent-

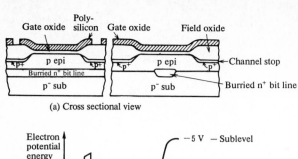

(a) Cross sectional view

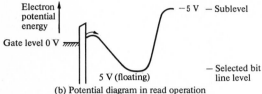

(b) Potential diagram in read operation

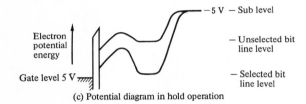

(c) Potential diagram in hold operation

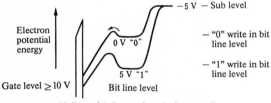

(d) Potential diagram in write in generation

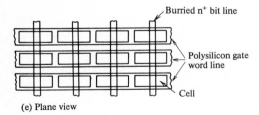

(e) Plane view

Fig.8.75. VCCD (Vertical Charge Coupled Device)

ly of the architecture of the cell array since it is nearly equal to the number of cells connected to each bit line.

The VCCD (Vertical Charge Coupled Device) is also a 2 control line cell, as shown in Figs.8.75a and c [8.72]. Namely, polysilicon word lines and n^+ buried bit lines are used. The data are stored in the surface potential well (Fig.8.75c). Read out, store and write in are made by choosing the gate levels

421

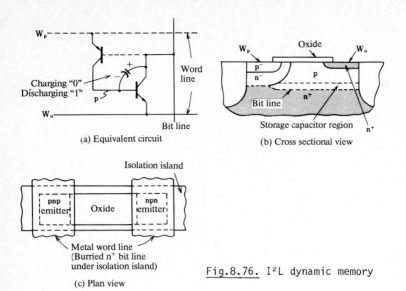

(a) Equivalent circuit

(b) Cross sectional view

(c) Plan view

Fig.8.76. I²L dynamic memory

and bit line levels as shown in Figs.8.75b-d. This cell can also realize a $4F^2$ cell area, but requires 3 word-line levels and its C_B/C_S ratio is difficult to be reduced since C_B is dominated by the bit-line junction capacitance.

Figure 8.76 shows a I^2L dynamic memory cell [8.73]. The data are stored in the capacitance C_S between base and collector. This cell consists of three devices including C_S, the two word lines W_n, W_p and a bit line. In the store mode, W_n and the bit line are set to 3 V, W_p is set to nearly zero value, and thereby the npn and pnp transistors are cut off. Thus, data are stored in C_S. In writing '1', C_S is charged by selecting the bit line level 3 V and the W_n level 0 V. In writing '0', C_S is discharged through the pnp transistor by selecting the W_p level 3 V and bit line level 2 V. In the read mode, the pnp transistor is cut off by selecting W_p and the bit line level 3 V. When C_S is charged, the npn transistor is cut off so that the bit line level does not change by setting W_n to level 0. When C_S is discharged, the npn transistor is on so that the amount of charge stored in C_S multiplied by β of the npn transistor is discharged from bit line to W_n line. This is equivalent to a storage of data in a capacitor of the value βC_S. The bit line level is sensed so that the '1' level is latched to 3 V and the '0' level to 2 V.

Restoring '1' is made by selecting the W_n level low, and restoring '0' by selecting W_n and W_p to 3 V and the bit line to 2 V. The sensing signal is determined by the $C_B/\beta C_S$ ratio which is $1/\beta$ smaller than compared to the 1-transistor memory. This cell can realize an $8F^2$ area, but the number of devices and control lines are not so small and the control is complicated.

422

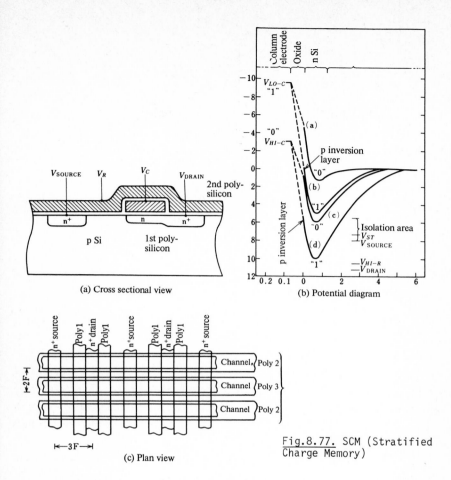

(a) Cross sectional view

(b) Potential diagram

(c) Plan view

Fig.8.77. SCM (Stratified Charge Memory)

Figure 8.77 shows the SCM (Stratified Charge Memory) principle [8.74]. The structure of the cell is like that of a two-transistor ROM. Transistor Q_1 makes the ROW selection and writes into transistor Q_2. Hole packets are dynamically stored in the surface potential well of the storing transistor Q_2. In reading, the transistor Q_2 operates as a buried n-channel MOS whose threshold voltage is controlled by the amount of holes stored in the surface channel. Operation of the cell is explained by referring to the potential diagram of Fig.8.77b.

In writing, transistor Q_1 operates in the accumulation mode by an appropriate choice of the V_R level. Writing '1', by setting the V_C level V_{Lo-C}(ONE), holes in the accumulation layer of Q_1 are injected into the surface potential well of Q_2 so that the potential curve b will result.

In writing '0', by setting the V_C level to V_{H1-C}(ZERO), holes are rejected so that the potential curve c will result. In the storing mode, by selecting

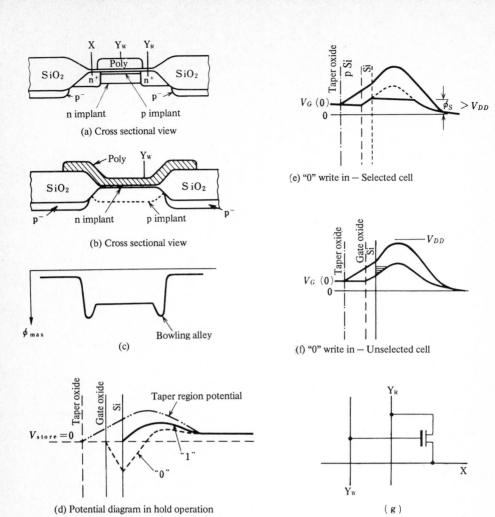

(a) Cross sectional view

(b) Cross sectional view

(c)

(d) Potential diagram in hold operation

(e) "0" write in — Selected cell

(f) "0" write in — Unselected cell

(g)

Fig.8.78. TI (Taper Isolated) RAM

the V_R level in such an intermediate level that the surface of Q_1 is depleted at level V_{ST}, holes injected into Q_2 are isolated from the p-Si substrate and stored independently of whether the V_C level is 'H1' or 'L0'.

When V_C is V_{Lo-C}, '0' corresponds to the curve a, while '1' corresponds to curve b. When V_C is V_{H1-C}, '0' corresponds to curve c and '1' to curve d. In reading, Q_1 is on by selecting the V_R level V_{H1-R}. When '1' is stored, then by selecting V_C to V_{H1-C}, Q_2 is on, as implied by the potential curve d, so that the electron currents pass from the drain to the source via the buried

424

channel of the transistor Q_2. When '0' is stored, Q_2 is off, as implied by the curve c, so that there is no current path between source and drain.

This cell is ROM like and operates in some sense like a 3-transistor memory so that reading is non-destructive and there is no limitation related to the amplitude of the output signal, a marked contrast to the one-transistor memory cell. In a $6F^2$ cell, one column is read in two cycles, but in the $8F^2$ cell in one cycle. When negative V_C bias is used, the X decoding circuit must be isolated from the substrate. However, the operation in positive V_C is possible by adding a shallow p layer in Q_2.

The TI (Taper Isolated) RAM (Fig.8.78) is a single-transistor ROM-like memory that utilizes the dynamic storage of holes and detects the conductivity modulation due to threshold shifts [8.75]. In the channel region, deep n and shallow ion implanted layers are formed (Fig.8.78a,b). In the taper region, only an n-implanted layer is formed by appropriate choice of the ion-implantation dose and energy. The surface potential along the direction of the channel width shows a hump in the taper region, as illustrated in Fig.8.78c. Therefore, along the direction of the channel length, there are so-called bowling alleys. They prevent stored holes from flowing out of the surface potential well. In the storing mode, by selecting the gate potential 0, the potential at the taper region will be positive, as compared to those of substrate and channel. This means that a barrier for the holes is formed. If the dose of ion implantation into the p-type region in the channel is large enough to make the surface potential negative when there are no stored holes, the potential curve should be like '0' in Fig.8.78d.

However, when there is a surface potential in the channel, the holes are confined there by the surrounding potential barrier provided by the bowling alley and source drain n^+ region. In this case, the threshold voltage of the buried n-channel MOS transistor makes negative shifts, and the potential in the channel region should be like '1' in Fig.8.78d. For selective writing, the gate electrode Y_w and the drain electrode X are used. In writing '0', the gate level is set 'high' so that the surface potential changes. If the drain voltage is lowered as $V_{DD} < \phi_s$, then the level of the n-type ion implanted layer is lowered and the hole potential well in the surface region is flattened, as shown in Fig.8.78e, so that surface holes are ejected into the p-type Si substrate. Then, drain voltage is made high and writing '0' is completed.

In Fig.8.78f, the potential curve of the unselected cell is shown. When the drain voltage is kept 'high', the holes in the surface region are kept stored independently o the gate bias voltage, and therefore the information stored in the unselected cell is never destructed. In writing '1', the gate level is lowered as far as the bowling alley in the surface region is kept positive. The drain voltage is applied in the same manner as in writing '0'. This cell is a ROM-like non-destructive read out memory. By utilizing the taper region, an ingenious one-transistor memory can be realized. However, since this transistor is of the depletion mode type, the conductivity modulation must be detected by the sense circuit, and the trade off between sensitivity and high density will remain as a conductivity ratio in a similar manner as the C_B/C_S ratio in the one-transistor one-capacitor dynamic RAM.

References

Chapter 1

1.1 G.E. Moore: Progress in digital integrated electronics. IEDM Technical Digest (1975) p.11
1.2 K. Ohta: J. Inst. Electron. Commun. Eng. Jpn. **60**-11, 1235 (1977)

Chapter 2

2.1 Y. Tarui, S. Denda, H. Baba, S. Miyauchi, K. Tanaka: Microelectron. Reliab. **8**, 101 (1969)
    H. Baba, S. Denda: Jpn. Utility Patent No.1012768 (1968)
2.2 E. Goto, T. Soma, M. Idesawa: J. Vac. Sci. Technol. **15**, 887 (1978)
2.3 E. Goto, T. Soma, M. Idesawa, T. Sasaki: AFIPS Nat. Comput. Conf. Expo. Conf. Proc. **47**, 1223 (1978)
2.4 H.C. Pfeiffer: IEEE Trans. Electr. Dev. ED-**26**, 663 (1979)
2.5 T.H.P. Chang: Proc. 7th Int. Conf. Electron and Ion Beam Sci. and Technol. (Electrochem. Soc., Princeton, NJ 1976) p.392
2.6 A. Iwata, M. Fujinami, Y. Uno, K. Yoshida, C. Munakata, S. Asai: 1978 National Convention Record of IECE Japan, Vol.2 (Inst. Electron. Commun. Eng. Jpn. Tokyo) p.91 [in Japanese]
2.7 G.L. Varnell, D.F. Spicer, J. Hebley, R. Robbins, C. Carpenter: J. Vac. Sci. Technol. **16**, 1787 (1979)
2.8 H.J. Binder, P. Hahmann: J. Vac. Sci. Technol. **16**, 1723 (1979)
2.9 M. Fujinami, T. Matsuda, K. Takamoto, H. Yoda, T. Ishiga, N. Saitou, T. Komoda: J. Vac. Sci. Technol. **19**, 941 (1981)
2.10 R.D. Moore, G.A. Caccoma, H.C. Pfeiffer, E.V. Weber, O.C. Woodard: J. Vac. Sci. Technol. **19**, 950 (1981)
2.11 M.G. R. Thomson, R.-J. Collier, D.R. Herriott: J. Vac. Sci. Technol. **15**, 891 (1978)
2.12 E.V. Weber, R.D. Moore: J. Vac. Sci. Technol. **16**, 1780 (1979)
2.13 E. Goto, T. Soma: Optik **48**, 255 (1977)
2.14 H.K.V. Lotsch: J. Appl. Phys. **38**, 3423 (1967)
2.15 W. Schumann, M. Dubas: *Holographic Interferometry*, Springer Ser. Opt. Sci., Vol.16 (Springer, Berlin, Heidelberg, New York 1979)
2.16 Y.I. Ostrovsky, M.M. Butusov, G. Ostrovskaya: *Interferometry by Holography*, Springer Ser. Opt. Sci., Vol.20 (Springer, Berlin, Heidelberg, New York 1980)
2.17 K. Hoh, N. Sugiyama, Y. Tarui: IEEE Trans. ED-**26**, 1363 (1979)
2.18 A. Sugata (ed.): *Denshi Ion Beam Handbook* [*Electron and Ion Beam Handbook*] Nihon Gakujutsushinkokai Dai 132 Iinkai (Nikkan Kogyo Shinbusha, Tokyo 1975) p.21 [in Japanese]

2.19 C.W. Oatley: *The Scanning Electron Microscope* (Cambridge University Press, Cambridge 1972) [Japanese transl.: *Sosagata Denshi Kenbikyo* (Koronasha, Tokyo 1979) p.26]
2.20 P.W. Hawkes (ed.): *Properties of Magnetic Lenses*, Topics Curr. Phys., Vol.18 (Springer, Berlin, Heidelberg, New York 1981)
2.21 W. Knauer: Optik **54**, 211 (1979)
2.22 S.F. Vogel: Rev. Sci. Instrum. **41**, 585 (1970)
2.23 A.N. Broers: Scanning Electron Microsc. **1**, 1 (1979)
2.24 O. Komoda: Denshisochi Kenkyukai EDD-**73-58**, 1 (1973) [in Japanese]
2.25 R.H. Fowler, L.W. Nordheim: Proc. Roy. Soc. A**119**, 173 (1928)
2.26 A.V. Crew, D.N. Eggenberger, J. Wall, L.M. Welter: Rev. Sci. Instrum. **39**, 576 (1968)
2.27 E.E. Windsor: Proc. IEE **116**, 348 (1969)
2.28 W.P. Dyke, F.M. Charbonnier, R.W. Strayer, R.L. Floyd, J.P. Barbour, J.K. Trolan: J. Appl. Phys. **31**, 790 (1960)
2.29 L.H. Veneklasen, B.M. Siegel: J. Appl. Phys. **43**, 1600 (1972)
2.30 J.W. Butler: 6th Int. Congr. Electron Microscopy **1**, 191 (1966)
2.31 L. Reimer: *Transmission Electron Microscopy*, Springer Ser. Opt. Sci., Vol.36 (Springer, Berlin, Heidelberg, New York 1983)
2.32 L.H. Veneklasen, B.M. Siegel: J. Appl. Phys. **43**, 4989 (1972)
2.33 R.J. Taylor, D.J. Swanson: System Appl. **36**, 1973 (1973)
2.34 Nikkei Electronics, No.162, 1977-6 (1977) p.32
2.35 H. Moss: Adv. Electr. Electr. Phys. Suppl. **3**, 97 (1968)
2.36 G. Liebmann: Proc. Phys. Soc. B**68**, 737 (1955)
2.37 F.H. Read: J. Phys. E**3**, 127 (1970)
2.38 A.D. Wilson, A. Kern, A.J. Speth, A.M. Patlach, P.R. Jaskar, T.W. Studwell, K.L. Keller: Proc. Symp. Electr. Ion Beam Sci. Technol., ed. by R. Bakish, (Electrochem. Soc., Princeton, NJ 1976) p.70
2.39 M. Fujinami, A. Shibayama, S. Moriya, T. Iwata, G. Tatsuno: Kenkyu Jitsu-yoka Hokoku, Electr. Comm. Lab. Tech. J. N.T.T. **27**(9), 193 (1978) [in Japanese]
2.40 Nikkei Electronics, 1977-6, 1032 (1977) [in Japanese]
2.41 K. Hoh, Y. Tarui: Oyo Butsuri **49**, 70 (1980) [in Japanese]
2.42 H.C. Pfeiffer, G.O. Longner: Proc. Symp. Electr. Ion Beam Sci. Technol. (1977) p.149
2.43 A.H. Eschenfelder: *Magnetic Bubble Technology*, 2nd ed., Springer Ser. Solid-State Sci., Vol.14 (Springer, Berlin, Heidelberg, New York 1981)
2.44 A.J. Speth, A.D. Wilson, A. Kern, T.H.P. Chang: J. Vac. Sci. Technol. **12**, 1235 (1976)
2.45 H.S. Yourke, E.V. Weber: Tech. Dgt. Int. Electr. Dev. Mtg. Sponc. Electr. Dev. IEEE (1976) p.431
2.46 D.R. Herriot, R.J. Collier, D.S. Alles, J.W. Strafford: IEEE Trans. ED-**22**, 385 (1975)
2.47 H. Baba, S. Denda: Jitsuyo-shinan Kokoku Pub. No. Sho47-34960 (1968) [in Japanese]
2.48 S. Okabe, T. Tabata, Y. Nakai: Oyo Butsuri **45**, 2 (1976) [in Japanese]
2.49 E.H. Snow, A.S. Grove, D.J. Fitzgerald: Proc. IEEE **55**, 1168 (1967)
2.50 H. Ryssel, H. Glawischnig (eds.): *Ion Implantation Techniques*, Springer Ser. Electrophys., Vol.10 (Springer, Berlin, Heidelberg, New York 1982)
2.51 J.M. Aitken, D.R. Young: J. Appl. Phys. **47**, 1196 (1976)
2.52 J.M. Aitken, D.R. Young, K. Pan: J. Appl. Phys. **49**, 3386 (1978)
2.53 J.M. Aitken: IEEE Trans. ED-**26**, 372 (1979)
2.54 E.P. Eernisse, C.B. Norris: J. Appl. Phys. **45**, 5196 (1974)
2.55 J.W. Corbett: Solid State Phys. Suppl. **7** (Academic, New York 1966) p.1
2.56 T. Wada: Oyo Butsuri **45**, 435 (1976) [in Japanese]
2.57 V.V. Vabilov, N.A. Ukhin: Radiation Effects in Semiconductors (Consultants Bureau, NY 1977) [translated from Russian]

2.58 H. Ryssel, H. Glawischnig (eds.): *Ion Implantation: Equipment and Techniques*, Spr. Ser. i. Electrophys., Vol. 11 (Springer, Berlin, Heidelberg, 1983)
2.59 N. Sugiyama, Y. Tarui: IEEE Trans. ED-**26**, 625 (1979)
2.60 N. Sugiyama, K. Saitoh: Computer Aided Des. **11**, 59 (1979)
2.61 T.H.P. Chang: J. Vac. Sci. Technol. **12**, 1271 (1975)
2.62 N. Sugiyama, K. Saitoh: Trans. IECE Jpn. E **63**, 198 (1980)
2.63 S. Hosaka, M. Ichihashi, H. Hayakawa, K. Asanami, S. Nishi, M. Migitaka: Proc. Microcircuit Eng., ed. by H. Beneking (Inst. of Semiconductor Electronics, Aachen 1979) p.7
2.64 J. Inst. Electr. Eng. Jpn. **99**, 776 (1979) [in Japanese]
2.65 M. Ichihashi, M. Mukohara, H. Hayakawa: Proc. 40th Jpn. Soc. Appl. Phys. Fall Meeting 1979, p.285 [in Japanese]
2.66 G.A. Wardly: J. Vac. Sci. Technol. **10**, 975 (1973)
2.67 S. Nishi: Proc. 18th SICE Meeting (Soc. Instr. Contr. Eng., Nagano 1979) p.233 [in Japanese]
2.68 K. Maio, N. Yokozawa, T. Sudo, A. Iwata: Proc. Jpn. Electric Soc. **551**, 689 (1977) [in Japanese]
2.69 A. Maekawa, M. Hirano, C. Munakata, T. Kozawa, H. Kouno: Japanese Patent No.799037 (1975)
2.70 S. Hosaka, H. Hayakawa, K. Asanami, M. Ichihashi, M. Migitaka: Proc. 40th Jpn. Soc. Appl. Phys. Fall Meeting 1979, p.289 [in Japanese]
2.71 S. Hosaka, M. Ichihashi, H. Hayakawa, M. Mukohara, S. Nishi, M. Migitaka: Proc. 27th Jpn. Soc. Appl. Phys. Spring Meeting 1980, p.356 [in Japanese]
2.72 S. Hosaka, M. Ichihashi, H. Hayakawa, S. Nishi, M. Migitaka: Jpn. J. Appl. Phys. **19**, 1797 (1980)
2.73 E. Goto, T. Soma, M. Idesawa: Patent Pend. in Japan "Electron Beam Scanning Method" No.50-127833 (Oct. 23, 1975)
2.74 H. Yasuda: Patent Pend. in Japan "Electron Beam Exposure System" No.50-158501 (Dec. 31, 1975)
2.75 T. Funayama, J. Kai, M. Nakamura: "An Alignment Technique in Variable-Shaped Beam Exposure", Nat. Conv. Rec. Inst. Electron. Commun. Eng. Jpn. (Inst. Electron. Commun. Eng. Jpn., Tokyo 1979) p.496 [in Japanese]
2.76 N. Yasutake, T. Funayama, M. Nakamura: "A Registration Technique in Variable-Shaped Beam Exposure", Nat. Conv. Rec. Inst. Electron. Commun. Eng. Jpn. (Inst. Electron. Commun. Eng. Jpn., Tokyo 1979) p.497 [in Japanese]
2.77 J. Kai, N. Yasutake, T. Funayama, M. Nakamura: "A Field Butting Technique in Variable-Shaped Beam Exposure", Nat. Conf. Rec. Inst. Electron. Commun. Eng. Jpn. (Inst. Electron. Commun. Eng. Jpn., Tokyo 1979) p.498 [in Japanese]

Chapter 3

3.1 D.L. Spears, H.I. Smith: Electron. Lett. **8**, 102 (1972)
3.2 D.L. Spears, H.I. Smith: Solid State Techn. **15**, 21 (1972)
3.3 D. Maydan, G.A. Coquin, J.R. Maldonado, S. Somekh, D.Y. Lou, G.N. Taylor: IEEE Trans. ED-**22**, 429 (1975)
3.4 J.R. Maldonado, M.E. Poulsen, T.E. Saunders, F. Vratny, A. Zacharias: J. Vac. Sci. Techn. **16**, 1942 (1979)
3.5 K. Okada, M. Nakamura: Preprints, 26th Spring Meeting, Jpn. Soc. Appl. Phys. (1979) p.294 [in Japanese]
3.6 K. Okada: J. Vac. Sci. Techn. **17**, 1233 (1980)

3.7 P.J. Mallozzi, H.M. Epstein, R.G. Jung, D.C. Applebaum, B.P. Fairand, W.J. Gallagher, R.L. Uecker, M.C. Muckerheide: J. Appl. Phys. **45**, 1891 (1974)
3.8 R.A. McCorkle, H.J. Vollmer: Rev. Sci. Instrum. **48**, 1055 (1977)
3.9 E. Spiller, D.E. Eastman, R. Feder, W.D. Grobman, W. Gudat, J. Topalian: J. Appl. Phys. **47**, 5450 (1976)
3.10 D.L. Spears, H.I. Smith, E. Stern: "X-Ray Replication of Scanning Electron Microscope Generated Patterns," in Proc. 5th Int. Conf. Electron and Ion Beam Sci. and Technology (1972) p.80
3.11 K. Suzuki, J. Matsui, T. Kadota, T. Ono: Jpn. J. Appl. Phys. **17**, 1447 (1978)
3.12 D.C. Flanders, H.I. Smith: J. Vac. Sci. Technol. **15**, 995 (1978)
3.13 D. Hofer, J. Powers, W.D. Grobman: J. Vac. Sci. Technol. **16**, 1968 (1979)
3.14 Y. Shirai, T. Nakamura, K. Hideshima, K. Okada, K. Sugizaki, M. Nakamura: A high throughput X-ray lithography system. Profess. Group S.S.D., Inst. Electr. Commun. Eng. Jpn. Vol.SSD **79-62**, 1 (1979) [in Japanese]
3.15 K. Yoshida, M. Asakawa, T. Saitoh: Mask alignment using vibration method. Nat. Conv. Rec. IECE Jpn. 8-238 (1979) [in Japanese]
3.16 S. Austin, H.I. Smith, D.C. Flanders: J. Vac. Sci. Technol. **15**, 984 (1978)
3.17 T. Saitoh, H. Watanabe, Y. Ohtsuka: Operation characteristics of the alignment mechanism using electrodynamic driver. Nat. Conv. Rec. Soc. Precision Eng. **359** (1976) [in Japanese]
3.18 T. Nakamura, K. Sugizaki: Mask alignment using piezoelectric transducer. Nat. Conv. Rec. IECE Jpn. No.433 (1979) [in Japanese]
3.19 H.L. Stover, F.L. Hause, D. McGreevy: Solid State Technol. **22**, 95 (1979)
3.20 S. Nakayama, S. Yamazaki. Y. Ohtsuka, S. Sasayama, M. Isshiki, S. Yoshida: X-ray lithography system. Nat. Conv. Rec. IPME Jpn. (1979) p.419 [in Japanese]
3.21 T. Asai, S. Ito, T. Eto, M. Migitaka: "A 1 : 4 Demagnifying Electron Projection System," in Proc. 11th Conf. on Solid State Devices (Tokyo 1980) p.47
3.22 M.B. Heritage: J. Vac. Sci. Technol. **12**, 1135 (1975)
3.23 B. Lischke, K. Anger, W. Münchmeyer, A. Oelmann, J. Frosien, R. Schmitt, M. Sturm: "Investigations About High Performance Electron-Microprojection System," in Proc. Symp. Electr. Ion Beam Sci. Technol. (1978) p.160
3.24 J. Frosien, B. Lischke, K. Anger: J. Vac. Sci. Technol. **16**, 1827 (1979)
3.25 M.B. Heritage, P.E. Stuckert, V. Dimilia: "A Solution to the Mask Stencil Problem in Electron Projection Microfabrication," in Proc. Symp. Electr. Ion Beam Sci. Technol. (1976) p.348
3.26 A. Meyer, A. Politycki, E. Fuchs: "Transmission Master for Demagnifying Electron Projection," in Proc. Symp. Electr. Ion Beam Sci. Technol. (1976) p.332
3.27 H. Bohlen, J. Greschner, W. Kulcke, P. Nehmiz: "Electron Beam Step and Repeat Proximity Printing," Electrochemical Soc. Spring Meeting, Seattle, Washington (1978) Abstract No.406
3.28 T.W. O'Keeffe, J. Vine, R.M. Handy: Solid State Electron. **12**, 841 (1969)
3.29 R. Speidel, M. Mayr. Proc. Microcircuit Engineering '79 (Aachen, Sept. 1979) p.43
3.30 G.A. Wardly: J. Vac. Sci. Technol. **12**, 1313 (1975)
3.31 G.A. Wardly: Rev. Sci. Instrum. **44**, 1506 (1973)
3.32 J.P. Scott: J. Appl. Phys. **46**, 661 (1975)
3.33 J.P. Scott: Philips Techn. Rev. **37**, 347 (1977)
3.34 W.R. Livesay: Solid State Technol. 21 (July 1974)
3.35 B. Fay: In Proc. 6th Int. Conf. Electron and Ion Beam Sci. and Technol. (San Francisco, CA, May 1974) p.527
3.36 J.P. Scott: In Proc. 6th Int. Conf. Electron and Ion Beam Sci. and Technol. (San Francisco, CA, May 1974) p.123

3.37 R. Ward: J. Vac. Sci. Technol. **16**, 1830 (1979)

3.38 W.R. Livesay: J. Vac. Sci. Technol. **15**, 1022 (1978)

3.39 I. Mori, T. Shinozaki, S. Sano: In Proc. Microcircuit Engineering '80 (Amsterdam, September 1980) p.145

3.40 S. Imamura, K. Harada, S. Sugawara: "Negative Resist Material CMS Resistant to Dry Etching Process (III)-Microfabrication for Deep UV Lithography," in Proc. 17th Symp. on Semiconductors and Integrated Circuits Technol. (1979) p.90 [in Japanese]

3.41 S. Iwamatsu, K. Asanami: "Development of Deep UV Projection Alignment System," in Proc. 17th Symp. on Semiconductors and Integrated Circuits Technol. (1979) p.78 [in Japanese]

3.42 Y. Mimura: Jpn. J. Appl. Phys. **17**, 541 (1978)

3.43 A. Charlesby: *Atomic Radiation and Polymers* (Pergamon, New York 1962) Chaps.9,10

3.44 K. Shinohara, K. Kashiwabara: *Atomic Radiation and Polymers* (Maki Books Ltd. 1968) [in Japanese]

3.45 T. Tada: "Molecular Orbital Studies of the Electron Beam Positive Resists," in Extended Abstracts of Electrochemical Society, Fall Meeting (1978) p.475

3.46 H. Kato, T. Tada, Y. Shimazaki, M. Koda: "Positive Electron and X-Ray Resist with High Sensitivity and Good Thermal Stability," in Proc. 17th Symp. on Semiconductors and Integrated Circuits Technol. (1979) p.54 [in Japanese]

3.47 H. Saeki, M. Kohda: "Poly(Methyl Methacrylate-co-t-Butyl Methacrylate) as a Highly Sensitive Positive Electron Beam Resist," in Proc. 17th Symp. on Semiconductors and Integrated Circuits Technol. (1979) p.48 [in Japanese]

3.48 H. Kato: "Dry Etch Resistant Negative Resist 'High Molecular Weight Polystyrene'", in Extended Abstracts of 158th Electrochemical Society Meeting (1980) p.831

3.49 Y. Shimazaki, H. Kato: "Highly Sensitive, Positive X-Ray Resist-ZnI_2 Blended PMMA", Preprint of 39th Meeting of Applied Physics, 3p-E-1 (1978) [in Japanese]

3.50 Y. Shimazaki: "Highly Sensitive, Positive X-Ray Resist," Preprint of 40th Fall Meeting of Chemical Society of Japan, 3C23 (1979) [in Japanese]

Chapter 4

4.1 *Institute of Electronics and Communications Engineers of Japan: Electronics and Communications Handbook* (Ohmsha Ltd., 1979) p.1476 [in Japanese]

4.2 K. Mizukami, Y. Hisamoto, Y. Wada, M. Migitaka: "S/N of Mask Pattern Detection Signal by Using Electron Beam", in 40th Meeting of Japan Soc. Appl. Phys. Technol. Dig. 2a-A-3, 284 (1979) [in Japanese]

4.3 M. Migitaka, K. Mizukami: "Evaluations for VLSI Conductive Photomasks" in Electronics Device Meeting, Inst. of Electron and Commun. Eng. Japan, EDD-78-115, 1 (1978) [in Japanese]

4.4 K. Mizukami, M. Migitaka: "Conductive Photo Masks", in 26th Meeting of Japan Soc. Appl. Phys. Technol. Dig. 29P-S-14, 301 (1979) [in Japanese]

4.5 M. Migitaka: "Conductive Photo Masks HSP Mask Pattern Inspection by Using Electron Beam", in Special Lecture in Konishiroku Photo Co. Ltd. (1980) p.1 [in Japanese]

4.6 K. Mizukami, M. Migitaka: "Relation to Electron Beam Diameter and Resolution Limit", in 41th Meeting of Japan Soc. Appl. Phys. Technol. Dig. L14, 17 (October 1980) [in Japanese]

4.7   K. Tsuda: "Trends for Photomasks Inspection System", Television Magazine
      32, 472 (1978) [in Japanese]
4.8   M. Mese, T. Uno, S. Ikeda, M. Ejiri; "Defects Extraction Method for Com-
      plicated Pattern," Trans. Inst. Electric Soc. **94**-C (5), 89 (1974) [in
      Japanese]
4.9   "LSI Photomasks Auto Inspection System for Detecting Minimum 0.3 μm
      Defects", Nikkei Electronics **164**, 34 (1977) [in Japanese]
4.10 M. Minami, H. Sekizawa: "IC Photomasks Inspection by Double Colour Method
      and S/N of Filtering", in 23rd Spring Meeting Japan Soc. Appl. Phys.
      Technol. Dig. Vol.289-B-10, 113 (1976) [in Japanese]
4.11 "To Find Out Photomasks Defect by Laser Beam Diffraction," Nikkei Elec-
      tronics **112**, 29 (1975) [in Japanese]
4.12 J.H. Bruning, M. Feldman, T.S. Kinsel, E.K. Sitting, R.L. Townsend:
      IEEE Trans. ED-**22**(7), 487 (1975)
4.13 Y. Wada, Y. Hisamoto, K. Mizukami, M. Migitaka: "Fundamental Evaluation
      for Mask Defects Inspection by Using Electron Beam", in 27th Spring
      Meeting Japan Soc. Appl. Phys. Phys. Techn. Dig. Vol.4a-N-9, 353 (1980)
      [in Japanese]
4.14 Y. Wada, Y. Hisamoto, K. Mizukami, M. Migitaka:""Mask Defects Inspection
      System by Using Electron Beam", Institute of Electron. Commun. Eng.
      Japan S55 Annual Meeting, Techn. Dig. Vol.433, 2 (1980) [in Japanese]
4.15 S. Yoshida: "Optical Technique in Micro-Lithography", Ryoko Technical
      Rev. **16**, 192 (1979)
4.16 Nikon Brochure for Laser Interferometric X-Y Measuring Machine Model 2I
      (1979)
4.17 K. Mizukami, Y. Hisamoto, Y. Wada, M. Migitaka: "Automated Linewidth
      Measuring System for Photomasks by Using Electron Beam", Proc. Micro-
      electronics Measurement Technol. Seminar 2 (1980) p.81
4.18 Y. Hisamoto, K. Mizukami, Y. Wada, M. Migitaka: "Mask Pattern Width Mea-
      surement by Using Electron Beam", in 40th Autumn Meeting, Japan Soc.
      Appl. Phys. Techn. Dig. 2a-A-5, 285 (1979) [in Japanese]
4.19 M. Mizukami, Y. Hisamoto, Y. Wada, M. Migitaka: "Mask Pattern Width
      Measurement by Using Electron Beam", to be published in the Transact.
      of the Institute of Electron. Commun. Eng. Japan **79**, 55 (1979) [in
      Japanese]
4.20 Y. Wada, Y. Hisamoto, K. Mizukami, M. Migitaka: "Studies of an Electron-
      Beam Mask Defect Inspection and Technology", J. Vac. Sci. Tech. **19**, 1
      (1981)
4.21 Y. Wada, Y. Hisamoto, K. Mizukami, M. Migitaka: "Investigation of
      Electron-Beam Mask-Defects Inspection System", Trans. IECE Vol.63,
      CNo.12 (1980) [in Japanese]

Chapter 5

5.1   H. Inuzuka, M. Takabayashi: *Semiconductor Materials and Their Single
      Crystal Production* (Nikkankogyo, Tokyo 1965) [in Japanese]
5.2   Semiconductor Materials Working Group of the Electrochemical Society
      of Japan (ed.): *Semiconductor Materials* (Asakura Press, Tokyo 1967)
      Chaps.1,3,5,7 [in Japanese]
5.3   Editing Board of Solid State Physics (ed.): *Progress of Semiconductor
      Technology* (Maki Press, Tokyo 1969) [in Japanese]
5.4   R.R. Haberecht, E.L. Kern (eds.): *Semiconductor Silicon 1969* (The Elec-
      trochemical Society Inc., New York 1969)
5.5   H.R. Huff, R.R. Burgess (eds.): *Semiconductor Silicon 1973* (The Elec-
      trochemical Society Inc., Princeton 1973)

5.6 H.R. Huff, E. Sirtl (eds.): *Semiconductor Silicon 1977* (The Electrochemical Society Inc., Princeton 1977)
5.7 J. Bourgoin, M. Lannoo: *Point Defects in Semiconductors II*, Springer
Ser. Solid-State Sci., Vol.35 (Springer, Berlin, Heidelberg, New York
1983)
5.8 A.A. Chernov: *Modern Crystallography III*, Springer Ser. Solid-State Sci.,
Vol.36 (Springer, Berlin, Heidelberg, New York, Tokyo 1984)
5.9a W. Kaiser, P.H. Keck: J. Appl. Phys. **28**, 882 (1957)
5.9b R.C. Newman, J.B. Willis: J. Phys. Chem. Solids **26**, 373 (1965)
5.10 R.J. Bell: *Introductory Fourier Transform Spectroscopy* (Academic,
New York 1972)
5.11 ASTM: Infrared Absorption Analysis of Impurities in Single Crystal
Semiconductor Material, *1977 Annual Book of ASTM Standards*, Part 43,
F120-75, F121-76, F123-74 (Academic, New York 1977)
5.12 M.L. Thewalt: Bound Multiexciton Complexes, *Physics of Semiconductors
1978*, Conf. Ser. **43**, 605 (1979)
5.13 M. Tajima: Appl. Phys. Lett. **32**, 719 (1978)
5.14 M. Tajima, A. Yusa: "Characterization of NTD Silicon Crystals by the
Photoluminescence Technique" in *Neutron-Transmutation-Doped Silicon*,
ed. by J. Guldberg (Plenum, New York 1981) p.377
5.15 A. Benninghoven, C.A. Evans, Jr., R.A. Powell, R. Shimizu, H.A. Storms
(eds.): *Secondary Ion Mass Spectrometry SIMS-II*, Springer Ser. Chem.
Phys., Vol.9 (Springer, Berlin, Heidelberg, New York 1979)
5.16 A. Benninghoven, J. Giber, J. László, M. Riedel, H.W. Werner (eds.):
Secondary Ion Mass Spectrometry SIMS-III, Springer Ser. Chem. Phys.,
Vol.19 (Springer, Berlin, Heidelberg, New York 1982)
5.17 A. Benninghoven, J. Okano, R. Shimizu, H.W. Werner (eds.): *Secondary
Ion Mass Spectrometry SIMS-IV*, Springer Ser. Chem. Phys., Vol.36
(Springer, Berlin, Heidelberg, New York, Tokyo 1984)
5.18 H. Oechsner (ed.): *Thin-Film and Depth-Profile Analysis*, Topics Curr.
Phys., Vol.37 (Springer, Berlin, Heidelberg, New York, Tokyo 1984)
5.19 R. Ueda, J.B. Mullin (eds.): *Crystal Growth and Characterization* (North-
Holland, Amsterdam 1974) Chap.10, p.107
5.20 J.C. Brice: *The Growth of Crystals from Liquids* (North-Holland, Amsterdam 1973) Chaps.4,7
5.21 A. Murgai, H.C. Gatos, A.F. Witt: *Semiconductor Silicon 1977*, ed. by
H.R. Huff, E. Sirtl (The Electrochemical Society Inc., Princeton 1977)
p.72
5.22 C.S. Fuller, R.A. Logan: J. Appl. Phys. **28**, 1427 (1957)
5.23 W. Kaiser, H.L. Frisch, H. Reiss: Phys. Rev. **112**, 1546 (1958)
5.24 A. Kanamori: Appl. Phys. Lett. **34**, 287 (1979)
5.25 P. Capper, A.W. Jones, E.J. Wallhouse, J.G. Wilkes: J. Appl. Phys. **48**,
1646 (1977)
5.26 A. Kanamori, M. Kanamori: J. Appl. Phys. **50**, 8095 (1979)
5.27 M. Tajima, A. Kanamori, T. Iizuka: Jpn. J. Appl. Phys. **18**, 1403 (1979)
5.28 Shin'Ichiro Takasu: Ohyobutsuri **49**, 83 (1980) [in Japanese]
5.29 Shin'Ichiro Takasu: Research Report of Ohyodenshibussei Working Group
No.382, 1 (1980) [in Japanese]
5.30 K. Morizane, P.S. Gleim: J. Appl. Phys. **40**, 4104 (1969)
5.31 E. Kosper, H. Clauss: Wiss. Ber. AEG-Telefunken **48**, 183 (1975)
5.32 S.M. Hu: J. Appl. Phys. **40**, 4413 (1969)
5.33 E.W. Hearn, E.H. Tekaat, G.H. Schwuttke: Microel. Reliab. **15**, 61 (1976)
5.34 T. Oku: Research Report of Ohyodenshibussei Working Group, No.382, 9
(1980) [in Japanese]
5.35 H. Otsuka, N. Yoshihiro, T. Oku, S. Takasu: ECS Extended Abstr. 79-2,
1366 (1979)
5.36 J.R. Patel, A.R. Chaudhuri: J. Appl. Phys. **34**, 2788 (1963)
5.37 S.M. Hu, S.P. Klepner, R.O. Schwenker, D.K. Seto: J. Appl. Phys. **47**,
4098 (1976)

5.38 N. Yoshihiro, H. Otsuka: 39th Meeting of Oyo-Butsuri-Gakukai Abstr. No.4a-M-5, p.260 (1978) [in Japanese]
5.39 S.M. Hu, W.J. Patrick: J. Appl. Phys. **46**, 1869 (1975)
5.40 Y. Aoki, Y. Hayafuji, S. Kawado: Role of Oxygen in Silicon on Defect Generation and MOS Characteristics, Spring Meeting of ECS, Extended Abstr. p.207 (1978)
5.41 S.M. Hu: Appl. Phys. Lett. **31**, 53 (1977)
5.42 N. Yoshihiro, H. Otsuka, T. Oku, Shin'Ichiro Takasu: J. Electrochem. Soc. **126**, 456C (1979)
5.43 J.R. Patel, A.R. Chaudhuri: J. Appl. Phys. **33**, 2223 (1962)
5.44 J.R. Patel: Disc. Faraday Soc. **38**, 201 (1964)
5.45 D. Sylwestrowicz: Philos. Mag. **7**, 1825 (1962)
5.46 T. Abe, S. Maruyama: Denki-Kagaku **35**, 149 (1967) [in Japanese]
5.47 D.I. Pomerantz: J. Electrochem. Soc. **119**, 255 (1972)
5.48 S.M. Hu: J. Vac. Sci. Tech. **14**, 17 (1977)
5.49 Y. Sugita: Oyo Butsuri **46**, 1056 (1977) [in Japanese]
5.50 G.H. Schwuttke: Microel. Reliab. **9**, 397 (1970)
5.51 C.J. Varker, K.V. Ravi: J. Appl. Phys. **45**, 272 (1974)
5.52 G.A. Rozgonyi, R.A. Kushner: J. Electrochem. Soc. **123**, 570 (1976)
5.53 Y. Matsushita, S. Kishino, M. Kanamori: Jpn. J. Appl. Phys. **19**, L63 (1980)
5.54 S. Kishino, M. Kanamori, N. Yoshihiro, M. Tajima, T. Iizuka: J. Appl. Phys. **50**, 8240 (1979)
5.55 S. Kishino, Y. Matsushita, M. Kanamori, T. Iizuka: Jpn. J. Appl. Phys. **21**, 1 (1982)
5.56 S. Kishino, Y. Matsushita, M. Kanamori: Appl. Phys. Lett. **35**(3), 213 (1979)
5.57 J.R. Patel: J. Appl. Phys. **40**, 3903 (1973)
5.58 S. Kishino, Y. Matsushita, M. Kanamori: Appl. Phys. Lett. **35**, 213 (1979)
5.59 H.J. Queisser (ed.): *X-Ray Optics*, Topics Appl. Phys., Vol.22 (Springer, Berlin, Heidelberg, New York 1977)
5.60 J.R. Patel, B.W. Batterman: J. Appl. Phys. **34**, 2716 (1963)
5.61 S. Kishino: J. Electron. Mat. **7**, 727 (1978)
5.62 J.R. Patel: J. Appl. Cryst. **8**, 186 (1975)
5.63 S. Kishino: Proc. 15th Int. Conf. Phys. Semicond. Kyoto 1980, J. Phys. Soc. Jpn. **49**, Suppl.A, 49 (1980)
5.64 L. Reimer: *Transmission Electron Microscopy*, Springer Ser. Opt. Sci., Vol.36 (Springer, Berlin, Heidelberg, New York, Tokyo 1984)
5.65 P.B. Hirsch, A. Howie, R.B. Nicholson, D.W. Pashley, M.J. Whelan: *Electron Microscopy of Thin Crystals* (Butterworths, London 1965)
5.66 M.J. Whelan: In *Modern Diffraction and Imaging Technique in Material Sciences*, ed. by S. Amelinckx, R. Gevers, G. Remaut, J. Van Landuxt, (North-Holland, Amsterdam 1970) p.35
5.67 Z.G. Pinsker: *Dynamical Scattering of X-Rays in Crystals*, Springer Ser. Solid-State Sci., Vol.3 (Springer, Berlin, Heidelberg, New York 1978)
5.68 Shih-Lin Chang: *Multiple Diffraction of X-Rays in Crystals*, Springer Ser. Solid-State Sci., Vol.50 (Springer, Berlin, Heidelberg, New York, Tokyo 1984)
5.69 W.M. Stobbs: In *Electron Microscopy in Materials Science*, ed. by E. Ruedl, U. Valdrè (Commission of the European Communities, Luxembourg 1978) p.591
5.70 Y. Matsushita, S. Aoki, S. Kishino: J. Crystallogr. Soc. Jpn. **21**, 358 (1979) [in Japanese]
5.71 M. Tajima, A. Kanamori, T. Iizuka: Jpn. J. Appl. Phys. **18**, 1401 (1979)
5.72 M. Tajima, S. Kishino, T. Iizuka: Jpn. J. Appl. Phys. **18**, 1403 (1979)
5.73 M. Tajima, S. Kishino, M. Kanamori, T. Iizuka: J. Appl. Phys. **51**, 2247 (1980);

M. Tajima, T. Masui, T. Abe, T. Iizuka: *Semiconductor Silicon 1981*, ed. by H.R. Huff, R. Kriegler, Y. Takeishi (Electrochem. Soc. 1981) p.72

5.74 H. Föll, U. Gösele, B.O. Kolbesen: J. Cryst. Growth **40**, 90 (1977)

5.75 J. Chikawa, S. Shirai: Proc. 10th Conf. Solid-State Devices Tokyo 1978, Jpn. J. Appl. Phys. **18**, Suppl.18-1, 153 (1979)

5.76 S. Kishino: Denshi Tsushin Gakkaishi **63**, 852 (1979) [in Japanese]

5.77 N. Inoue, J. Osaka, K. Wada: Oyo Butsuri **48**, 1126 (1979) [in Japanese]

5.78 S. Kishino, Y. Matsushita, M. Kanamori, T. Iizuka: Jpn. J. Appl. Phys. **21**, 1 (1982)

5.79 J. Burke: *The Kinetics of Phase Transformations in Metals* (Pergamon, London 1965) Chaps.6,7

5.80 S. Shirai: Appl. Phys. Lett. **36**, 156 (1980)

5.81 J.W. Mayer, L. Eriksson, J.A. Davies: *Ion Implantation in Semiconductors* (Academic, New York 1970)

5.82 G. Dearnaley, J.H. Freeman, R.S. Nelson, J. Stephen: *Ion Implantation* (North-Holland, Amsterdam 1973)

5.83 S. Namba (ed.): *Ion Chunyu Gijutsu* (Kogyo Chosakai, Tokyo 1975) [in Japanese]

5.84 K. Itoh, M. Tsurushima, M. Yatsuta, I. Ohdomari: *Ion Implantation* (Shokodo, Tokyo 1976) [in Japanese]

5.85 F. Chernow, J.A. Borders, D.K. Brice (eds.): *Ion Implantation in Semiconductors* (Plenum, New York 1977)

5.86 S. Namba (ed.): *Ion Implantation* (Plenum, New York 1975)

5.87 F.L. Vook: Radiation Damage and Defects in Semiconductors. Inst. Phys. Conf. Ser. **16**, 60 (1973)

5.88 K. Yagi, M. Tamura, Y. Yanagi, K. Inamiwa, T. Tokuyama: Jpn. J. Appl. Phys. **19**, Suppl.19-1, 61 (1980)

5.89 H. Ryssel, H. Glawischnig (eds.): *Ion Implantation Techniques*, Springer Ser. Electrophys., Vol.10 (Springer, Berlin, Heidelberg, New York 1982)

5.90 H. Ryssel, H. Glawischnig (eds.): *Ion Implantation: Equipment and Techniques*, Springer Ser. Electrophys., Vol.11 (Springer, Berlin, Heidelberg, New York 1983)

5.91 T.Y. Tan, E.E. Gardner, W.K. Tice: Appl. Phys. Lett. **30**, 175 (1977)

5.92 K. Yamamoto, S. Kishino, Y. Matsushita, T. Iizuka: Appl. Phys. Lett. **36**, 195 (1980)

5.93 K. Nagasawa, Y. Matsushita, S. Kishino: Appl. Phys. Lett. **37**, 622 (1980)

5.94 S. Kishino, K. Nagasawa, T. Iizuka: Jpn. J. Appl. Phys. **19**, L827 (1980)

5.95 K. Ploog, K. Graf: *Molecular Beam Epitaxy of III-V Compound, A Comprehensive Bibliography 1958-1983* (Springer, Berlin, Heidelberg, New York 1983)

5.96 J.R. Deines, A. Spiro: Low-Pressure Silicon Epitaxy, Electrochem. Soc. Spring Meeting, Extended Abstr.62 (1974)

5.97 M. Ogirima, H. Saida, M. Suzuki, M. Maki: J. Electrochem. Soc. **125**, 1879 (1978)

5.98 R.E. Logar, R.B. Herring, M.T. Wauk: Reduced Pressure Silicon Epitaxy in a Cylindrical Geometry Reactor, Electrochem. Soc. Fall Meeting, Extended Abstr. 225 (1978)

5.99 M. Ogirima, H. Saida, M. Suzuki, M. Maki: J. Electrochem. Soc. **124**, 903 (1977)

5.100 M.J.P. Duchemin, M.M. Bonnet, M.F. Koelsch: J. Electrochem. Soc. **125**, 637 (1978)

5.101 M. Nomura, Y. Kohno: Research Report of Electronic Materials Working Group, Jpn. Electr. Eng. Soc. EFM-79-9 (1979)

5.102 Y. Ota: J. Electrochem. Soc. **124**, 1795 (1977)

5.103 G.E. Becker, J.C. Bean: J. Appl. Phys. **48**, 3396 (1977)

5.104 T. Sakamoto, T. Takahashi, E. Suzuki, A. Shoji, H. Kawanami: Report of Professional Group on Semiconductor and Transistor, I.E.C.E.J., SSD 78-92, Feb. 1979 [in Japanese]
5.105 Y. Komiya, K. Shirai, A. Kanamori, T. Sakamoto: Si Molecular Beam Epitaxy, Symposium S5-2, National Convention of I.E.E.J., Spring 1979 [in Japanese]
5.106 K.H. Ebert, P. Deuflhard, W. Jäger (eds.): *Modelling of Chemical Reaction Systems*, Springer Ser. Chem. Phys., Vol.18 (Springer, Berlin, Heidelberg, New York 1981)
5.107 R. Vanselow, R. Howe (eds.): *Chemistry and Physics of Solid Surfaces V*, Springer Ser. Chem. Phys., Vol.35 (Springer, Berlin, Heidelberg, New York, Tokyo 1984)
5.108 A. Benninghoven (ed.): *Ion Formation from Organic Solids*, Springer Ser. Chem. Phys., Vol.25 (Springer, Berlin, Heidelberg, New York 1983)
5.109 Y. Ota: Proc. 11th Conf. Solid State Device Tokyo 1979, Jpn. J. Appl. Phys. **19**, Suppl. 19-1, 637 (1980)
5.110 T. Takahashi, Y. Komiya: Private communication
5.111 J.C. Bean: Appl. Phys. Lett. **36**, 741 (1980)

Chapter 6

6.1 N. Tsubouchi, M. Miyoshi, H. Abe: Jpn. J. Appl. Phys. **17**, Suppl.17-1, 223 (1978)
6.2 N. Tsubouchi, H. Miyoshi, H. Abe, T. Enomoto: IEEE Trans. ED-**26**, 618 (1979)
6.3 K. Tsukamoto, Y. Akasaka, H. Miyoshi, N. Tsubouchi, Y. Horiba, K. Kijima, H. Nakata: "High Pressure Oxidation for Isolation of High Speed Bipolar Devices," IEDM Tech. Dig. 340 (1979)
6.4 S.M. Irving: Kodak Photoresist Seminar Proc. 1968 Ed., **2**, 26 (1968)
6.5 H. Abe, H. Matsui, K. Demizu, H. Komiya: Denki Kagaku **41**, 544 (1973) [in Japanese]
6.6 R.A.H. Heinecke: Solid State Electron. **18**, 1146 (1975)
6.7 R.A.H. Heinecke: Solid State Electron. **19**, 1039 (1976)
6.8 H.F. Winters: J. Appl. Phys. **49**, 5165 (1978)
6.9 H. Itakura, H. Toyoda, H. Komiya: 26th Joint Meeting of Applied Physics, Japan, Meeting Abstract 30a-H-5 (1979) [in Japanese]
6.10 J.A. Bondur: J. Vac. Sci. Technol. **13**, 1023 (1976)
6.11 S. Matsuo: Jpn. J. Appl. Phys. **16**, 175 (1977)
6.12 S. Matsuo: Jpn. J. Appl. Phys. **17**, 235 (1978)
6.13 H. Toyoda, H. Itakura, H. Komiya: Jpn. J. Appl. Phys. **20**, 667 (1981)
6.14 H. Toyoda, H. Komiya, H. Itakura: J. Electron. Mat. **9**, 569 (1980)
6.15 H. Itakura, H. Komiya, H. Toyoda: Jpn. J. Appl. Phys. **19**, 1429 (1980)
6.16 H. Komiya, M. Tobinaga: unpublished
6.17 C.J. Mogab: J. Electrochem. Soc. **124**, 1262 (1977)
6.18 G.C. Schwartz, L.B. Rothman, T.J. Schopen: J. Electrochem. Soc. **126**, 464 (1979)
6.19 P.D. Parry, A.F. Rodde: Solid State Techn. **4**, 125 (1979)
6.20 S. Matsuo: 40th Meeting of Jpn. Soc. Appl. Phys., Meeting Abstract 1p-A-7 (1979) [in Japanese]
6.21 P.M. Schaible, W.C. Metzger, J.P. Anderson: J. Vac. Sci. Techn. **15**, 334 (1978)
6.22 A. Yasuoka, H. Nagata, H. Harada, T. Enomoto: IECE Jpn. Techn. Rep. SSD-77-30 (1977) [in Japanese]
6.23 R. Reichelderfer, D. Vogel, R.L. Bersin: 152nd ECS Meeting, Abstract **151** (1977)

6.24 H. Kinoshita, M. Iiri: Proc. 15th Semiconductor and Integrated Circuit Technology Symp. (1978) p.60 [in Japanese]

6.25 K. Ueki, H. Komatsu, T. Mizutani, S. Iida: 26th Joint Meeting of Appl. Phys. Japan, Meeting Abstract 30p-S-6 (1979) [in Japanese]

6.26 T. Yamazaki, Y. Horiike, M. Shibagaki: 40th Meeting of Japan Soc. Appl. Phys. Meeting Abstract 1p-A-2 (1979) [in Japanese]

6.27 T. Tokuyama, N. Natsuaki, M. Miyao: Nikkei Electronics, June 11, 116 (1979) [in Japanese]

6.28 A.E. Bell: RCA Rev. **40**, 295 (1979)

6.29 A. Gat, J.F. Gibbons, T.J. Magee, J. Peng, V.H. Deline, P. Williams, C.A. Evans: Appl. Phys. Lett. **32**, 276 (1978)

6.30 P. Baeri, S.U. Campisano, G. Foti, E. Rimini: Appl. Phys. Lett. **33**, 137 (1978)

6.31 M. Koyanagi, H. Tamura, M. Miyao, N. Hashimoto, T. Tokuyama: Appl. Phys. Lett. **35**, 621 (1979);
S. Iwamatsu, M. Ogawa: 157th Electrochem. Soc. Meeting, May 1980, Abstract No.174

6.32 K. Uda, T. Takahashi, Y. Komiya, H. Koyama, Y. Miura: 27th Joint Conf. Appl. Phys., April 1980, Abstract p.460 [in Japanese]

6.33 M.W. Geis, D.C. Flanders, D.A. Antoniadis, H.I. Smith: IEDM Techn. Digest (1979) p.210

6.34 K.F. Lee, J.F. Gibbons, K.C. Saraswat, T.I. Kamins: Appl. Phys. Lett. **35**, 173 (1979)

6.35 M. Tamura, H. Tamura, T. Tokuyama: Jpn. J. Appl. Phys. **19**, L23 (1980)

6.36 J.M. Poate, H.J. Leamy, T.T. Sheng: Appl. Phys. Lett. **33**, 918 (1978)

6.37 T. Shibata, J.F. Gibbons, T.W. Sigmon: Appl. Phys. Lett. **36**, 566 (1980)

6.38 Y. Miura, M. Saito, K. Hoh: 156th Electrochem. Soc. Meeting, Oct. 1979 RNP 680

6.39 H.S. Carslaw, J.C. Jaeger: *Conduction of Heat in Solids* (Oxford University Press, Oxford 1950I Chap.11

6.40 A. Lietoila, J.F. Gibbons: Appl. Phys. Lett. **34**, 332 (1979)

6.41 J.A. van Vechten: In *Laser and Electron Beam Processing of Materials*, ed. by C.W. White, P.S. Peercy (Academic, New York 1980) p.53

6.42 D.H. Auston, C.M. Surko, T.N.C. Venkatesan, R.E. Slusher, J.A. Golovchenko: Appl. Phys. Lett. **33**, 437 (1978)

6.43 H.A. Bomke, H.L. Berkowitz, M. Harmatz, S. Kronenberg, R. Lux: Appl. Phys. Lett. **33**, 955 (1978)

6.44 A.C. Greenwald, A.R. Kirkpatrick, R.G. Little, J.A. Minnucci: J. Appl. Phys. **50**, 783 (1979)

6.45 N.M. Johnson, J.L. Regolini, D.J. Bartelink, J.F. Gibbons, K.N. Ratnakumar: Appl. Phys. Lett. **36**, 425 (1980)

6.46 R.T. Hodgson, J.E.E. Baglin, R. Pal, J.M. Neri, D.A. Hammer: Appl. Phys. Lett. **37**, 187 (1980)

6.47 J.L. Vossen, W. Kern: *Thin Film Process* (Academic, New York 1978)

6.48 W. Kern, R.S. Rosler: J. Vac. Sci. Techn. **14**, 1082 (1977)

6.49 C. Chang: J. Electrochem. Soc. **123**, 1245 (1976)

6.50 K.L. Chopra: *Thin Film Phenomena* (Krieger, Melbourne 1979)

6.51 R.W. Wilson, L.E. Terry. J. Vac. Sci. Techn. **13**, 157 (1976)

6.52 T. Mamaki: *Fundamentals of Thin Film Deposition* (Nikkan Kogyo, Tokyo 1977) [in Japanese]

6.53 S. Yanagisawa, T. Fukuyama: J. Electrochem. Soc. **127**, 1120 (1980)

6.54 T. Fukuyama, S. Yanagisawa: Jpn. J. Appl. Phys. **18**, 987 (1979)

6.55 J. Hrbek: Thin Solid Films **42**, 185 (1977)

6.56 J. Black: IEEE Trans. ED-**26**, 360 (1969)

6.57 N. Hosokawa, R. Matsuzaki, A. Tasuo: Jpn. J. Appl. Phys. Suppl.2, Pt.1, 435 (1974)

6.58 A. Yasuoka, K. Nagata, H. Harada, T. Enomoto: Professional Group S.S.D. Inst. Electr. Commun. Eng. Jpn. SSD-77-30 (1977) [in Japanese]

6.59 W.D. Grobman, H.E. Luhn, T.P. Donohue, A.J. Speth, A. Wilson, M. Hatzakis, T.H.P. Chang: IEEE Trans. ED-**26**, 360 (1979)
6.60 H. Shibata, H. Iwasaki: Professional Group S.S.D. Inst. Electr. Commun. Eng. Jpn. Vol. SSD 79-66, 1 (1979) [in Japanese]
6.61 H. Shibata, H. Iwasaki, K. Yamada, T. Oku, Y. Tarui: IEEE Trans. ED-**26**, 604 (1979)
6.62 S. Yanagisawa, T. Fukuyama: In Abstract of the 16th Semiconductor and LSI Symposium (1979) p.18 [in Japanese]
6.63 B.L. Crowder, S. Zirinsky: IEEE Trans. ED-**26**, 369 (1979)
6.64 M.M. Atalla, E. Tannenbaum, E.J. Scheibner: Bell Syst. Techn. J. **38**, 749 (1959)
6.65 R. Williams, M.H. Woods: J. Appl. Phys. **44**, 1026 (1973)
6.66 M.M. Shahin: Photogr. Sci. Eng. **15**, 322 (1971)
6.67 W.C. Johnson: IEEE Trans. NS-**22**, 2144 (1975)
6.68 S. Iwamatsu, Y. Tarui: J. Electrochem. Soc. **126**, 1078 (1979)
6.69 E.H. Nicollian, A. Goetzberger, C.N. Bergund: Appl. Phys. Lett. **15**, 174 (1969)
6.70 T.H. Ning, C.M. Osburn, H.N. Yu: J. Electron. Mat. **6**, 65 (1977)
6.71 Y. Miura, K. Yamabe, Y. Komiya, Y. Tarui: J. Electrochem. Soc. **127**, 191 (1980)
6.72 A. Kanagawa: In *Contamination Control for Air Conditioning*, ed. by K. Hayakawa, M. Tsukiji (Soft Science, Tokyo 1974) p.285 [in Japanese]
6.73 For example, Type ASAS-X made by Particle Measurement Systems, Inc.
6.74 L.M. Ephrath: J. Electrochem. Soc. **126**, 1419 (1979)
6.75 S. Iida, H. Komatsu, S. Matsuo, M. Asakawa: 38th Meeting of Jpn. Soc. Appl. Phys., Meeting Abstract 13p-W-14 (1977) [in Japanese]
6.76 H. Nagata, M. Denda, A. Yasuoka, T. Enomoto: IECE Jpn. Techn. Rep. SSD78-28 SSD-78-28 (1978) [in Japanese]
6.77 H.W. Lehmann, R. Winter: J. Vac. Sci. Techn. **15**, 319 (1978)
6.78 Y. Kuroki, K. Mori, K. Sugibuchi: Proc. 14th Semiconductor and Integrated Circuit Techn. Symposium (1978) p.12 [in Japanese]
6.79 T. Kure: Proc. 1st Dry Process Symposium (1979) p.31 [in Japanese]
6.80 M. Shibagaki, Y. Horiike: 27th Joint Meeting of Appl. Phys., Japan, Meeting Abstract 3a-N-8 (1980) [in Japanese]
6.81 R.A. Gdula: 156th ECS Meeting, Abstract 608 (1979)
6.82 R.G. Poulsen: Proc. IEDM 205 (1976)

Chapter 7

7.1 Z. Tanaka, T. Yamamura: "Outline of Automatic Test for LSI Mask Pattern Data, Nikkei Electronics (April 1980) p.80 [in Japanese]
7.2 H. Ohno, H. Kawanishi, H. Yoshizawa, A. Kishimoto, Y. Fujinami: Logic Circuit Interconnection Check System PALMS-1 System Block and Check Procedure. Nat. Conv. Rec. Inst. Electron. Commun. Eng. Jpn. No.409 (March 1979) [in Japense]
7.3 A. Kishimoto, Y. Fujinami: Logic Circuit Interconnection Check System PALMS-(2)-Figure Processing. Nat. Conv. Rec. Inst. Electron. Commun. Eng. Jpn. No.410 (March 1979) [in Japanese]
7.4 K. Eguchi, H. Fukuda, T. Takahashi, T. Ozawa, J. Sakami: LSI Circuit Interconnection Check System IVS-Function Outline and Fundamental Composition. Nat. Conv. Rec. Inst. Electron. Commun. Eng. Jpn. No.58 (March 1979) [in Japanese]
7.5 K. Tansho, H. Yoshimura, N. Ohwada, T. Nishide, H. Honma: On a Circuit Characteristics Estimation Program from LSI Mask Pattern. Profess. Group on Semicond. Semicon. Device Inst. Electron. Commun. Eng. Jpn. Vol.SSD-78 No.12 (December 1978)

7.6　M. Takashima, T. Mitsuhashi, K. Yoshida: Circuit Diagram Plot Program in Mask Pattern Analysis System. Nat. Conv. Rec. Inst. Electron. Commun. Eng. Jpn. No.402 (March 1979)

7.7　B.T. Preas, B.W. Lindsay, C.W. Gwyn: "Automatic Circuit Analysis Based on Mask Information", in 13th Design Automation Conference (June 1976) p.309

7.8　L. Szántó: Comput. Aided Des. **10**, 135 (1978)

7.9　K. Fujiki: Mask Data Inspection for IC Operation by THOM System. Profess. Group on Semicond. Semicon. Device Inst. Electron. Commun. Eng. Jpn. Vol.SSD-80, No.12 (May 1980)

7.10　J.R. Yoder: Appl. Opt. **7**, 1791 (1968)

7.11　O.C. Wells: *Scanning Electron Microscopy* (McGraw-Hill, New York 1974)

7.12　P.R. Thornton: *Scanning Electron Microscopy* (Chapman and Hall, 1968);

7.13　T.R. Touw, P.A. Herman, G.V. Lukianoff: Practical Techniques for Application of Voltage Contrast to Diagnosis of Integrated Circuits, IITRI/SEM, 177-182 (1977)

7.14　K. Nakahara, Y. Tarui, S. Kawashiro, N. Narukami, Y. Hayashi: Bull. Electrotechn. Lab. **40**, 380 (1976)

7.15　K.G. Gopinathan, A. Gopinath: J. Phys. E**11**, 229 (1978)

7.16　H. Fujioka, T. Hosokawa, Y. Kanda, K. Ura: Submicron Electron Beam Probe to Measure Signal Waveform at Arbitrarily Specified Positions on MHZ IC, IITRI/SEM/1978, **1**, 755 (1978)

7.17　A. Gopinath, K.G. Gopinathan, P.R. Thomas: Voltage Contrast: A Review, IITRI/SEM/1978 **1**, 375 (1978)

7.18　E. Wolfgang, R. Lindner, P. Fazekas, H.P. Feuerbaum: Electron-beam testing of VLSI circuits. IEEE J. SC-**14**, 471 (1979)

7.19　H.P. Feuerbaum: VLSI Testing Using the Electron Probe, IITRI/SEM/1979, **1**, 285 (1979)

7.20　E. Menzel, E. Kubalek: Electron beam test system for VLSI circuit inspection, IITRI/SEM/1979, **1**, 297 (1979)

7.21　R.E. McMahon: Electronics **44**, 92 (1971)

7.22　S. Iwamatsu, S. Shimizu: Proc. 16th Joint of Automatic Control Conf. (1973) p.431 [in Japanese]

7.23　C.N. Potter, D.E. Sawyer: Rev. Sci. Instrum. **39**, 180 (1968)

7.24　D.E. Sawyer, D.W. Berning, D.C. Lewis: Solid State Technol. **20**(6), 37 (1977)

7.25　D.E. Sawyer, D.W. Berning: NBS Special Publication, 400-24 (1977)

7.26　D.E. Sawyer, D.W. Berning: Proc. IEEE **64**, 393 (1976)

7.27　D.E. Sawyer, D.W. Berning: IEDM Tech. Digest 111 (1975)

7.28　M. Nagase: Proc. Seventeenth Symp. on Semiconductors and Integrated Circuits Technology (1979) p.18 [in Japanese]

7.29　M. Nagase: In Prof. Group on Semiconductor Devices Inst. Electron. Commun. Eng. Jpn. SSD-**79** (56) 9 (1979) [in Japanese]

7.30　M. Nagase: Microel. Reliab. **20**, 717 (1980)

7.31　M. Nagase: Denshi-Zairyo Special Publication: Remarkable Basic Technology for VLSI (1980) p.122 [in Japanese]

7.32　G.J. van Gurp: Appl. Phys. Lett. **11**, 476 (1971)

7.33　J.R. Black: "The Reaction of Al with Vitreous Silica," in 15th Annual Proceedings Reliability Physics (1977) p.257

7.34　W.H. Class: Solid State Tech. **22** (6), 61 (1979)

7.35　K. Fujiki: "Dividing Algorithm for Mask Pattern into Trapezoids", Trans. IECE **63**-C (11), 745 (1980)

7.36　K. Fujiki: "Network Recognition for VLSI Mask Pattern Data," Trans. IECE J**64**-C (2), 76 (1981)

8.1 R.H. Dennard: IEEE J. SC-**9**, 256 (1974)
8.2 D.F. Barbe (ed.): *Very Large Scale Integration*, Springer Ser. Electrophys., Vol.5, 2nd ed. (Springer, Berlin, Heidelberg, New York 1982)
8.3 Y. Tarui, K. Hoh: J. Inst. Electron. Commun. Eng. Jpn. **62**, 364 (1979)
8.4 A.S. Grove: In *Physics and Technology of Semiconductor Devices* (Wiley, New York 1967) p.193
8.5 J. Bourgoin, M. Lannoo: *Point Defect in Semiconductors II*, Springer Ser. Solid-State Sci., Vol.35 (Springer, Berlin, Heidelberg, New York 1983)
8.6 M. Lenzlinger, H.H. Snow: J. Appl. Phys. **40**, 278 (1969)
8.7 B. Hoeneisen, C.A. Mead: Solid State Electron. **15**, 819 (1972)
8.8 I.M. Bateman: Solid-State Electron. **17**, 539 (1974)
8.9 R.W. Keyes: Proc. IEEE **63**, 740 (1975)
8.10 R.W. Keyes: IEEE J. SC-**10**, 245 (1975)
8.11 F.M. D'Heurle: Proc. IEEE **59**, 1409 (1971)
8.12 F.M. D'Heurle, R. Rosenberg: Phys. Thin Films **13**, 257 (1973)
8.13 I.A. Blech: Thin Solid Films **13**, 117 (1972)
8.14 C.B. Norris, J.F. Gibbons: IEEE Trans. ED-**14**, 38 (1967)
8.15 C.Y. Duh, J.L. Moll: IEEE Trans. ED-**14**, 46 (1967)
8.16 P.A. Tove, G. Anderson, G. Ericsson, R. Lidholt: IEEE Trans. ED-**17**, 407 (1970)
8.17 T.H. Ning, P.W. Cook, R.H. Dennard, C. Osburn, S.E. Schuster, H.N. Yu: IEEE J. SC-**14**, 268 (1979)
8.18 J.T. Wallmark: IEEE Trans. ED-**26**, 135 (1979)
8.19 L. Brancomb: Science **203**, 143 (1979)
8.20 J.E. Selleck, R.A. Kenyon, D.P. Gaffney, F.W. Wiedman, A. Bhattacharyya: Dig. Techn. Paper ISSCC 220 (1980)
8.21 Electronics Design **11**, 44 (1979)
8.22 G.E. Moore: IEDM Techn. Dig. 11 (1975)
8.23 H.H. Berger, S.K. Wiedmann: IEEE J. SC-**7**, 340 (1972)
8.24 P. Richmann, J.A. Hayes: Electronics 111 (1972)
8.25 F. Faggin, T. Klein: Solid State Electron. **13**, 1125 (1970)
8.26 C.N. Ahlquist, J.R. Breivogel, J.T. Koo, J.L. McCollum, W.G. Oldham, A.L. Renninger: IEEE J. SC-**11**, 570 (1976)
8.27 J.H. Friedrich: Techn. Dig. ISSCC 104 (1968)
8.28 R.H. Dannard: U.S. Patent 3 387 286
8.29 E. Kooi, J.G. van Lierop, J.A. Appels: J. Electrochem. Soc. **123**, 1117 (1976)
8.30 J. Nishizawa, T. Terasaki, J. Shibata: IEEE Trans. ED-**22**, 185 (1975)
8.31 D.D. Tang, T.H. Ning, S.K. Wiedmann, R.D. Isaac, G.C. Feth, H.N. Yu: IEDM Techn. Dig. 201 (1979)
8.32 N. Akiyama, Y. Yatsurugi, Y. Endo, Z. Imayoshi, T. Nozaki: Appl. Phys. Lett. **22**, 630 (1973)
8.33 K. Onodera, T. Sekigawa, K. Shimizu, Y. Tarui: Nat. Conf. Rec. Inst. Electron. Commun. Eng. Jpn. S2-**14**, 2 (1980) [in Japanese]
8.34 K. Onodera: Electrochem. Soc. Meeting '80 (St. Louis), Recent News Paper (May 1980)
8.35 K. Onodera, T. Sekigawa, Y. Tarui: Prof. Group S.S.D., Inst. Electron. Commun. Eng. Jpn. SSD79-**62 70**, 61 (1979) [in Japanese]
8.36 T. Sekigawa, Y. Hayashi, Y. Tarui: Trans. Inst. Electron. Commun. Eng. Jpn. **58**-C, 509 (1975) [in Japanese]
8.37 Y. Hayashi, T. Sekigawa, Y. Tarui: Jpn. J. Appl. Phys. Suppl. **16**, 163 (1977)
8.38 H. Shibata, H. Iwasaki, T. Oku, Y. Tarui: Techn. Dig. IEDM 395 (1977)

8.39 H. Shibata, H. Iwasaki, K. Yamada, T. Oku, Y. Tarui: IEEE Trans. ED-**26**, 604 (1979)
8.40 H. Shibata: Technical Meeting on ED (IEEJ) 80/1, 97 (1980) [in Japanese]
8.41 M. Ogawa, S. Iwamatsu: Techn. Dig. of 26th Joint Conf. on Appl. Phys. 433 (1979) [in Japanese]
8.42 H.B. Pogge: Recent Newspaper Abstracts of 155th Electrochem. Soc. Meeting 114 RNP (1979)
8.43 K.F. Lee, J.F. Gibbons, K.C. Saraswat: Appl. Phys. Lett. **35**, 173 (1979)
8.44 S. Taguchi, E. Sugino, H. Tango: Techn. Dig. of 26th Joint Conf. of Appl. Phys. 455 (1979) [in Japanese]
8.45 M.E. Roulet, P. Schob, I. Golecki, M.A. Nicolet: Electron. Lett. **15**, 527 (1979)
8.46 S. Iwamatsu: Abstract of 27th Joint Conf. on Appl. Phys. 1a-E-2 (1979) [in Japanese]
8.47 E. Preuss, H. Schlötter: Inst. Phys. Conf. Ser. No.40 (ESSDERC 77), 7 (1978)
8.48 Y. Nishi, H. Hara: Jpn. J. Appl. Phys. **17**, Suppl.17-1, 27 (1978)
8.49 S.A. Evans: IEEE Trans. ED-**26**, 396 (1979)
8.50 D.D. Tang, T.H. Ning, S.K. Wiedmann, R.D. Isaac, G.C. Feth, H.N. Yu: IEDM 201 (1979)
8.51 S. Horiba, J. Nishizawa: Digest Techn. Papers (ISSCC) 64 (1980)
8.52 R.C. Eden, B.M. Welch, R. Zucca, S.I. Long: IEEE J. Solid State Circuits, SC-**14**, 221 (1979)
8.53 T. Mimura, S. Hiyamizu, T. Fujii, K. Nanbu: Jpn. J. Appl. Phys. **19**, L 225 (1980)
8.54 M.S. Shur, L.F. Eastman: IEEE Trans. ED-**26**, 1677 (1979)
8.55 I. Matisoo: IBM J. Res. Dev. **24**, 113 (1980)
8.56 T.R. Gheewala: IBM J. Res. Dev. **24**, 130 (1980)
8.57 S.M. Faris, W.H. Henkels, E.A. Valsamakis, H.H. Zappe: IBM J. Res. Dev. **24**, 143 (1980)
8.58 H. Ohta: IEEE Trans. ED-**29**, 2027 (1980)
8.59 K. Okada, K. Aomura, T. Nakamura, H. Shiba: IEEE Trans. ED-**26**, 385 (1979)
8.60 S.P. Gauer: IEEE Trans. ED-**26**, 349 (1979)
8.61 S. Asai: IEDM Digest 185 (1976)
8.62 E.Z. Hamdy, M.I. Elmasry: IEDM 576 (1979)
8.63 M.T. Elliot, M.R. Splinter, A.B. Jones, J. Reekstin: IEEE Trans. ED-**26**, 469 (1979)
8.64 R.H. Dennard, F.H. Gaensslen, E.J. Walker, P.W. Cook: IEEE Trans. ED-**26**, 325 (1979)
8.65 P.W. Cook, S.E. Schuster, J.T. Parrish, V.DiLonardo, D.R. Freedman: IEEE Trans. ED-**26**, 333 (1979)
8.66 K.N. Ratnakumar, J.D. Meindl: ISSCC, Digest Techn. Papers 72 (1980)
8.67 L.R. Weisberg: IEDM Digest 684 (1979)
8.68 D.A. Patterson, C.H. Séquin: IEEE Trans. Comp. C-**29**, 108 (1980)
8.69 Y. Matsue, H. Yamamoto, K. Kobayashi, T. Wada, M. Tameda, T. Okuda, Y. Inagaki: ISSCC 232 (1980)
8.70 T. Mano, K. Takeya, T. Watanabe, K. Kiuchi, T. Ogawa, K. Hirata: ISSCC 234 (1980)
8.71 H.S. Lee, W.D. Pricer: IEDM 15 (1976)
8.72 A. Mohsen: ISSCC 152 (1979)
8.73 W.B. Sander: ISSCC 182 (1976)
8.74 D. Erb: ISSCC Dig. Techn. Papers 24 (1978)
8.75 P.K. Chatterjee, G.W. Taylor, M. Walsh: IEDM, Late News (1978)
8.76 K. Ohta, K. Yamada, M. Saitoh, K. Shimizu, Y. Tarui: IEEE Trans. ED-**27**, 1352 (1980)

Subject Index

Temperature rise 357

Thermal cycling 402

Thermal donor 244

Thermal electron 16

Thermal field emission 27,84

Thermal history 237,238,239,248,
249,252

Thermal warpage 230,232

Thermally induced microdefect 235,
246,249

TI (taper isolated) RAM 425

Transistor transistor logic 402

Transmission electron microscopy
(TEM) 230,240

Triode gun 17

TTL (transistor transistor logic)
402

Tungsten single-crystal wire 25

Tunnel junction 396

Two-chip comparison 188

Two-dimensional information 174

Vacancy 51

Vacuum evaporation 305

Variable rectangular beam 15

Variable shape electron-beam
system 9

Variable-shaped beam 10,97

Variable-shaped-beam lithography
97

VCCD (vertical charge coupled
device) 421

Vector-scan delineator 81

Vector scan machine 38

Vector scanning 10

Velocity saturation 375

Vertical charge coupled device
421

Very large scale integration 1

VHSIC project 407

VLSI (very large scale integration)
1

VLSI cooperative laboratories 7

VLSI tester 346

Vogel's mechanism 21

Voltage measurement 335

V-shaped groove mark 46

Wafer bow 215

Wafer warpage 54,199

Wafer-warpage correction 54

Walking test 345

Warpage 196,215,216,217,224,227

Warped-wafer correction 76

Weak-inversion MOS transistor 412

Weak-inversion region 412

Weak link 396

Wehnelt electrode 17

WIMT (weak-inversion MOS transistor)
412

Work stage 40

Writing position compensation
method 48

X-ray diffraction 239,246

X-ray lithography 135

X-ray resist 170

X-ray section topograph 211

X-ray topography 239